Number Theory: An Applied Approach

Number Theory: An Applied Approach

Edited by
Rowan Payne

MURPHY & MOORE
www.murphy-moorepublishing.com

Published by Murphy & Moore Publishing,
1 Rockefeller Plaza,
New York City, NY 10020, USA

ISBN: 978-1-63987-403-3

Cataloging-in-Publication Data

Number theory : an applied approach / edited by Rowan Payne.
p. cm.
Includes bibliographical references and index.
ISBN 978-1-63987-403-3
1. Number theory. 2. Algebra. 3. Mathematics. I. Payne, Rowan.
QA241 .N863 2022
512.7--dc23

For information on all Murphy & Moore Publications
visit our website at www.murphy-moorepublishing.com

MURPHY & MOORE

Contents

Preface

The main aim of this book is to educate learners and enhance their research focus by presenting diverse topics covering this vast field. This is an advanced book which compiles significant studies by distinguished experts in the area of analysis. This book addresses successive solutions to the challenges arising in the area of application, along with it; the book provides scope for future developments.

Number theory is a branch of pure mathematics that deals with the study of natural numbers and integers. It focuses on the set of positive whole numbers which are called the set of natural numbers. The number theory helps in discovering interesting relationships between different sorts of numbers and in proving their legitimacy. It includes both experimental and theoretical aspects. The experimental aspect generates new questions and discovers their answer. The theoretical aspect tries to devise an argument that provides a conclusive answer to the questions. This book covers in detail some existent theories and innovative concepts revolving around number theory. Most of the topics introduced herein cover new techniques and applications of number theory. This book is appropriate for students seeking detailed information in this area as well as for experts.

It was a great honour to edit this book, though there were challenges, as it involved a lot of communication and networking between me and the editorial team. However, the end result was this all-inclusive book covering diverse themes in the field.

Finally, it is important to acknowledge the efforts of the contributors for their excellent chapters, through which a wide variety of issues have been addressed. I would also like to thank my colleagues for their valuable feedback during the making of this book.

Editor

1

COMPLEMENTARY FAMILIES OF THE FIBONACCI-LUCAS RELATIONS

Ivica Martinjak
Faculty of Science, University of Zagreb, Zagreb, Croatia
imartinjak@phy.hr

Helmut Prodinger
Department of Mathematics, University of Stellenbosch, Stellenbosch, South Africa
hproding@sun.ac.za

Abstract

In this paper we present two families of Fibonacci-Lucas identities, with the Sury's identity being the best known representative of one of the families. While these results can be proved by means of the basic identity relating Fibonacci and Lucas sequences we also provide a bijective proof. Both families are then treated by generating functions.

1. Introduction

The Fibonacci and Lucas sequences of numbers are defined as

$$F_{n+2} = F_{n+1} + F_n, \; F_0 = 0, \; F_1 = 1, \tag{1}$$
$$L_{n+2} = L_{n+1} + L_n, \; L_0 = 2, \; L_1 = 1, \tag{2}$$

where $n \in \mathbb{N}_0$, and denoted by $(F_n)_{n\geq 0}$ and $(L_n)_{n\geq 0}$, respectively. Equivalently, these sequences could be defined as the only solutions (x, y), $x = L_n$, $y = F_n$ of the Diophantine equation

$$x^2 - 5y^2 = 4(-1)^n.$$

For natural numbers k, n we consider the number of ways to tile a $k \times n$ rectangle with certain tiles. The area of a 1×1 rectangle is usually called a *cell*. Tiles that are 1×1 rectangles are called *squares* and tiles of dimension 1×2 are called *dominoes*. In particular, when $k = 1$ a rectangle to tile is called *n-board*.

Among many interpretations of the Fibonacci and Lucas numbers, here we use the fact that the number f_n of n-board tilings with squares and dominoes is equal to F_{n+1}, while the number l_n of the same type of tilings of a labeled circular n-board (labeled in the clockwise direction) is equal to L_n. Namely, an n-board tiling begins either with a square or with a domino, meaning that the numbers f_n obey the same recurrence relation as the Fibonacci numbers. Having in mind that the sequence $(f_n)_{n\geq 0}$ begins with $f_1 = 1$, $f_2 = 2$ we get

$$f_n = F_{n+1}. \tag{3}$$

Similar reasoning proves that the number of tilings of a labeled circular board of the length n is equal to the n-th Lucas number,

$$l_n = L_n. \tag{4}$$

A labeled circular board of length n we shall call an *n-bracelet.* We say that a bracelet is *out of phase* when a single domino covers cells n and 1, and *in phase* otherwise. Consequently there are more ways to tile an n-bracelet than an n-board. More precisely, n-bracelet tilings are equinumerous to the sum of tilings of an n-board and $(n-2)$-board,

$$l_n = f_n + f_{n-2}. \tag{5}$$

This follows from the fact that in phase tilings can be unfolded into n-board tilings while out of phase tilings can be straightened into $(n-2)$-board tilings since a single domino is fixed on cells n and 1. Once having (3) and (4), the statement of Lemma 1 follows immediately from relation (5).

Lemma 1. *The nth Lucas number is equal to the sum of $(n-1)^{\text{st}}$ and $(n+1)^{\text{st}}$ Fibonacci number:*

$$L_n = F_{n-1} + F_{n+1}. \tag{6}$$

Lemma 1 presents the most elementary identity involving both the Fibonacci and Lucas sequence. In what follows we use it in order to prove further identities. It is worth mentioning that there are numerous identities known for these sequences. Many of them one can find in the classic reference [9]. An introduction to Fibonacci polynomials can be seen in [5] while recent results on this subject one can find in [1].

2. Colored Tilings and the Product $m^n F_{n+1}$

A tiling either of a board or bracelet of length n with two kinds of tiles, squares or oriented dominoes, in which each cell can be painted with any of m available colors we shall call an (n, m)-tiling. Thus, the orientation of a domino is defined by colors of cells it covers, meaning that for m available colors we have m^2 types of dominoes. Note that within the (n, m)-tilings one can equivalently use dominoes that paint both cells with the same colors and have m possible directions. The third alternative to define such a tiling is to use dominoes in m^2 colors that are different from the colors of squares (see Figure 1). In what follows we shall use the first definition.

We let $c_1, c_2, \ldots, c_m$ denote colors in such a tiling. In particular, when $m = 2$ we choose white and black as colors c_1 and c_2, respectively. In case $m = 3$ we choose white, black and gray for c_1, c_2, c_3, respectively.

One can easily check that there are 8 tilings of a 2-board in two colors, having two types of squares and four types of dominoes (Figure 1). Furthermore, there are 24 such tilings of a 3-board, 80 tilings of a 4-board, etc. In general, there are $2^n f_n$ such tilings of an n-board. The numbers $1, 3, 18, 81, \ldots, 3^n f_n, \ldots$ also have a combinatorial interpretation. They count the ways to tile an n-board with 3 types of squares and 9 types of dominoes. We generalize these facts in the following lemma.

Lemma 2. *The product $m^n f_n$ represents the number of colored n-board tilings with two kinds of tiles, squares and oriented dominoes.*

Proof. There are f_n uncolored tilings of an n-board with squares and dominoes. According to the definition of oriented dominoes, each cell of a tiling can be painted with only one of the m available colors. Thus, each of f_n tilings gives m^n tilings when we use m types of squares and m^2 types of dominoes. □

The same argument proves that $m^n L_n$ represents the number of colored n-bracelet tilings with squares in m colors and m^2 types of dominoes.

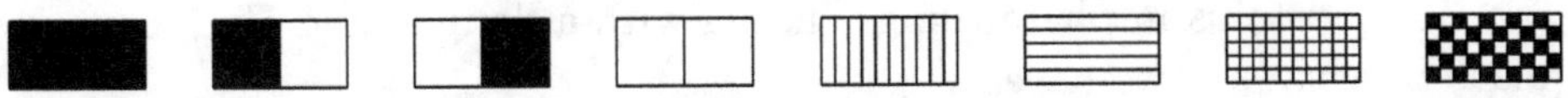

Figure 1: The eight $(2, 2)$-tilings of a board of length 2. In general there are $m^n F_{n+1}$ tilings with squares in m colors and dominoes in m^2 colors of a board of length n.

There are a few ways to prove the Fibonacci-Lucas identity (7). It can be proved using Binet's formula [8], and by means of generating functions as well [6]. Here we continue to call it Sury's identity, as it is already done in [6, 7].

Lemma 3. *For the Fibonacci sequence $(F_n)_{n\geq 0}$ and the Lucas sequence $(L_n)_{n\geq 0}$*

$$\sum_{k=0}^{n} 2^k L_k = 2^{n+1} F_{n+1}. \tag{7}$$

Proof. We present an induction proof. We use Lemma 1 and relation (4) to immediately get

$$F_{n+1} + L_{n+1} = 2F_{n+2},$$

which will be used to complete the inductive step. The relation (7) obviously holds true for $n = 1$, since $L_0 + 2L_1 = 4F_2$. Furtermore, we have

$$\sum_{k=0}^{n} 2^k L_k + 2^{n+1} L_{n+1} = 2^{n+1} F_{n+1} + 2^{n+1} L_{n+1}$$

$$\sum_{k=0}^{n+1} 2^k L_k = 2^{n+1}(F_{n+1} + L_{n+1})$$

$$= 2^{n+2} F_{n+2}$$

which completes the proof. □

We let $\mathcal{A}_{n,m}$ denote the set of all (n, m)-board tilings. Let $t_{n,w}$ be the unique all-white squares tiling. Then the other tiling must end with a series of white squares (which could be empty) preceded by a domino or a non-white square. Let $\mathcal{A}^d_{n,m}$ and $\mathcal{A}^s_{n,m}$, resprectively, denote the sets of these tilings. Thus,

$$\mathcal{A}_{n,m} = \mathcal{A}^s_{n,m} \cup \mathcal{A}^d_{n,m} \cup \{t_{n,w}\}.$$

We let $\mathcal{B}_{n,m}$ denote the set of all (n, m)-bracelet tilings. It is worth to emphasize that within $(n, 2)$-bracelet tilings the n-th cell can be tiled by

- a square, in colors c_1, c_2,
- a domino in phase, and a domino out of phase.

We let $\mathcal{B}^{c_i}_{n,m}$ denote the set of (n, m)-bracelet tilings ending with a square of color c_i, the $\mathcal{B}^p_{n,m}$ contains in phase tilings ending with a domino and $\mathcal{B}^o_{n,m}$ contains out of phase tilings. Now we have

$$\mathcal{B}_{n,m} = \bigcup_{i=1}^{m} \mathcal{B}^{c_i}_{n,m} \cup \mathcal{B}^p_{n,m} \cup \mathcal{B}^o_{n,m}.$$

It is worth to note that equalities

$$|\mathcal{B}^{c_1}_{n,m}| = |\mathcal{B}^{c_1}_{n,m}| = \cdots = |\mathcal{B}^{c_m}_{n,m}|$$
$$|\mathcal{B}^p_{n,m}| = |\mathcal{B}^o_{n,m}|$$

obviously holds true.

There is also an elegant bijective proof of Lemma 3 [2]. It is based on a 1-to-2 correspondence between the set $\mathcal{B}_{k,2}$ of bracelet tilings and the set $\mathcal{A}_{k,2}$ of board tilings, $1 \leq k \leq n$. Thus, in both sets tiles are colored in two colors (white and black) meaning that there are two types of squares and four types of dominoes. We let $\mathcal{A}^{c_i}_{n,m}$ denote the set of m-colored n-board tilings ending with a square of color c_i, $i = 1, \ldots, m$.

More precisely, we establish a 1-to-1 correspondence between the sets $\mathcal{A}^{c_2}_{k,2}$ and $\mathcal{B}^{c_1}_{k,2}$ as well as between the sets $\mathcal{A}^{d}_{k,2}$ and $\mathcal{B}^{p}_{n,2}$, $1 \le k \le n$, by the operations *i)* removing tiles from $k+1$ through n and *ii)* gluing cells k and 1 together.

In the same manner we generate bracelets ending with a black square and out of phase bracelets. Two remaining tilings $t_{n,w}$ consisting from all-white squares are mapped to two 0-bracelets. Thus, twice the number of $(n,2)$-board tilings is needed to establish correspondence with bracelets of length at most n that are tiled with two types of squares and four types of dominoes. Having in mind $|\mathcal{A}_{n,2}| = 2^n f_n$, we have

$$\sum_{k=0}^{n} 2^k L_k = 2 \cdot 2^n f_n = 2^{n+1} F_{n+1},$$

which completes the proof.

3. The Main Result

The next result, Theorem 1, gives an extension of the relation (7). It provides the answer whether there is an identity involving the product $3^n l_n$, respectively $3^n f_n$ (which appears in some other contexts [4]). We prove it combinatorially while a proof by means of Lemma 1 is also possible.

Theorem 1. *For the Fibonacci sequence $(F_n)_{n\ge0}$ and the Lucas sequence $(L_n)_{n\ge0}$*

$$\sum_{k=0}^{n} 3^k (L_k + F_{k+1}) = 3^{n+1} F_{n+1}. \tag{8}$$

Proof. Note that within $(n,3)$-bracelet tilings the n-th cell can be tiled by

- a square, in one of the colors c_1, c_2, c_3,
- a domino in phase, and a domino out of phase.

According to the previouis notations, the sets $\mathcal{B}^{c_1}_{n,3}$, $\mathcal{B}^{c_2}_{n,3}$ and $\mathcal{B}^{c_3}_{n,3}$ denote those tilings ending with a square in a color c_1, c_2, c_3, respectively, $\mathcal{B}^{p}_{n,3}$ denote tilings ending with a domino in phase and $\mathcal{B}^{o}_{n,3}$ denote tilings with a domino out of phase.

The set $\mathcal{A}_{n,3}$ of $(n,3)$-board tilings we divide into disjoint subsets $\mathcal{A}^{s}_{n,3}$, $\mathcal{A}^{d}_{n,3}$ and $\{t_{n,w}\}$,

$$\mathcal{A}_{n,3} = \mathcal{A}^{s}_{n,3} \cup \mathcal{A}^{d}_{n,3} \cup \{t_{n,w}\}.$$

In the set $\mathcal{A}^{s}_{n,3}$, there are tilings consisting of a nonwhite square covering cell k and white squares on cells $k+1$ through n, $1 \le k \le n$. The set $\mathcal{A}^{d}_{n,3}$ contains tilings having a domino on cell k while white squares cover cells $k+1$ through n. Now, removing tiles from $k+1$ through n and gluing cells k and 1 together, from $\mathcal{A}^s_{n,3}$ we obtain $(k,3)$-bracelet tilings ending with gray and those tilings ending with a black square. Thus,

$$|\mathcal{A}^s_{n,3}| = |\mathcal{B}^{c_2}_{n,3}| + |\mathcal{B}^{c_3}_{n,3}|.$$

By the same operations, from the set $\mathcal{A}^d_{n,3}$ we get $(k,3)$-bracelet tilings ending with a domino in phase,

$$|\mathcal{A}^d_{n,3}| = |\mathcal{B}^p_{n,3}|.$$

The board tiling $t_{n,w}$ consisting of all-white squares is mapped to the 0-bracelet.

In the same way we establish a correspondence between $(n,3)$-board tilings and out of phase $(k,3)$-bracelet tilings together with twice the number of $(k,3)$-bracelets ending with a white square,

$$|\mathcal{A}^s_{n,3}| = |\mathcal{B}^{c_1}_{n,3}| + |\mathcal{B}^{c_3}_{n,3}|,$$
$$|\mathcal{A}^d_{n,3}| = |\mathcal{B}^o_{n,3}|.$$

The extra set $\mathcal{B}^{c_3}_{n,3}$ of $(k,3)$-bracelets ending with black square can be unfolded into $(k-1,3)$-board tilings (according to the arguments we use when proving Lemma 1). This means that relation

$$\sum_{i=1}^{3} |\mathcal{B}^{c_i}_{n,3}| + |\mathcal{B}^p_{n,3}| + |\mathcal{B}^o_{n,3}| + 1 + \sum_{k=0}^{n-1} 3^k F_{k+1} = 2|\mathcal{A}_{n,3}|$$

holds true. Furthermore, we have

$$\sum_{k=0}^{n} 3^k L_k + \sum_{k=0}^{n-1} 3^k F_{k+1} = 2 \cdot 3^n F_{n+1}$$

and finally

$$\sum_{k=0}^{n} 3^k L_k + \sum_{k=0}^{n} 3^k F_{k+1} = 3^{n+1} F_{n+1},$$

which completes the proof. □

For example, when $n = 3$ then

$$|\mathcal{B}^{c_i}_{n,3}| = 18 + 3 + 1 = 22, \quad i = 1, 2, 3,$$
$$|\mathcal{B}^p_{n,3}| = |\mathcal{B}^o_{n,3}| = 27 + 9 = 36$$

and

$$|\mathcal{A}^s_{n,3}| = 36 + 6 + 2 = 44,$$
$$|\mathcal{A}^d_{n,3}| = 27 + 9 = 36.$$

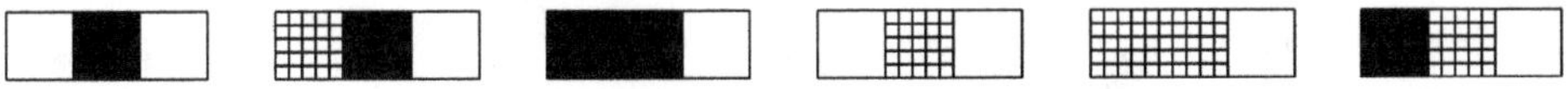

Figure 2: The six tilings in the set $\mathcal{A}^{s}_{3,3}$, for the case $k = 2$.

Figure 2 illustrates this instance showing $(3,3)$-board tilings in the set $\mathcal{A}^{s}_{3,3}$ for which $k = 2$.

Clearly, there is a 1-to-1 correspondence between (n, m)-boards with a domino as the last nonwhite tile and either in phase tilings or out of phase tilings. There is also a 1-to-1 correspondence between boards with a square as the last nonwhite tile and (k, m)-bracelets, $1 \leq k \leq n$ ending with a square in color c_i, $1 \leq i \leq m-1$. In order to establish a 1-to-2 correspondence we have to add $m-2$ sets of $(k-1, m)$-boards to the set of (k, m)-bracelets. This reasoning proves Theorem 2. The identity for the next case $m = 4$ is as follows:

$$\sum_{k=0}^{n} 4^k (L_k + 2F_{k+1}) = 4^{n+1} F_{n+1}. \tag{9}$$

Theorem 2. *For the Fibonacci sequence $(F_n)_{n\geq 0}$ and the Lucas sequence $(L_n)_{n\geq 0}$ of numbers, with $m \geq 2$*

$$\sum_{k=0}^{n} m^k [L_k + (m-2)F_{k+1}] = m^{n+1} F_{n+1}. \tag{10}$$

In Theorem 3 we give a further extension of Theorem 3, to any m. We prove it by applying Lemma 1.

Theorem 3. *The alternating sum of products m^{n-k}, $0 \leq k \leq n$ with Fibonacci and Lucas numbers is equal to either positive or negative value of $(n+1)$-th Fibonacci number,*

$$\sum_{k=0}^{n} (-m)^{n-k} [L_{k+1} + (m-2)F_k] = F_{n+1}. \tag{11}$$

where $m \geq 2$.

Proof. We have

$$\begin{aligned}
&m^n L_1 - m^{n-1} L_2 + m^{n-2} L_3 - \cdots + (-1)^n L_{n+1} \\
&\quad = m^n (F_0 + F_2) - m^{n-1}(F_1 + F_3) + m^{n-2}(F_2 + F_4) - \cdots + (-1)^n (F_n + F_{n+2}) \\
&\quad = m^n F_1 - m^{n-1}(mF_1 + F_2) + m^{n-2}(mF_2 + F_3) - \cdots + (-1)^n (mF_n + F_{n+1}) \\
&\quad = m^n F_1 - m^n F_1 - m^{n-1} F_2 + m^{n-1} F_2 + m^{n-2} F_3 - m^{n-2} F_3 - \cdots + (-1)^n F_{n+1} \\
&\quad = (-1)^n F_{n+1}.
\end{aligned}$$

□

Corollary 1. *The alternating sum of products* $2^{n-k}L_{k+1}$, $0 \le k \le n$ *is equal to either a positive or negative value of the* $(n+1)^{\text{st}}$ *Fibonacci number:*

$$\sum_{k=0}^{n}(-1)^k 2^{n-k} L_{k+1} = (-1)^n F_{n+1}. \tag{12}$$

Thus, Theorem 2 and Theorem 3 give complementary families of the Fibonacci-Lucas identities, with the Sury identity as the best known representative of (10).

4. A Generating Functions Approach

Sums as the one in (8) or (10) can be systematically evaluated using *generating functions.* We recall that

$$\sum_{n\ge 0} F_n z^n = \frac{z}{1-z-z^2} \quad \text{and} \quad \sum_{n\ge 0} L_n z^n = \frac{2-z}{1-z-z^2}.$$

Consequently we have

$$\begin{aligned}\sum_{n\ge 0}\Big(\sum_{0\le k\le n} m^k F_k\Big)z^n &= \frac{1}{1-z}\frac{mz}{1-mz-m^2z^2}\\ &= \frac{m}{m^2+m-1}\frac{1+m^2z}{1-mz-m^2z^2} - \frac{m}{m^2+m-1}\frac{1}{1-z}.\end{aligned}$$

Comparing coefficients, this leads to the explicit formula

$$\sum_{0\le k\le n} m^k F_k = \frac{m^{n+1}}{m^2+m-1}[F_{n+1}+mF_n] - \frac{m}{m^2+m-1}.$$

Likewise,

$$\begin{aligned}\sum_{n\ge 0}\Big(\sum_{0\le k\le n} m^k L_k\Big)z^n &= \frac{1}{1-z}\frac{2-mz}{1-mz-m^2z^2}\\ &= \frac{m}{m^2+m-1}\frac{(2m+1)-m(m-2)z}{1-mz-m^2z^2} + \frac{m-2}{m^2+m-1}\frac{1}{1-z}\end{aligned}$$

and

$$\sum_{0\le k\le n} m^k L_k = \frac{m^{n+1}}{m^2+m-1}[(2m+1)F_{n+1}-(m-2)F_n] + \frac{m-2}{m^2+m-1}.$$

This evaluates the formula (10):

$$\begin{aligned}\sum_{0\le k\le n} m^k[L_k+(m-2)F_{k+1}] &= \sum_{0\le k\le n} m^k L_k + \frac{m-2}{m}\sum_{1\le k\le n+1} m^k F_k\\ &= \frac{m^{n+1}}{m^2+m-1}[(2m+1)F_{n+1}-(m-2)F_n] + \frac{m-2}{m^2+m-1}\\ &+ \frac{m-2}{m}\frac{m^{n+1}}{m^2+m-1}[F_{n+1}+mF_n] - \frac{m-2}{m}\frac{m}{m^2+m-1} + \frac{m-2}{m}m^{n+1}F_{n+1}\\ &= m^{n+1}F_{n+1}.\end{aligned}$$

Alternating sums, as in (11), can be handled as well: Since $L_{k+1}=F_k+F_{k+2}=2F_k+F_{k+1}$ we find

$$L_{k+1}+(m-2)F_k = F_{k+1}+mF_k.$$

It follows that

$$\begin{aligned}\sum_{n=0}^{\infty}\Bigg(\sum_{k=0}^{n}(-m)^{n-k}[L_{k+1}+(m-2)F_k]\Bigg)z^n &= \sum_{n=0}^{\infty}\Bigg(\sum_{k=0}^{n}(-m)^{n-k}[F_{k+1}+mF_k]\Bigg)z^n\\ &= \left[\sum_{i=0}^{\infty}(-mz)^i\right]\left[\sum_{j=0}^{\infty}(F_{j+1}+mF_j)z^j\right]\\ &= \frac{1}{1+mz}\;\;\frac{1}{1-z-z^2}+\frac{mz}{1-z-z^2}\Bigg)\\ &= \frac{1}{1-z-z^2}\\ &= \sum_{n=0}^{\infty}F_{n+1}z^n.\end{aligned}$$

Similarly,

$$\begin{aligned}\sum_{n\ge 0}\Bigg(\sum_{0\le k\le n}(-1)^k m^{n-k}L_k\Bigg)z^n &= \sum_{k,l\ge 0}(-1)^k L_k z^k m^l z^l\\ &= \frac{1}{1-mz}\frac{2+z}{1+z-z^2}\\ &= \frac{1}{m^2+m-1}\frac{(m-2)-z(2m+1)}{1+z-z^2} - \frac{m(2m+1)}{m^2+m-1}\frac{1}{1-mz},\end{aligned}$$

which leads to

$$\sum_{0\le k\le n}(-1)^k m^{n-k}L_k = \frac{(-1)^{n+1}}{m^2+m-1}[(m-2)F_{n+1}-(2m+1)F_n] + \frac{m(2m+1)}{m^2+m-1}m^n.$$

Formulæ (11) and (12) follow from these two in a straightforward way.

Acknowledgment The authors thank the referee for careful reading, help and valuable suggestions that improved the final version of the paper.

References

[1] T. Amdeberhan, X. Chen and V. H. Moll, B. E. Sagan, Generalized Fibonacci polynomials and Fibonomial coefficients, *Ann. Comb.* **18** (2014), 541–562.

[2] A. T. Benjamin and J. J. Quinn, *Proofs That Really Count*, Mathematical Association of America, Washington DC, 2003.

[3] A. T. Benjamin, S. S. Plott and J. A. Sellers, Tiling of recent sum identities involving Pell numbers, *Ann. Comb.* **12** (2008), 271–278.

[4] T. Došlić and I. Zubac, Counting maximal matchings in linear polymers, *Ars Math. Contemp.* **11** (2016), 255–276.

[5] R. L. Graham, D. E. Knuth and O. Patashnik, *Concrete Mathematics*, Addison-Wesley, 1994.

[6] H. Kwong, An alternate proof of Sury's Fibonacci-Lucas relation, *Amer. Math. Monthly* **6** (2014), 514–514.

[7] I. Martinjak, Two extensions of the Sury's identity, *Amer. Math. Monthly*, **123 (9)** (2016), p. 919.

[8] B. Sury, A polynomial parent to a Fibonacci-Lucas relation, *Amer. Math. Monthly* **3** (2014), 236–236.

[9] S. Vajda, *Fibonacci and Lucas Numbers, and the Golden Section*, John Wiley & Sons, Inc., New York, 1989.

2

INVARIANT MEASURES FOR NEW CLASSES OF PIECEWISE FRACTIONAL LINEAR MAPS

Fritz Schweiger
Department of Mathematics, University of Salzburg, Salzburg, Austria
fritz.schweiger@sbg.ac.at

Abstract
This paper continues the investigations of invariant measures for the so-called Moebius maps T. Various modifications of these maps are considered. The left-most and the right-most branch do not map their domain onto the interval $[0,1]$, but since $T0 = T1$, the union of their images is $[0,1]$.

1. Introduction

In recent years the ergodic theory of non-invertible maps has been of interest. A possible frame work for these maps is the concept of *fibred systems* [8]. A map $T : B \to B$ is called a *fibred system* if there is a partition $\{B(k) : k \in I\}$ of B with the index set I that is finite or countable, such that the restriction of T to $B(k)$ is injective. This restriction is called a *branch* of T. The inverse map of a branch is denoted by V_k and its Jacobian by ω_k. A short account of the historical development of the related f-expansions can be found in [5].

A lot of work has already done on the ergodicity and the existence of invariant measures for such maps. Note that the existence of a finite invariant measure can be used to calculate probabilities by the use of ergodic theorems (see [8]). In many cases the existence of an invariant measure can be proved by the verification of some sufficient conditions. Much less is known about the shape of an invariant density. Lebesgue measure is invariant for g-adic expansions. Due to an old conjecture of Gauss, the invariant density for continued fractions is known, namely $h(x) = \frac{1}{1+x}$. One extension of continued fractions to higher dimensions is the Jacobi-Perron algorithm. For $n = 2$ the following map

$$T(x, y) = (\frac{y}{x} - a, \frac{1}{x} - b),\ a = a(x) = \lfloor \frac{y}{x} \rfloor,\ b = b(x) = \lfloor \frac{1}{x} \rfloor$$

is ergodic and admits a finite invariant measure, but the shape of the density is not known.

Another tricky example is the Bolyai algorithm which is related to the map

$$T : [1,2] \to [1,2],\ Tx = x^2, 1 \leq x < \sqrt{2};$$

$$Tx = x^2 - 1, \sqrt{2} \leq x < \sqrt{3};\ Tx = x^2 - 2,\ \sqrt{3} \leq x < 2.$$

The shape of its invariant density is not known [2, 6]. Furthermore, we mention a/b-expansions [1, 3].

In this note modified Moebius maps are considered (in [7], the name Moebius maps has been proposed for fractional linear maps $T : [0,1] \to [0,1]$ with three branches). The left-most and the right-most branch do not map their domain of definition onto $[0,1]$, but since $T0 = T1$ the union of their images build $[0,1]$. The graphs of these maps have suggested the name *butterfly maps.*
Important tools are the dual Moebius map and the jump transformation (see [8]). We denote the *(natural) dual map* (or dual algorithm with respect to the kernel $K(x,y) = (1+xy)^{-2}$) of T by T^*. This is a Moebius map, the branches of which are formed by the transposed matrices. The dual map T^* is called *differentiably isomorphic* to T if there is a bijective map $\psi(t) = \frac{b+dt}{a+bt}$ such that $\psi \circ T = T^* \circ \psi$. Otherwise we call T^* an *exceptional* dual. Since T can be written piecewise by 2×2 -matrices, one uses the matrix equations

$$\begin{pmatrix} a & b \\ b & d \end{pmatrix} T = T^* \begin{pmatrix} a & b \\ b & d \end{pmatrix}$$

to calculate a map ψ. If the dual map is defined on a set B^*, then the invariant density for T can be written down as

$$h(x) = \int_{B^*} \frac{dy}{(1+xy)^2}.$$

It can happen that a constant map $\psi(t) = K$ satisfies $\psi \circ T = T^* \circ \psi$. Then $B^* = \{K\}$ and

$$h(x) = \frac{1}{(1+Kx)^2}.$$

The case $K = 0$ means that Lebesgue measure is invariant.

2. Maps With a Fixed Partition

A point ζ is called a *central fixed point* of the butterfly map if $T\zeta = \zeta$ and $T0 = T1 = \zeta$.

Theorem 1. *The invariant density can be written down in an explicit form for butterfly maps with three branches on a fixed partition and a central fixed point if* *there is polynomial $P(X, Y)$ such that the parameters μ and ν of the map satisfy $P(\mu, \nu) = 0$.*

Proof. (1) We first consider specific maps with three decreasing branches and the partition $0, \frac{1}{4}, \frac{1}{2}, 1$. The choice of a fixed partition reduces the number of variables and is sufficient to show the method of proof.

We list the three branches of T

$$Tx = \frac{1-4x}{3+\delta x}, \; 0 \le x < \frac{1}{4}$$

$$Tx = \frac{1-2x}{-1+6x}, \; \frac{1}{4} \le x < \frac{1}{2}$$

$$Tx = \frac{5+\eta-4x}{3-\eta+4\eta x}, \; \frac{1}{2} \le x \le 1$$

and their inverse maps

$$V_0 x = \frac{1-3x}{4+\delta x}, \; V_1 x = \frac{1+x}{2+6x}, \; V_2 x = \frac{5+\eta+(\eta-3)x}{4+4\eta x}.$$

The essential conditions are

$$T(\frac{1}{3}) = \frac{1}{3}$$

and

$$T0 = T1 = \frac{1}{3}.$$

This means that the central fixed point is $\zeta = \frac{1}{3}$. To avoid poles in the intervals of definition, we require $\delta > -3$ and $\eta > -1$. Note that the condition that $\frac{1}{3}$ is a fixed point implies that the map V_1 is uniquely determined.

We are looking for a map $\psi(t) = \frac{b+dt}{a+bt}$ such that $\psi \circ T = T^* \circ \psi$ holds. This leads to three linear equations in a, b, and d. One finds a non-trivial solution if the condition

$$84 + \delta + 12\eta = \delta\eta$$

is satisfied. The map ψ can be calculated as

$$\psi(t) = \frac{-6+\delta+(42-\delta)t}{6+(\delta-6)t}.$$

If we look at $216+6\delta-\delta^2$, we see that ψ is increasing for $-1 < \delta \le 18$ and decreasing for $\delta > 18$.

The map T^* essentially is of the same type on the interval with endpoints $\psi(0) = \frac{\delta-6}{6}$ and $\psi(1) = \frac{36}{\delta}$. The crucial fixed point is $\psi(\frac{1}{3}) = 2$.

If $\delta \neq 18$ the invariant density has two parts h_I and h_{II}. The function h_I is the density on $[0, \frac{1}{3}]$ and h_{II} the density on $[\frac{1}{3}, 1]$. They satisfy the following conditions:

$$h_I(x) = h_I(V_0x)\omega_0(x) + h_{II}(V_1x)\omega_1(x),\ 0 \leq x < \frac{1}{3},$$

$$h_{II}(x) = h_I(V_1x)\omega_1(x) + h_{II}(V_2x)\omega_2(x),\ \frac{1}{3} \leq x < 1.$$

The map T^* on the interval with endpoints $\psi(0)$ and $\psi(1)$ is a dual algorithm. Let D_I be the interval with endpoints 2 and $\psi(0)$, and D_{II} be the interval with endpoints 2 and $\psi(1)$. Then we see that

$$((T^*)^{-1}D_I \cap B^*(0)) \cup ((T^*)^{-1}D_{II} \cap B^*(1)) = D_I$$

and

$$((T^*)^{-1}D_I \cap B^*(1)) \cup ((T^*)^{-1}D_{II} \cap B^*(2)) = D_{II}.$$

Therefore, the device of the dual algorithm gives the densities as

$$h_I(x) = \int_{D_I} \frac{dy}{(1+xy)^2} \text{ and } h_{II}(x) = \int_{D_{II}} \frac{dy}{(1+xy)^2}.$$

For $\delta = 18$ the set B^* shrinks to the point 2. Therefore the invariant density is given by $h(x) = \frac{1}{(1+2x)^2}$. Note that the second root of $216 + 6\delta - \delta^2 = 0$, namely $\delta = -12$, does not give an allowed value for the map V_0.

(2) Now we look shortly at three increasing branches with the same partition. The essential conditions are again

$$T(\frac{1}{3}) = \frac{1}{3}$$

and

$$T(0) = T(1) = \frac{1}{3}.$$

We find

$$V_0x = \frac{1-3x}{-8-\delta+\delta x},\ V_1x = \frac{1+x}{4},\ V_2x = \frac{1+(3+\eta)x}{2+\eta x}.$$

If $2\eta = \delta$, the map ψ can be calculated as

$$\psi(t) = \frac{\delta - 3\delta x}{-8-\delta+\delta x}.$$

The invariant densities

$$h_I(x) = |\int_0^{\psi(1)} \frac{dy}{(1+xy)^2}|,\ h_{II}(x) = |\int_{\psi(0)}^0 \frac{dy}{(1+xy)^2}|$$

satisfy the invariance conditions

$$h_I(x) = h_I(V_1x)\omega_1(x) + h_{II}(V_2x)\omega_2(x),\ 0 \le x < \frac{1}{3}$$

$$h_{II}(x) = h_I(V_0x)\omega_0(x) + h_{II}(V_1x)\omega_1(x),\ \frac{1}{3} \le x < 1.$$

(3) The first mixture of (1) and (2) is not really interesting. Its inverse branches are given by

$$V_0x = \frac{1-3x}{4+\delta x},\ V_1x = \frac{1+x}{4},\ V_2x = \frac{5+\eta+(\eta-3)x}{4+4\eta x}.$$

One sees that this map has two ergodic components, namely the intervals $[0, \frac{1}{3}]$ and $[\frac{1}{3}, 1]$. The dual algorithm of the first map is given by the map ψ on $[0, \frac{1}{3}]$

$$\psi(t) = \frac{\delta - 3\delta t}{4+\delta t},$$

and the the dual algorithm of the second map is given by by the map ψ on $[\frac{1}{3}, 1]$

$$\psi(t) = \frac{\eta - 3\eta t}{-\eta - 2 + \eta t}.$$

If $4\eta + \eta\delta + 2\delta = 0$ then we obtain the same function ψ.

(4) The second mixture has a little different behavior. Its inverse branches are given by

$$V_0x = \frac{1-3x}{-8+\delta+\delta x},\ V_1x = \frac{1+x}{2+6x},\ V_2x = \frac{1+(3+\eta)x}{2+\eta x}.$$

The map T^2 has two ergodic components, namely the intervals $[0, \frac{1}{3}]$ and $[\frac{1}{3}, 1]$. Under the condition $6 + 2\delta = 3\eta$, the map

$$\psi(t) = \frac{6-\delta+(30+7\delta)t}{6+\delta+(6-\delta)t}$$

allows us to write down an invariant density for T.

We take a short look at T^2 on the interval $[0, \frac{1}{3}]$. It has four branches

$$V_{01}x = \frac{1-3x}{16+\delta+(48+5\delta)x},\ V_{02}x = \frac{1+(9+2\eta)x}{16+\delta+(-3\delta+8\eta)x},$$

$$V_{12}x = \frac{3+(3+2\eta)x}{10+(18+8\eta)x},\ V_{11}x = \frac{3+7x}{10+18x}$$

with the partition 0, $\frac{1}{16+\delta}$, $\frac{1}{4}$, $\frac{3}{10}$, and $\frac{1}{3}$. □

In the proof of theorem 1 we constructed invariant densities for case (1) and (2) with the help of an appropriate function ψ. The next theorem shows that no exceptional dual maps exist.

Theorem 2. *There is no exceptional dual in case (1) or (2) .*

Proof. (1) To construct a dual map on an interval B^*, we have to find points α and β as its endpoints and ζ as a central fixed point such that

$$\alpha = V_0^*\zeta,\ V_0^*\alpha = V_1^*\beta,\ V_1^*\zeta = \zeta,\ V_1^*\alpha = V_2^*\beta,\ V_2^*\zeta = \beta.$$

The map $y \mapsto \frac{6+y}{2+y}$ has two fixed points $\zeta = 2$ and $\zeta = -3$.

(1a) If $\zeta = 2$ then $\alpha = \frac{\delta-6}{6}$ and $\beta = \frac{-3+3\eta}{7+\eta}$. The equation $V_0^*\alpha = V_1^*\beta$ gives

$$\eta\delta = \delta + 12\eta + 84.$$

This is exactly the condition which was given for the existence of a differentiably isomorphic dual. The equation $V_1^*\alpha = V_2^*\beta$ leads to

$$\frac{30+\delta}{6+\delta} = \frac{7\eta^2+16\eta+9}{3\eta^2+16\eta+13} = \frac{(7\eta+9)(\eta+1)}{(3\eta+13)(\eta+1)} = \frac{7\eta+9}{3\eta+13}.$$

This equation gives the same condition. Note that $\eta = -1$ is not an admissible value for the parameter η. Therefore this dual is not exceptional.

(1b) If $\zeta = -3$, then $\alpha = \delta + 9$ and $\beta = \frac{9+\eta}{-11-3\eta}$. The equation $V_1^*\alpha = V_2^*\beta$ leads to

$$27\eta\delta + 356\eta + 83\delta + 1092 = 0.$$

The equation $V_1^*\alpha = V_2^*\beta$ gives the relation

$$\delta\eta + 8\eta = \delta + 24.$$

These equations represent hyperbolas with parallel asymptotic lines. The common points are $(\delta, \eta) = (-12, -3)$ and $(\delta, \eta) = (-\frac{144}{11}, -\frac{15}{7})$ which do not correspond to a suitable map T.

(2) We proceed in a similar way as before. The map $y \mapsto \frac{y}{4+y}$ has two fixed points $\zeta = 0$ and $\zeta = -3$.

(2a) If $\zeta = 0$ then $\alpha = \frac{\delta}{-8-\delta}$ and $\beta = \frac{\eta}{2}$. We first use the equation $V_0^*\beta = V_1^*\alpha$ which leads to the relation

$$2\eta\delta - \delta^2 - 12\delta + 24\eta = (\delta - 2\eta)(-\delta - 12).$$

Since $\delta = -12$ is not allowed we see that $\delta = 2\eta$. The condition $V_1^*\beta = V_2^*\alpha$ implies

$$\delta + 6\delta - 12\eta - 2\eta^2 = (-2\eta + \delta)(\eta + 6) = 0.$$

Therefore we find $\delta = 2\eta$ and this dual is not exceptional.

(2b) The value $\zeta = -3$ gives $\alpha = \frac{\delta+9}{-\delta+11}$ and $\beta = 9 + 2\eta$. $V_0^*\beta = V_1^*\alpha$ gives

$$\delta^2 - 8\eta\delta - 27\delta - 96\eta - 468 = 0,$$

and $V_1^*\beta = V_2^*\alpha$ implies

$$2\eta^2 - 4\eta\delta - 27\eta - 24\delta - 234 = 0.$$

Then

$$\begin{aligned} 0 &= \delta^2 - 8\eta\delta - 27\delta - 96\eta - 468 - 2(2\eta^2 - 4\eta\delta - 27\eta - 24\delta - 234) \\ &= \delta^2 - 4\eta^2 + 21\delta - 42\eta = (\delta - 2\eta)(\delta + 2\eta + 21). \end{aligned}$$

If we insert $\delta = 2\eta$ and $\delta = -2\eta - 21$ into one of the equations, we eventually find the four common points of these hyperbolas: $(-12, -6)$, $(-13, -\frac{13}{2})$, $(-9, -6)$, and $(-12, -\frac{9}{2})$ which do not represent admissible parameters. □

3. Maps with a Variable Partition

In this section we consider a fractional linear map $g(x)$ such that $Tx = g(x) \bmod 1$ is a butterfly map with three branches. The partition of the unit interval then depends on the given map $g(x)$.

Theorem 3. *The invariant density can be written down in an explicit form for butterfly maps with three branches, given by a fractional linear map and a central fixed point.*

Proof. (1) We start with a decreasing map

$$g(x) = \frac{A + Bx}{C + Dx}$$

and the conditions

$$g(1) = \beta,\ g(0) = \beta + 2, 0 < \beta < 1$$

and

$$\frac{A + B\beta}{C + D\beta} - 1 = \beta.$$

Then we choose $C = 1$ and the parameter $\delta = D$. We calculate $A = \beta + 2$, $B = -\frac{\delta+4}{\delta+2}$, and $\beta = \frac{1}{2+\delta}$, which implies $-1 < \delta$. The map $Tx = g(x) \bmod 1$ then is given by the three branches

$$Tx = \frac{1 - (2\delta^2 + 5\delta + 4)x}{2 + \delta + (2\delta + \delta^2)x},\ 0 \le x < \frac{1}{4 + 5\delta + 2\delta^2},$$

$$Tx = \frac{3+\delta-(\delta^2+3\delta+4)x}{2+\delta+(2\delta+\delta^2)x}, \frac{1}{4+5\delta+2\delta^2} \le x < \frac{3+\delta}{4+3\delta+\delta^2},$$

$$Tx = \frac{5+2\delta-(\delta+4)x}{2+\delta+(2\delta+\delta^2)x}, \frac{3+\delta}{4+3\delta+\delta^2} \le x < 1.$$

In this case, for $\delta > -1$, a map ψ with $\psi \circ T = T^* \circ \psi$ can be found. A calculation shows

$$\psi(t) = \frac{\delta+\delta^2 t}{2\delta+3+\delta t}.$$

Then the invariant density can be calculated in the usual way. Note that β is the central fixed point of T.

(2) We may also start with an increasing map

$$g(x) = \frac{A+Bx}{C+Dx}$$

and the conditions

$$g(0) = \beta,\ g(1) = \beta+2, 0 < \beta < 1$$

and

$$\frac{A+B\beta}{C+D\beta} - 1 = \beta.$$

In a similar way as before we find

$$Tx = \frac{1+(4+7\delta+2\delta^2)x}{2+\delta+(2\delta+\delta^2)x}, 0 \le x < \frac{1+\delta}{4+5\delta+\delta^2},$$

$$Tx = \frac{-1-\delta+(4+5\delta+\delta^2)x}{2+\delta+(2\delta+\delta^2)x}, \frac{1+\delta}{4+5\delta+\delta^2} \le x < \frac{3+2\delta}{4+3\delta},$$

$$Tx = \frac{-3-2\delta+(4+3\delta)x}{2+\delta+(2\delta+\delta^2)x}, \frac{3+2\delta}{4+3\delta} \le x < 1.$$

Again, we find a function ψ with $\psi \circ T = T^* \circ \psi$. It is

$$\psi(t) = \frac{\delta+\delta^2 t}{-2\delta-1+\delta t}$$

which serves to calculate an invariant density. Again, β is the central fixed point of T. □

4. A Further Variation

We consider butterfly maps with three branches, but the point $T0 = T1$ is a point of partition. These three branches will be labeled by the parameters λ, μ, and ν.

Theorem 4. *For butterfly maps with three increasing branches such that the point $T0 = T1$ is a point of partition, an explicit form of the invariant density can be given if the three parameters satisfy $P(\lambda, \mu, \nu) = 0$ for a polynomial $P(X, Y, Z)$.*

Proof. Here we use the partition 0, $\frac{1}{3}$, $\frac{2}{3}$, and 1.

(1) For sake of completeness we first look at the case of decreasing branches. More precisely, let

$$V(\lambda)(\frac{2}{3}) = 0,\ V(\lambda)(0) = \frac{1}{3},$$
$$V(\mu)(1) = \frac{1}{3},\ V(\mu)(0) = \frac{2}{3},$$
$$V(\nu)(1) = \frac{2}{3},\ V(\nu)(\frac{2}{3}) = 1.$$

But this case is not really interesting because the interval $[\frac{2}{3}, 1]$ is an absorbing region.

(2) Therefore we now consider the increasing case:

$$V(\lambda)(\frac{2}{3}) = 0,\ V(\lambda)(1) = \frac{1}{3},$$
$$V(\mu)(0) = \frac{1}{3},\ V(\mu)(1) = \frac{2}{3},$$
$$V(\nu)(0) = \frac{2}{3},\ V(\nu)(\frac{2}{3}) = 1.$$

Then we give these branches as follows:

$$V(\lambda)x = \frac{-2+3x}{3+\lambda-\lambda x},\ V(\mu)x = \frac{1+(1+2\mu)x}{3+3\mu x},\ V(\nu)x = \frac{4+(3-2\nu)x}{6-2\nu x}.$$

The parameters λ, μ, and ν have to obey some conditions to avoid certain types of fixed points, e. g., the map $V(\mu, \nu)$ has the fixed point $\frac{2}{3}$ which will not affect the existence of an invariant measure on $[0, 1]$ and the fixed point $\frac{5+4\mu}{2\nu-3\mu+2\mu\nu}$. This fixed point must be outside of the interval $[\frac{5+4\mu}{9+6\mu}, \frac{2}{3}[$. The reason for this will become clear in the course of our construction.

If our map admits an invariant density, it has two pieces which satisfy the following equations:

$$(1a)\ \ h_I(x) = h_I(V(\mu)x)\omega(\mu; x) + h_{II}(V(\nu)x)\omega(\nu; x),\ 0 \leq x < \frac{2}{3},$$

$$(1b)\ \ h_{II}(x) = h_I(V(\lambda)x)\omega(\lambda; x) + h_I(V(\mu)x)\omega(\mu; x),\ \frac{2}{3} \leq x < 1.$$

We find an invariant density by using the device of jump transformations and dual algorithm (see [8]). We introduce the jump transformation S by setting

$$Sx = T^2x,\ x \in B(\lambda,\nu) = B(\lambda),\ Sx = Tx, x \in B(\mu) \cup B(\nu).$$

Let g_I, g_{II} be the invariant density on $[0, \frac{2}{3}]$ and $[\frac{2}{3}, 1]$, respectively. Then we find the equations

$$(2a)\ g_I(x) = g_I(V(\lambda,\nu)x)\omega(\lambda,\nu;x) + g_I(V(\mu)x)\omega(\mu;x) + g_{II}(V(\nu)x)\omega(\nu;x),$$

$$(2b)\ g_{II}(x) = g_I(V(\mu)x)\omega(\mu;x).$$

We insert the expression for g_{II} into equation $(2a)$ and obtain

$$(3) g_I(x) = g_I(V(\lambda,\nu)x)\omega(\lambda,\nu;x) + g_I(V(\mu)x)\omega(\mu;x) + g_I(V(\mu,\nu)x)\omega(\mu,\nu;x).$$

This equation can be seen as a Kuzmin equation for a map $R : [0, \frac{2}{3}] \to [0, \frac{2}{3}]$ with the partition 0, $\frac{1}{3}$, $\frac{5+4\mu}{9+6\mu}$, and $\frac{2}{3}$. The map R is a map with three branches which have been studied in [7]. The inverse branches are given as

$$V(\lambda,\nu)x = \frac{(9-2\nu)x}{18+2\lambda-(6\nu+3\lambda)x},$$

$$V(\mu)x = \frac{1+(1+2\mu)x}{3+3\mu x},$$

$$V(\mu,\nu)x = \frac{10+8\mu+(3-4\nu+6\mu-4\mu\nu)x}{18+12\mu+(-6\nu+9\mu-6\mu\nu)x}.$$

According to [7], we can write down an explicit form of the invariant density if the condition

$$\det\begin{pmatrix} -6\nu-3\lambda & -9-2\nu-2\lambda & 0 \\ 3\mu & -2+2\mu & 1 \\ -6\nu+9\mu-6\nu\mu & -15+4\nu-6\mu-4\mu\nu & 10+8\mu \end{pmatrix} = 0$$

is satisfied.

Then one can construct a map $\psi(t) = \frac{a+bt}{c+dt}$ such that $\psi \circ R = R^* \circ \psi$ where R^* denotes the *dual map* to R.

Another possibility is to check if an *exceptional dual* for the map R can be constructed (see [7], [4]). One has to find an interval with endpoints α and β (including the case ∞) such that the following conditions are satisfied:

$$V(\lambda,\nu)^*\alpha = \alpha,\ V(\lambda,\nu)^*\beta = V(\mu,\nu)^*\alpha,\ V(\lambda,\nu)^*\beta = V(\mu)^*\alpha,\ V(\mu)^*\beta = \beta.$$

According to the theory of jump transformations, we find

$$h_I(x) = g_I(x),\ h_{II}(x) = g_{II}(x) + g_I(V(\lambda)x)\omega(\lambda;x).$$

Now it is easy to verify the equations (1a) and (1b):

$$h_I(V(\mu)x)\omega(\mu;x) + h_{II}(V(\nu)x)\omega(\nu;x) =$$

$$h_I(V(\mu)x)\omega(\mu;x) + h_I(V(\mu,\nu)x)\omega(\mu,\nu;x) + h_I(V(\lambda,\nu)x)\omega(\lambda,\nu;x) = h_I(x),$$

$$h_I(V(\lambda)x)\omega(\lambda;x) + h_I(V(\mu)x)\omega(\mu;x) = g_I(V(\lambda)x)\omega(\lambda;x) + g_{II}(x) = h_{II}(x).$$

□

Examples. These examples illustrate the case of increasing branches of theorem 4.

(a)

$$\lambda = -\frac{13}{5}, \mu = 0, \nu = \frac{1}{2}$$

$$V(\lambda)x = \frac{-10+15x}{2+13x}, \ V(\mu)x = \frac{1+x}{3}, \ V(\nu)x = \frac{4+2x}{6-x}$$

$$V(\lambda,\nu)x = \frac{5x}{8+3x}, \ V(\mu)x = \frac{1+x}{3}, \ V(\mu,\nu)x = \frac{10+x}{18-3x}$$

$$\psi(t) = \frac{1-2t}{1+t}$$

$$g_I(x) = h_I(x) = \frac{1}{5-x} + \frac{1}{1+x}, \ g_{II}(x) = \frac{1}{14-x} + \frac{1}{4+x},$$

$$h_{II}(x) = \frac{1}{14-x} + \frac{1}{4+x} - \frac{5}{2+5x} + \frac{7}{-2+7x}.$$

(b)

$$\lambda = -3, \mu = -\frac{3}{4}, \nu = \frac{3}{2}$$

$$V(\lambda)x = \frac{-2+3x}{3x}, \ V(\mu)x = \frac{4-2x}{12-9x}, \ V(\nu)x = \frac{4}{6-3x}$$

$$V(\lambda,\nu)x = \frac{x}{2}, \ V(\mu)x = \frac{4-2x}{12-9x}, \ V(\mu,\nu)x = \frac{4-3x}{9-9x}$$

$$\psi(t) = -t$$

$$g_I(x) = h_I(x) = \frac{3}{2-3x}, \ g_{II}(x) = \frac{1}{1-x} - \frac{3}{4-3x},$$

$$h_{II}(x) = \frac{1}{1-x} - \frac{3}{4-3x} + \frac{1}{2-x} + \frac{1}{x}.$$

(c) In this example we use an exceptional dual!

$$\text{Let } \gamma = \frac{-1+\sqrt{7}}{2} \text{ and } \lambda = -\frac{-3+8\gamma}{3}, \mu = \frac{1}{2}, \nu = \frac{3-8\gamma}{6}.$$

$$V(\lambda)x = \frac{-6+9x}{6+8\gamma+(3-8\gamma)x}, \; V(\mu)x = \frac{2+4x}{6+3x}, \; V(\nu)x = \frac{12+(6+8\gamma)x}{18+(-3+8\gamma)x}$$

$$V(\lambda,\nu)x = \frac{x}{2}, \; V(\mu)x = \frac{2+4x}{6+3x}, \; V(\mu,\nu)x = \frac{14+(3+8\gamma)x}{24+12\gamma x}$$

The exceptional dual is given on $[0,\gamma]$ by the maps

$$V(\lambda,\nu)^*y = \frac{y}{2}, \; V(\mu,\nu)^*y = \frac{12\gamma+(3+8\gamma)y}{24+14y}, \; V(\mu)^*y = \frac{3+4y}{6+2y}$$

$$g_I(x) = h_I(x) = \frac{\gamma}{1+\gamma x}, \; g_{II}(x) = \frac{3+4\gamma}{6+2\gamma+(3+4\gamma)x} - \frac{1}{2+x}$$

$$h_{II}(x) = \frac{3+4\gamma}{6+2\gamma+(3+4\gamma)x} - \frac{3-8\gamma}{6+8\gamma+(3-8\gamma)x}.$$

(d) In this example we use again an exceptional dual!

$$\lambda = -\frac{3}{2}, \mu = 0, \nu = -3$$

$$V(\lambda)x = \frac{-4+6x}{3+3x}, \; V(\mu)x = \frac{1+x}{3}, \; V(\nu)x = \frac{4+9x}{6+6x}$$

$$V(\lambda,\nu)x = \frac{2x}{2+3x}, \; V(\mu)x = \frac{1+x}{3}, \; V(\mu,\nu)x = \frac{10+15x}{18+18x}$$

The exceptional dual is given on $[0,\infty[$ by the maps

$$V(\mu)^*y = \frac{y}{3+y}, V(\mu,\nu)^*y = \frac{18+15y}{18+10y}, \; V(\lambda,\nu)^*y = \frac{3+2y}{2}$$

$$g_I(x) = h_I(x) = \frac{1}{x}, \; g_{II}(x) = \frac{1}{1+x}, \; h_{II}(x) = \frac{3}{-2+3x}.$$

5. Outlook: Maps With Four Branches

We start with the increasing map $g(x) = \frac{1+(6+7\mu)x}{2+2\mu x}$, $\mu > -1$ and consider the map $Tx = g(x) \bmod 1$. Note that $T0 = \frac{1}{2} = T1$. This map leads to a fibred system T with the partition

$$0 < \frac{1}{6+5\mu} < \frac{3}{6+3\mu} = \frac{1}{2+\mu} < \frac{5}{6+\mu} < 1.$$

The map T is given piecewise as

$$T_k x = g(x) - k, \;\; k = 0,1,2,3.$$

If $\mu = 0$, then T is piecewise linear and $h(x) = 1$ (Lebesgue measure) is the invariant density.

Theorem 5. *For $\mu > -1$ and $\mu \neq 0$ it is not possible to construct an exceptional dual.*

Proof. We try to find $\alpha, \xi, \theta, \eta, \beta$, and μ such that

$$T_0^*\alpha = \xi = T_1^*\xi,\ T_0^*\xi = \beta = T_1^*\theta,\ T_2^*\theta = \alpha = T_3^*\eta,\ T_2^*\eta = \eta = T_3^*\beta.$$

From

$$\frac{2\mu + (6+7\mu)\xi}{2+\xi} = \beta = \frac{2\mu + (6+5\mu)\theta}{2-\theta},$$

one calculates the relation

$$\theta = \frac{\xi}{1+\xi},$$

and from

$$\frac{2\mu + (6+3\mu)\theta}{2-3\theta} = \alpha = \frac{2\mu + (6+\mu)\eta}{2-5\eta},$$

the relation

$$\theta = \frac{\eta}{1-\eta}.$$

The fixed points ξ and η satisfy the quadratic equations

$$(Q_1)\quad \xi^2 + (4+5\mu)\xi + 2\mu = 0,$$

$$(Q_2)\quad 3\eta^2 + (4+3\mu)\eta + 2\mu = 0.$$

If we insert $\xi = \frac{\eta}{1-2\eta}$ into equation (Q_1) we find

$$(-7-2\mu)\eta^2 + (4-3\mu)\eta + 2\mu = 0.$$

If the quadratic equations for η are irreducible, we see that $3 = -7 - 2\mu$ and $4 + 3\mu = 4 - 3\mu$, a contradiction.

If these equations are reducible then

$$\eta = -\frac{2\mu^2 + 10\mu}{3\mu^2 + 10\mu + 20}.$$

We insert $\eta = \frac{\xi}{1+2\xi}$ into equation (Q_2) and obtain

$$(11+14\mu)\xi^2 + (4+11\mu)\xi + 2\mu = 0.$$

Since these equations now are reducible we obtain

$$\xi = -\frac{14\mu^2 + 10\mu}{35\mu^2 + 50\mu + 20}.$$

We substitute this value into the relation $\eta = \frac{\xi}{1+2\xi}$ and obtain

$$\eta = -\frac{14\mu^2 + 10\mu}{7\mu^2 + 30\mu + 20}.$$

We equate both values for η and observe $\mu \neq 0$. Then we get the equation

$$7\mu^2 + 10\mu + 10 = 0,$$

which has no real solution. □

We also consider the decreasing map $g(x) = \frac{7+(\mu-6)x}{2+2\mu x}$, $\mu > -1$ and study shortly the map $Tx = g(x) \bmod 1$. This map leads to a fibred system T with the partition

$$0 < \frac{1}{6+5\mu} < \frac{3}{6+3\mu} = \frac{1}{2+\mu} < \frac{5}{6+\mu} < 1.$$

The map T is given piecewise as

$$T_k x = g(x) - k, \; k = 0, 1, 2, 3.$$

If $\mu = 0$ then T is piecewise linear and again $h(x) = 1$ (Lebesgue measure) is the invariant density.

Theorem 6. *For $\mu > -1$ and $\mu \neq 0$ it is not possible to construct an exceptional dual.*

Proof. In a similar way as before we try to find $\alpha, \xi, \theta, \eta, \beta$, and μ such that

$$T_2^*\theta = \alpha = T_3^*\xi, \, T_1^*\theta = \beta = T_0^*\eta, \, T_3^*\alpha = \eta = T_2^*\xi, \, T_1^*\eta = \xi = T_0^*\beta.$$

As before we find $\xi = \frac{\theta}{1+\theta}$ and $\eta = \frac{\theta}{1-\theta}$. We use

$$W_-(\xi) = \frac{\xi}{1-2\xi} = \eta$$

and

$$W_+(\eta) = \frac{\eta}{1+2\eta} = \xi.$$

The maps $T_0^* \circ T_0^* \circ W_-$ and $W_+ \circ T_3^* \circ T_3^*$ both have ξ as a fixed point. The corresponding linear maps have the eigenvalue equations

$$\lambda^2 - \lambda(40 + 32\mu - 3\mu^2) + 144(1+\mu)^2 = 0,$$

$$\lambda^2 - \lambda(40 + 48\mu + 5\mu^2) + 144(1+\mu)^2 = 0.$$

The comparison of these equations leads to $\mu = 0$ (the linear case) or $\mu = -2$, which is not an admissible parameter.
The same result can be derived from a comparison of the maps $W_- \circ T_0^* \circ T_0^*$ and $T_3^* \circ T_3^* \circ W_+$. □

We consider two further examples of maps with four branches. An invariant density can be found for some special values of parameters by constructing an *exceptional dual* (this is a dual map T^* which is not differentiably isomorphic).

(a) We consider the partition $0 < \frac{1}{4} < \frac{1}{2} < \frac{3}{4} < 1$ and the map T with the four inverse branches

$$V_\lambda(x) = \frac{1-2x}{4+\lambda x},\ V_\mu(x) = \frac{2+(2+\mu)x}{8+2\mu x},$$

$$V_\nu(x) = \frac{3-\nu x}{4+(2-2\nu)x},\ V_\tau(x) = \frac{3+(2+\tau)x}{4+\tau x}.$$

The map T admits the blocks

$$\lambda\lambda, \lambda\mu; \mu\lambda, \mu\mu, \mu\nu, \mu\tau; \nu\lambda, \nu\mu, \nu\nu, \nu\tau; \tau\lambda, \tau\mu.$$

A dual map T^* must admit the blocks

$$\lambda\lambda, \lambda\mu, \lambda\nu, \lambda\tau; \mu\lambda, \mu\mu, \mu\nu, \mu\tau; \nu\mu, \nu\nu; \tau\mu, \tau\nu.$$

Therefore no map $\psi(t) = \frac{c+dt}{a+bt}$ can exist such that $\psi \circ T = T^* \circ \psi$. We look for a dual with partition $\alpha < \xi < \theta < \eta < \beta$. One has to solve eight equations for nine parameters:

$$\alpha = \frac{\lambda-2\beta}{4+\beta},\ \frac{\lambda-2\alpha}{4+\alpha} = \xi = \frac{2\mu+(2+\mu)\alpha}{8+2\alpha},\ \frac{2\mu+(2+\mu)\beta}{8+2\beta} = \theta = \frac{2-2\nu-\nu\eta}{4+3\eta},$$

$$\frac{2-2\nu-\nu\xi}{4+3\xi} = \eta = \frac{\tau+(2+\tau)\xi}{4+3\xi},\ \frac{\tau+(2+\tau)\eta}{4+3\eta} = \beta.$$

If $4+\alpha \neq 0$ and $4+3\xi \neq 0$ one gets the equations

$$\alpha = \frac{2\lambda-2\mu}{6+\mu},\ \beta = \frac{-2\lambda+8\mu+\lambda\mu}{2\lambda+12},$$

$$\theta = \frac{8\mu^2+\lambda\mu^2+4\lambda\mu+40\mu-4\lambda}{2\lambda\mu+16\mu+12\lambda+96},\ 2-2\nu-\nu\xi = \tau+2\xi+\tau\xi.$$

We solve one special case, namely $\theta = 0$. We then get

$$\beta = -\frac{2\mu}{2+\mu},\ \eta = \frac{2-2\nu}{\nu}.$$

These equations lead to

$$\alpha = \frac{2\lambda+\lambda\mu+4\mu}{8+2\mu},\ \xi = \frac{-8+10\nu+2\nu^2}{6-6\nu+\nu^2},\ \tau = \frac{28-36\nu+8\nu^2}{-2+4\nu-\nu^2}.$$

From

$$\alpha = \frac{2\lambda+\lambda\mu+4\mu}{8+2\mu} = \frac{2\lambda-2\mu}{6+\mu},$$

we get

$$\lambda = \frac{40\mu + 8\mu^2}{4 - 4\mu - \mu^2}.$$

Now we choose $\mu = -1$ and obtain with $\nu = \frac{9-\sqrt{33}}{4}$ (a root of $2\nu^2 - 9\nu + 6 = 0$) the values

$$\alpha = -\frac{10}{7},\ \xi = -\frac{2}{3}, \theta = 0,\ \eta = \frac{-3+\sqrt{33}}{6},\ \beta = 2,$$

$$\lambda = -\frac{32}{7},\ \mu = -1,\ \nu = \frac{9-\sqrt{33}}{4},\ \tau = 1+\sqrt{33}.$$

The invariant density then is

$$h_I(x) = \frac{2}{1+2x} + \frac{10}{7-10x} \text{ for } 0 \le x < \frac{1}{2},$$

$$h_{II}(x) = \frac{2-2\nu}{\nu+(2-2\nu)x} + \frac{2}{3-2x} \text{ for } \frac{1}{2} < x \le 1.$$

We outline the connection with the jump map S which avoids the non-full cylinders labeled λ and τ. If g denotes the invariant density for S then

$$h_I(x) = g(x) + g(V_\tau x)\omega(\tau; x) + g(V_\lambda x)\omega(\lambda; x) + g(V_{\lambda\lambda}x)\omega(\lambda\lambda; x),$$

$$+g(V_{\tau\lambda}x\omega(\tau\lambda; \omega(\tau\lambda; x) + g(V_{\lambda\lambda\lambda}x)\omega(\lambda\lambda\lambda; x) + g(V_{\tau\lambda\lambda}x)\omega(\tau\lambda\lambda; x) + \ldots$$

$$h_{II}(x) = g(x).$$

(b) We consider again the partition $0 < \frac{1}{4} < \frac{1}{2} < \frac{3}{4} < 1$ and the map T with the four inverse branches

$$V_\lambda(x) = \frac{-1+2x}{4+\lambda-\lambda x},\ V_\mu(x) = \frac{2+2\mu x}{8+2\mu x},$$

$$V_\nu(x) = \frac{4+(2+3\nu)x}{8+4\nu x},\ V_\tau(x) = \frac{3+(2+\tau)x}{4+\tau x}.$$

This map admits the blocks

$$\lambda\nu, \lambda\tau; \mu\lambda, \mu\mu, \mu\nu, \mu\tau; \nu\lambda, \nu\mu, \nu\nu, \nu\tau; \tau\lambda, \tau\mu.$$

Then a dual map T^* must admit the blocks

$$\lambda\mu, \lambda\nu, \lambda\tau; \mu\mu, \mu\nu, \mu\tau; \nu\lambda, \nu\mu, \nu\nu; \tau\lambda, \tau\mu, \tau\nu.$$

Again, no map $\psi(t) = \frac{c+dt}{a+bt}$ can exist such that $\psi \circ T = T^* \circ \psi$. We look for a dual with partition $\alpha < \xi < \theta < \eta < \beta$. One has to solve eight equations for nine parameters:

$$\alpha = \frac{-\lambda+2\xi}{4\lambda-\xi}, \frac{-\lambda+2\beta}{4\lambda-\beta} = \xi = \frac{2\mu+(2+\mu)\xi}{8+2\xi}, \frac{2\mu+(2+\mu)\beta}{8+2\beta} = \theta = \frac{4\nu+(2+3\nu)\alpha}{8+4\alpha},$$

$$\frac{4\nu+(2+3\nu)\eta}{8+4\eta}=\eta=\frac{\tau+(2+\tau)\alpha}{4+3\alpha},\ \frac{\tau+(2+\tau)\eta}{4+3\eta}=\beta.$$

The two fixed points ξ and η provide a good start! The equations which relate to them are:

$$2\xi^2+\xi(6-\mu)-2\mu=0,$$

$$4\eta^2+\eta(6-3\nu)-4\nu=0.$$

We want to keep $\alpha=0$ and $\beta=1$ and find $\lambda=2\xi$, $\tau=4\eta$, and $\mu=\nu=1$. Then we have

$$\xi=\frac{-5+\sqrt{41}}{4},\ \eta=\frac{-3+\sqrt{73}}{8}.$$

The invariant density should satisfy

$$h_I(x)=h_I(V_\mu x)\omega(\mu;x)+h_{II}(V_\nu x)\omega(\nu;x)+h_{II}(V_\tau x)(\omega(\tau;x),\ 0\le x<\frac{1}{2},$$

$$h_I(x)=h_I(V_\lambda x)\omega(\lambda;x)+h_I(V_\mu x)\omega(\mu;x)+h_{II}(V_\nu x)\omega(\nu;x),\ \frac{1}{2}<x\le 1$$

and can be given as

$$h_I(x)=\frac{1-\xi}{(1+x)(1+\xi x)},\ h_{II}(x)=\frac{\eta}{1+\eta x}.$$

Again, we outline the connection with the jump map S which avoids the non-full cylinders labeled λ and τ. If g denotes the invariant density for S then

$$h_I(x)=g(x)+g(V_\tau x)\omega(\tau;x)+g(V_{\lambda\tau}x)\omega(\lambda\tau;x)+g(V_{\tau\lambda\tau}x)\omega(\tau\lambda\tau;x)+...$$

$$h_{II}(x)=g(x)+g(V_\lambda x)\omega(\lambda;x)+g(V_{\tau\lambda}x)\omega(\tau\lambda;x)+g(V_{\lambda\tau\lambda}x)\omega(\lambda\tau\lambda;x)+....$$

Acknowledgment. My thanks go to the referee whose remarks contributed essential improvements to my paper.

References

[1] K. Dajani, C. Kraaikamp, and N. D. S. Langeveld, Continued fraction expansions with variable numerators, *Ramanujan J.* **37** (2015), no. 3, 617–639.

[2] O. Jenkinson and M. Pollicott, Ergodic properties of the Bolyai-Rényi expansion, *Indag. Math. (N.S.)* **11** (2000), 399–418.

[3] F. Schweiger, Invariant measures for continued fractions with variable numerators, *Integers* **17** (2017), #A56.

[4] F. Schweiger, Invariant measures for Moebius maps with three branches, *J. Number Theory* **184** (2018), 206–215.

[5] F. Schweiger, *Continued Fractions and Their Generalizations: A Short History of f-Expansions*, Boston, Massachusetts, Docent Press, 2017.

[6] F. Schweiger, Invariant measures of piecewise fractional linear maps and piecewise quadratic maps, *Int. J. Number Theory* **14** (2018), 1559-1572.

[7] F. Schweiger, Differentiable equivalence of fractional linear maps, *Dynamics & Stochastics*, 237–247. IMS Lecture Notes Monogr. Ser., **48**, 2006.

[8] F. Schweiger, *Ergodic Theory of Fibred Systems and Metric Number Theory*, Oxford Science Publications, New York, The Clarendon Press, Oxford University Press, 1995.

3

DETERMINANTAL EXPRESSIONS FOR BERNOULLI POLYNOMIALS

Takashi Agoh

Department of Mathematics, Tokyo University of Science, Noda, Chiba, Japan

agoh_takashi@ma.noda.tus.ac.jp

Abstract

In this paper, we study determinantal expressions for Bernoulli polynomials by making use of certain polynomial analogues of the Saalschütz-Gelfand identity for Bernoulli numbers. As a result, we establish quite new determinantal expressions that provide us with various properties of these polynomials.

1. Introduction

The Bernoulli numbers B_n and polynomials $B_n(x)$, $n = 0, 1, 2, \ldots$, are defined by the generating functions

$$\begin{aligned}\mathcal{F}(t) &:= \frac{t}{e^t - 1} = \sum_{n=0}^{\infty} \frac{B_n t^n}{n!} \quad (|t| < 2\pi);\\ \mathcal{F}(t, x) &:= \frac{te^{xt}}{e^t - 1} = \sum_{n=0}^{\infty} \frac{B_n(x) t^n}{n!} \quad (|t| < 2\pi),\end{aligned} \tag{1.1}$$

respectively. They have numerous important applications in various branches of mathematics, most notably in number theory, combinatorics, numerical and asymptotic analysis, the theory of difference equations, and others.

It is easy to find that $B_0 = 1$, $B_1 = -1/2$, $B_2 = 1/6$, $B_3 = 0$, $B_4 = -1/30$, $\ldots$, and these numbers satisfy $B_{2n+1} = 0$ and $(-1)^{n-1} B_{2n} > 0$ for all $n \geq 1$. Further, since $\mathcal{F}(t, x) = \mathcal{F}(t) e^{xt}$, we see that $B_n(0) = B_n$, and $B_n(x)$ is expressed as

$$B_n(x) = \sum_{i=0}^{n} \binom{n}{i} B_i x^{n-i} \quad (n \geq 0). \tag{1.2}$$

Using this expression, it can be shown that for $n \geq 1$,

$$\text{(i)} \quad \frac{d}{dx}B_n(x) = nB_{n-1}(x); \quad \text{(ii)} \quad \int_0^x B_{n-1}(t)dt = \frac{B_n(x) - B_n}{n}. \tag{1.3}$$

Since $B_0(x) = 1$, the first property (i) in (1.3) tells us that the Bernoulli polynomials form an Appell sequence.

Many kinds of recurrence relations for these numbers and polynomials are known (see, e.g., [19, 14, 12, 8]). Among them, the most basic linear ones are

$$\begin{aligned} &\text{(i)} \qquad \sum_{i=0}^{n-1} \binom{n}{i} \frac{B_{i+1}}{i+1} + \frac{1}{n+1} = 0 \quad (n \geq 1); \\ &\text{(ii)} \qquad \sum_{i=0}^{n-1} \binom{n}{i} \frac{B_{i+1}(x)}{i+1} + \frac{1}{n+1} = x^n \quad (n \geq 1). \end{aligned} \tag{1.4}$$

These identities are easily obtained by expanding both sides of each of the functional relations $\mathcal{F}(t)(e^t - 1) = t$ and $\mathcal{F}(t, x)(e^t - 1) = te^{xt}$ into the Maclaurin power series and then comparing the coefficients of t^{n+1} on both sides, respectively.

The identity (1.4) (i) involves all the preceding Bernoulli numbers up to B_n. In contrast to this, the following *shortened* (or *incomplete*) recurrence relation was discovered by Saalschütz [18], and later by M. B. Gelfand [11]:

$$\begin{aligned} &\sum_{i=0}^{k} \binom{k}{i} \frac{B_{m+1+i}}{m+1+i} + (-1)^{k+m} \sum_{j=0}^{m} \binom{m}{j} \frac{B_{k+1+j}}{k+1+j} \\ &\qquad = \frac{(-1)^{m+1}}{k+m+1} \binom{k+m}{k}^{-1}, \end{aligned} \tag{1.5}$$

which is valid for arbitrary integers $k, m \geq 0$. An elementary proof and a brief historical overview can be found in [1, 2, 3]. The most remarkable feature of (1.5) is that the first $\min\{k, m\}$ Bernoulli numbers are completely missing.

Such type recurrences must go back to von Ettingshausen [10] in 1827 and later Stern [21] in 1878. Indeed, they first discovered the following surprising identity that involves only the second half of Bernoulli numbers up to B_{2m+1}:

$$\sum_{i=0}^{m+1} \binom{m+1}{i} (m+i+1) B_{m+i} = 0, \quad m \geq 0. \tag{1.6}$$

As will be mentioned details in the next Section 2, both (1.5) and (1.6) can be extended to Bernoulli polynomials and they play important roles in this paper.

There are many interesting approaches to searching for determinantal expressions for Bernoulli polynomials (see, e.g., [6, 5, 7, 16, 17] and the references therein). For instance, by making use of the fundamental difference equation

$$B_k(x+1) - B_k(x) = kx^{k-1} \quad (k \geq 1), \tag{1.7}$$

which is easily shown from the relation $\mathcal{F}(t.x+1)-\mathcal{F}(t,x)=te^{xt}$, we can deduce for $n \geq 1$,

$$\frac{B_n(x)}{n!} = \det \begin{bmatrix} 1 & 0 & 0 & \cdots & 0 & 1 \\ \frac{1}{2!} & 1 & 0 & \cdots & 0 & \frac{x}{1!} \\ \frac{1}{3!} & \frac{1}{2!} & 1 & \cdots & 0 & \frac{x^2}{2!} \\ \vdots & \vdots & \vdots & \ddots & \vdots & \vdots \\ \frac{1}{n!} & \frac{1}{(n-1)!} & \frac{1}{(n-2)!} & \cdots & 1 & \frac{x^{n-1}}{(n-1)!} \\ \frac{1}{(n+1)!} & \frac{1}{n!} & \frac{1}{(n-1)!} & \cdots & \frac{1}{2!} & \frac{x^n}{n!} \end{bmatrix}. \tag{1.8}$$

Indeed, writing (1.7) as

$$\sum_{i=0}^{k-1} \frac{B_i(x)}{i!(k-i)!} = \frac{x^{k-1}}{(k-1)!},$$

which is actually equivalent to (1.4) (ii) with $n=k-1$, consider a system of linear equations by taking $k=1,2,\ldots,n+1$. Then, according to Cramer's rule we can obtain (1.8) by using ordinary determinantal operations. See also [6, 7] for a slightly different proof of (1.8).

The main purpose of this paper is to study determinantal expressions for Bernoulli polynomials based on shortened recurrence relations. This paper is organized as follows: In Section 2 we first present a polynomial analogue of (1.5) (Lemma 2.1), and subsequently, applying it we deduce three kinds of recurrence relations (Lemma 2.2). Based on them, we establish, in Section 3, quite new determinantal expressions for Bernoulli polynomials. In Section 4, to confirm the effectiveness of our expressions, by making use of them we recover some of well-known basic properties of Bernoulli polynomials. In Section 5 we discuss special topics related to the golden ratio. We conclude this paper, in Section 6, with some additional remarks on determinantal expressions for Euler polynomials that have a intimate connection with those for Bernoulli polynomials.

We would like to point out beforehand that the methods we will use are quite elementary without use of intricate tools and it does not require any advanced knowledge other than basic linear algebra.

2. Recurrence Relations

The following identity is a polynomial analogue of Saalschütz-Gelfand's formula (1.5), in which the first $\min\{m,k\}$ Bernoulli polynomials are completely missing.

Lemma 2.1. *For integers $k, m \geq 0$ and $q \geq 1$ we have*

$$\begin{aligned}&\sum_{j=0}^{k}\binom{k}{j}\frac{B_{m+1+j}(x)}{m+1+j}q^{k-j}+\sum_{j=0}^{m}(-1)^{m+1+j}\binom{m}{j}\frac{B_{k+1+j}(x)}{k+1+j}q^{m-j}\\&\quad=\sum_{r=0}^{q-1}(x+r)^k(x+r-q)^m+\frac{(-1)^{m+1}}{k+m+1}\binom{k+m}{k}^{-1}q^{k+m+1}.\end{aligned}\tag{2.1}$$

It is clear that (2.1) reduces to (1.5) by setting $x = 0$ and $q = 1$. An elementary proof of (2.1) and more details surrounding can be found, e.g., in [1, 2].

As consequences of (2.1), we can deduce the following three kinds of shortened recurrence relations that (typically) involve only the second half of all the Bernoulli polynomials up to $B_{2k}(x)$ or $B_{2k-1}(x)$.

Lemma 2.2. *For an integer $k \geq 1$ we have*

$$\begin{aligned}&\text{(i)}\quad \sum_{i=0}^{\lfloor (k-1)/2\rfloor}\binom{k}{2i+1}\frac{B_{2(k-i)}(x)}{2(k-i)}=X_k(x);\\&\text{(ii)}\quad \sum_{i=0}^{\lfloor (k-1)/2\rfloor}\binom{k}{2i+1}B_{2(k-i)-1}(x)=Y_k(x);\\&\text{(iii)}\quad \sum_{i=0}^{\lfloor (k-1)/2\rfloor}\binom{k}{2i+1}(2(k-i)-1)B_{2(k-i-1)}(x)=Z_k(x),\end{aligned}\tag{2.2}$$

where

$$\begin{aligned}X_k(x):&=\frac{1}{2}(x^2-x)^k+\frac{(-1)^{k+1}}{2(2k+1)}\binom{2k}{k}^{-1};\\Y_k(x):&=\frac{k}{2}(2x-1)(x^2-x)^{k-1};\\Z_k(x):&=\frac{k}{2}\left\{(4k-2)x^2-(4k-2)x+k-1\right\}(x^2-x)^{k-2}\\&=\binom{2k}{2}(x^2-x)^{k-1}+\binom{k}{2}(x^2-x)^{k-2}.\end{aligned}\tag{2.3}$$

Proof. In particular, taking $m = k \geq 0$ and $q = 1$ in (2.1), we have

$$\sum_{j=0}^{k}\binom{k}{j}\left(1+(-1)^{k+1+j}\right)\frac{B_{k+1+j}(x)}{k+1+j}=(x^2-x)^k+\frac{(-1)^{k+1}}{2k+1}\binom{2k}{k}^{-1}.\tag{2.4}$$

Since all the terms on the left-hand side of this identity involving odd-index Bernoulli polynomials vanish, we easily see that (2.4) leads to (2.2) (i) after dividing by 2. In order to derive (2.2) (ii) and (iii), differentiate successively both sides of (2.2) (i) with respect to x based on (1.3) (i). □

As is easily seen, (2.2) (iii) is a polynomial analogue of von Ettingshausen-Stern's formula (1.6). Indeed, when $k > 2$, we can deduce (1.6) by setting $x = 0$.

3. Determinantal Expressions

In this section, based on shortened recurrence relations in (2.2) we establish new determinant expressions for Bernoulli polynomials. In addition, we discuss mutual relations between two adjacent even-index or odd-index Bernoulli polynomials.

For an integer $k \geq 1$, let us denote by $\boldsymbol{p}_k$ the row vector of length $\lfloor (k+1)/2 \rfloor$ constituted of alternate binomial coefficients, namely

$$\boldsymbol{p}_k := \begin{cases} \left(\binom{k}{0}, \binom{k}{2}, \ldots, \binom{k}{k-3}, \binom{k}{k-1}\right) & \text{if } k \text{ is odd;} \\ \left(\binom{k}{1}, \binom{k}{3}, \ldots, \binom{k}{k-3}, \binom{k}{k-1}\right) & \text{if } k \text{ is even.} \end{cases}$$

Using this vector for $k = 1, 2, \ldots, n$, consider the lower triangular matrix of order n such that

$$P_n := \begin{bmatrix} \boldsymbol{p}_1 & & & O \\ & \boldsymbol{p}_2 & & \\ & \boldsymbol{p}_3 & & \\ & & \boldsymbol{p}_4 & \\ & & \boldsymbol{p}_5 & \\ O & & & \ddots \\ & & & \boldsymbol{p}_n \end{bmatrix} \qquad (n \geq 1).$$

Since the diagonal entries are $\binom{1}{0}, \binom{2}{1}, \ldots, \binom{n}{n-1}$, we have $\det P_n = n!$.

Further, let $P_n(x)$ be the polynomial matrix, replacing the last column of P_n by $(X_1(x), \ldots, X_n(x))^{\mathrm{T}}$, where "T" means the transpose of a vector, namely

$$P_n(x) := \begin{bmatrix} \boldsymbol{p}_1 & & & & X_1(x) \\ & \boldsymbol{p}_2 & & & X_2(x) \\ & \boldsymbol{p}_3 & & O & X_3(x) \\ & & \boldsymbol{p}_4 & & \cdot \\ & & \boldsymbol{p}_5 & & \cdot \\ O & & & \ddots & \cdot \\ & & & \boldsymbol{p}_n^* & X_n(x) \end{bmatrix} \qquad (n \geq 1). \tag{3.1}$$

Here, we denote by $\boldsymbol{p}_n^*$ the row vector of length $\lfloor (n-1)/2 \rfloor$ obtained by removing the last entry $\binom{n}{n-1}$ from $\boldsymbol{p}_n$, that is,

$$\boldsymbol{p}_n^* := \begin{cases} \left(\binom{n}{0}, \binom{n}{2}, \dots, \binom{n}{n-3}\right) & \text{if } n \text{ is odd;} \\ \left(\binom{n}{1}, \binom{n}{3}, \dots, \binom{n}{n-3}\right) & \text{if } n \text{ is even.} \end{cases}$$

Similarly, let $Q_n(x)$ and $R_n(x)$ be the matrices, replacing the last column of $P_n(x)$ by $(Y_1(x), \dots, Y_n(x))^{\mathrm{T}}$ and $(Z_1(x), \dots, Z_n(x))^{\mathrm{T}}$, respectively.

Then, according to Cramer's rule, we obtain from Lemma 2.2 that for $n \geq 1$,

$$\begin{aligned} &\text{(i)} \quad \frac{B_{2n}(x)}{2n} = \frac{\det P_n(x)}{n!}; \quad \text{(ii)} \quad B_{2n-1}(x) = \frac{\det Q_n(x)}{n!}; \\ &\text{(iii)} \quad (2n-1)B_{2(n-1)}(x) = \frac{\det R_n(x)}{n!}. \end{aligned} \tag{3.2}$$

Now expand $\det P_n(x)$ along the first row. Since $\boldsymbol{p}_1 = (\binom{1}{0}) = (1)$ and the first column of $P_n(x)$ is $(1, O_{n-1})^{\mathrm{T}}$, where O_k means the zero vector of length k, we see that the $(1, n)$ minor of $P_n(x)$ completely vanishes. So that, denoting newly

$$\widetilde{P}_{n-1}(x) := \begin{bmatrix} \boldsymbol{p}_2 & & & & X_2(x) \\ \boldsymbol{p}_3 & & & & X_3(x) \\ & \boldsymbol{p}_4 & & O & \cdot \\ & \boldsymbol{p}_5 & & & \cdot \\ O & & \ddots & & \cdot \\ & & & \boldsymbol{p}_n^* & X_n(x) \end{bmatrix} \quad (n \geq 2), \tag{3.3}$$

we have $\det P_n(x) = \det \widetilde{P}_{n-1}(x)$. Similarly, denoting by $\widetilde{Q}_{n-1}(x)$ and $\widetilde{R}_{n-1}(x)$ the matrices replaced the last column of $\widetilde{P}_{n-1}(x)$ by

$$(Y_2(x), \dots, Y_n(x))^{\mathrm{T}} \quad \text{and} \quad (Z_2(x), \dots, Z_n(x))^{\mathrm{T}},$$

respectively, we have $\det Q_n(x) = \det \widetilde{Q}_{n-1}(x)$ and $\det R_n(x) = \det \widetilde{R}_{n-1}(x)$.

Therefore, from (3.2) we can deduce the following determinantal expressions:

Theorem 3.1. *We have $B_0(x) = Z_1(x)$, $B_1(x) = Y_1(x)$, and for $n \geq 2$,*

$$\begin{aligned} &\text{(i)} \quad B_{2n}(x) = \frac{2\det \widetilde{P}_{n-1}(x)}{(n-1)!}; \qquad \text{(ii)} \quad B_{2n-1}(x) = \frac{\det \widetilde{Q}_{n-1}(x)}{n!}; \\ &\text{(iii)} \quad B_{2(n-1)}(x) = \frac{\det \widetilde{R}_{n-1}(x)}{(2n-1)\cdot n!}. \end{aligned} \tag{3.4}$$

Since both (i) and (iii) in (3.4) are the expressions for Bernoulli polynomials with even-indices, it follows that

$$B_{2n}(x) = \frac{2\det \widetilde{P}_{n-1}(x)}{(n-1)!} = \frac{\det \widetilde{R}_n(x)}{(2n+1)\cdot(n+1)!} \quad (n \geq 2).$$

In addition, since $\frac{d}{dx}X_k(x) = Y_k(x)$ and $\frac{d}{dx}Y_k(x) = Z_k(x)$, we have

$$\frac{d}{dx}\det \widetilde{P}_{n-1}(x) = \det \widetilde{Q}_{n-1}(x) \quad\text{and}\quad \frac{d}{dx}\det \widetilde{Q}_{n-1}(x) = \det \widetilde{R}_{n-1}(x).$$

We next rephrase all the determinantal expressions in (3.4) in terms of $X_k(x)$, $Y_k(x)$ and $Z_k(x)$ as follows:

Theorem 3.2. *For $n \geq 2$, there exist uniquely $c_k \in \mathbb{Z}$, $k = 2, 3, \ldots, n$, such that*

$$\text{(i)}\quad B_{2n}(x) = \frac{2}{(n-1)!}\sum_{k=2}^{n} c_k X_k(x); \quad \text{(ii)}\quad B_{2n-1}(x) = \frac{1}{n!}\sum_{k=2}^{n} c_k Y_k(x);$$

$$\text{(iii)}\quad B_{2(n-1)}(x) = \frac{1}{(2n-1)\cdot n!}\sum_{k=2}^{n} c_k Z_k(x). \tag{3.5}$$

Proof. For an integer k with $2 \leq k \leq n$, let c_k be the $(k-1, n-1)$ cofactor of $\widetilde{P}_{n-1}(x)$. Then, it is clear that c_k is also the common $(k-1, n-1)$ cofactor of $\widetilde{Q}_{n-1}(x)$ and $\widetilde{R}_{n-1}(x)$. By expanding all the determinants appeared in (3.4) along their last columns, we can derive the expressions in (3.5). The uniqueness can be shown by the following inductive method. Suppose that there exists another integer $(n-1)$-tuple $(d_2, d_3, \ldots, d_n)$ satisfying (3.5) (i), that is,

$$\sum_{k=2}^{n} c_k X_k(x) = \sum_{k=2}^{n} d_k X_k(x).$$

Comparing the coefficients of the highest degree term x^{2n} on the both sides, we see that $c_n = d_n$; thus, the above relation reduces to the one replaced n by $n-1$. By the same argument, we can show that $c_{n-1} = d_{n-1}$. Continue such procedures until to arrive at $c_2 = d_2$. □

Since the sum parts on the right-hand sides of (i), (ii) and (iii) in (3.5) keep the same forms of linear combinations, it is possible to obtain $B_{2n-1}(x)$ and $B_{2n}(x)$ automatically by utilizing the linear form of $B_{2(n-1)}(x)$; and hence of $\det \widetilde{R}_{n-1}(x)$. We present below an easy example in the case when $n = 4$. At first, find the form of $\det \widetilde{R}_3(x)$. By a direct calculation one has

$$\det \widetilde{R}_3(x) = \det \begin{bmatrix} \binom{2}{1} & 0 & Z_2(x) \\ \binom{3}{0} & \binom{3}{2} & Z_3(x) \\ 0 & \binom{4}{1} & Z_4(x) \end{bmatrix} = 4Z_2(x) - 8Z_3(x) + 6Z_4(x), \tag{3.6}$$

and thus, $(c_2, c_3, c_4) = (4, -8, 6)$. Therefore, (3.6) yields from (3.5) (iii) that

$$\begin{aligned} B_6(x) &= \frac{1}{7\cdot 4!}(4Z_2(x) - 8Z_3(x) + 6Z_4(x)) \\ &= \frac{1}{42}\left\{42(x^2-x)^3 - 21(x^2-x)^2 + 1\right\} \\ &= \frac{1}{42}\left(42x^6 - 126x^5 + 105x^4 - 21x^2 + 1\right). \end{aligned}$$

By utilizing this form we automatically obtain from (3.5) (ii),

$$\begin{aligned} B_7(x) &= \frac{1}{4!}(4Y_2(x) - 8Y_3(x) + 6Y_4(x)) \\ &= \frac{1}{6}(2x-1)(x^2-x)\left\{3(x^2-x)^2 - 6(x^2-x) + 4\right\} \\ &= \frac{1}{6}\left(6x^7 - 21x^6 + 21x^5 - 7x^3 + x\right) \end{aligned}$$

and also from (3.5) (i),

$$\begin{aligned} B_8(x) &= \frac{2}{3!}(4X_2(x) - 8X_3(x) + 6X_4(x)) \\ &= \frac{1}{3}\left\{3(x^2-x)^4 - 4(x^2-x)^3 + 2(x^2-x)^2 + (6s_4 - 8s_3 + 4s_2)\right\} \\ &= \frac{1}{3}\left(3x^8 - 12x^7 + 14x^6 - 7x^4 + 2x^2 - 1/10\right), \end{aligned}$$

where s_k is the constant term of $X_k(x)$, i.e.,

$$s_k := X_k(0) = \frac{(-1)^{k+1}}{2(2k+1)}\binom{2k}{k}^{-1} \quad (k \geq 1). \tag{3.7}$$

We next search for mutual relations between two adjacent even-index or odd-index Bernoulli polynomials based on the expressions in (3.4). For this purpose, consider the matrix $\widetilde{P}_n(x)$ of order n and expand its determinant along the $(n-1)$th column (i.e., the second column from the last). Since this column vector is

$$\left(O_{n-2}, \binom{n}{n-1}, \binom{n+1}{n-2}\right)^{\mathrm{T}} = (O_{n-2}, n, (n+1)n(n-1)/6)^{\mathrm{T}},$$

we have

$$\det \widetilde{P}_n(x) = n \det L_{n-1}(x) - \frac{(n+1)n(n-1)}{6} \det \widetilde{P}_{n-1}(x), \tag{3.8}$$

where $L_{n-1}(x)$ is the matrix of order $n-1$ obtained by deleting the $(n-1)$th row and column from $\widetilde{P}_n(x)$, namely

$$L_{n-1}(x) := \begin{bmatrix} \boldsymbol{p}_2 & & & & X_2(x) \\ \boldsymbol{p}_3 & & O & & X_3(x) \\ & \boldsymbol{p}_4 & & & \vdots \\ & \boldsymbol{p}_5 & & & \\ & & \ddots & & \\ & O & & \boldsymbol{p}_{n-1} & X_{n-1}(x) \\ & & & \boldsymbol{p}_{n+1}^{**} & X_{n+1}(x) \end{bmatrix} \tag{3.9}$$

with

$$\boldsymbol{p}_{n+1}^{**} := \begin{cases} \left(\binom{n+1}{0}, \binom{n+1}{2}, \dots, \binom{n+1}{n-4}\right) & \text{if } n \text{ is even;} \\ \left(\binom{n+1}{1}, \binom{n+1}{3}, \dots, \binom{n+1}{n-4}\right) & \text{if } n \text{ is odd.} \end{cases} \tag{3.10}$$

Note that $\boldsymbol{p}_{n+1}^{**}$ is the row vector removed the last two entries $\binom{n+1}{n-2}$ and $\binom{n+1}{n}$ from $\boldsymbol{p}_{n+1}$. In particular, $\boldsymbol{p}_2^{**}$, $\boldsymbol{p}_3^{**}$ and $\boldsymbol{p}_4^{**}$ become the empty vectors of length 0.

Similar to the above, denote by $M_{n-1}(x)$ and $N_{n-1}(x)$ the submatrices of $\widetilde{Q}_n(x)$ and $\widetilde{R}_n(x)$ obtained by replacing the last column of $L_{n-1}(x)$ by

$$(Y_2(x), \dots, Y_{n-1}(x), Y_{n+1}(x))^{\mathrm{T}}; \quad (Z_2(x), \dots, Z_{n-1}(x), Z_{n+1}(x))^{\mathrm{T}},$$

respectively. Expanding $\det \widetilde{Q}_n(x)$ and $\det \widetilde{R}_n(x)$ along their $(n-1)$th columns, one can get the relations

$$\det \widetilde{Q}_n(x) = n \det M_{n-1}(x) - \frac{(n+1)n(n-1)}{6} \det \widetilde{Q}_{n-1}(x); \tag{3.11}$$

$$\det \widetilde{R}_n(x) = n \det N_{n-1}(x) - \frac{(n+1)n(n-1)}{6} \det \widetilde{R}_{n-1}(x). \tag{3.12}$$

Applying (3.8), (3.11) and (3.12) to the corresponding expressions in (3.4), we are able to obtain the following mutual relations between two adjacent even-index or odd-index Bernoulli polynomials.

Theorem 3.3. *For $n \geq 2$, we have*

$$\begin{aligned} &\text{(i)} \quad B_{2(n+1)}(x) + \frac{(n+1)(n-1)}{6} B_{2n}(x) = \frac{2 \det L_{n-1}(x)}{(n-1)!}; \\ &\text{(ii)} \quad B_{2n+1}(x) + \frac{n(n-1)}{6} B_{2n-1}(x) = \frac{n \det M_{n-1}(x)}{(n+1)!}; \\ &\text{(iii)} \quad B_{2n}(x) + \frac{n(n-1)(2n-1)}{6(2n+1)} B_{2(n-1)}(x) = \frac{n \det N_{n-1}(x)}{(2n+1) \cdot (n+1)!}. \end{aligned} \tag{3.13}$$

As is easily seen, using repeatedly these relations, it is possible to derive three types of recurrences for 'any' number of consecutive even-index or odd-index Bernoulli polynomials. For example, taking $n = k, k+1, \ldots, k+m-1$ in (3.13) (i) for any integers $k, m \geq 2$ and then adding up all of them, we get

$$\begin{aligned} B_{2(m+k)}(x) + \sum_{i=1}^{m-1} \frac{(k+m-i)^2+5}{6} B_{2(m+k-i)}(x) + \frac{k^2-1}{6} B_{2k}(x) \\ = 2\sum_{j=1}^{m} \frac{\det L_{k+m-1-j}(x)}{(k+m-1-j)!}. \end{aligned} \tag{3.14}$$

In the same way as the above, using (3.13) (ii) and (iii), slightly different types of recurrences of arbitrary length can be obtained.

If there exists a common zero θ of $B_{2(n+1)}(x)$ and $B_{2n}(x)$, then θ must satisfy $\det L_{n-1}(\theta) = \det N_n(\theta) = 0$ by (3.13) (i) and (iii). Similarly, we can say from (3.13) (ii) that if there exists a common zero ξ of $B_{2n+1}(x)$ and $B_{2n-1}(x)$, then $\det M_{n-1}(\xi) = 0$. Note that even if there exist such common zeros, they are never multiple zeros. Indeed, it has been proved by Dilcher [9] that Bernoulli polynomials have no multiple zeros. For the trivial common zeros $0, 1/2$ and 1 of $B_{2n+1}(x)$, see (h) in the next Section 4.

Using the above results on Bernoulli polynomials, we next discuss determinantal expressions for Bernoulli numbers.

Since $Y_1(0) = 1/2$ and $Y_k(0) = 0$ for $k \geq 2$, we see from (3.5) (ii) that $B_1 = -1/2$ and $B_{2n-1} = 0$ for $n \geq 2$. Next, by setting $x = 0$ in (3.4) (i) and (iii) we have for $n \geq 2$,

$$\text{(i)}\quad B_{2n} = \frac{2\det \widetilde{P}_{n-1}(0)}{(n-1)!}; \qquad \text{(ii)}\quad B_{2(n-1)} = \frac{\det \widetilde{R}_{n-1}(0)}{(2n-1)\cdot n!}. \tag{3.15}$$

Here, it is obvious that $\det \widetilde{R}_n(0) = 2n(n+1)(2n+1)\det \widetilde{P}_{n-1}(0)$.

To obtain more terse expressions, we now attempt to expand $\det \widetilde{P}_{n-1}(0)$ and $\det \widetilde{R}_{n-1}(0)$ along the first row and the last column, respectively. Setting $x = 0$ in (3.3) and using s_k defined in (3.7), we have

$$\widetilde{P}_{n-1}(0) = \begin{bmatrix} \boldsymbol{p}_2 & & & & s_2 \\ \boldsymbol{p}_3 & & & & s_3 \\ & \boldsymbol{p}_4 & & O & \cdot \\ & \boldsymbol{p}_5 & & & \cdot \\ O & & \ddots & & \cdot \\ & & & \boldsymbol{p}_n^* & s_n \end{bmatrix} \quad (n \geq 2). \tag{3.16}$$

First, expand $\det \widetilde{P}_{n-1}(0)$ along the first row in consideration of $\boldsymbol{p}_2 = (\binom{2}{1}) = (2)$

and $s_2 = -1/60$. Then we obtain the determinantal identity

$$\det \widetilde{P}_{n-1}(0) = 2\det C_{n-2} + \frac{(-1)^{n-1}}{60}\det D_{n-2} \quad (n \geq 2), \tag{3.17}$$

where C_{n-2} and D_{n-2} are the submatrices of $\widetilde{P}_{n-1}(0)$ defined by $C_0 := [0]$ and $D_0 := [1]$ by convention, and for $n \geq 3$, using the last entry $\binom{3}{2} = 3$ of $\boldsymbol{p}_3$,

$$C_{n-2} := \begin{bmatrix} 3 & & & s_3 \\ \boldsymbol{p}_4 & & O & s_4 \\ \boldsymbol{p}_5 & & & s_5 \\ & \boldsymbol{p}_6 & & \cdot \\ & \boldsymbol{p}_7 & & \cdot \\ O & & \ddots & \cdot \\ & & \boldsymbol{p}_n^* & s_n \end{bmatrix}; \quad D_{n-2} := \begin{bmatrix} \boldsymbol{p}_3 & & & \\ & \boldsymbol{p}_4 & & O \\ & \boldsymbol{p}_5 & & \\ & & \boldsymbol{p}_6 & \\ & & \boldsymbol{p}_7 & \\ O & & & \ddots \\ & & & \boldsymbol{p}_n^* \end{bmatrix},$$

respectively. In particular, $\det C_1 = \det[s_3] = 1/280$ and $\det D_1 = \det[\binom{3}{0}] = 1$. Next, expand $\det \widetilde{R}_{n-1}(0)$ along the last column $(1, O_{n-2})^{\mathrm{T}}$. Then, since $Z_2(0) = 1$ and $Z_k(0) = 0$ for $k \geq 3$, we get immediately

$$\det \widetilde{R}_{n-1}(0) = (-1)^n \det D_{n-2}. \tag{3.18}$$

Consequently, by applying (3.17) and (3.18) to (3.15), we can finally obtain the following two kinds of determinantal expressions for nonzero Bernoulli numbers.

Corollary 3.4. *For an integer $n \geq 2$ we have*

$$\begin{aligned} &\text{(i)} \quad B_{2n} = \frac{1}{(n-1)!}\left(4\det C_{n-2} + (-1)^{n-1}\det D_{n-2}/30\right); \\ &\text{(ii)} \quad B_{2(n-1)} = \frac{(-1)^n \det D_{n-2}}{(2n-1)\cdot n!}. \end{aligned} \tag{3.19}$$

Subsequently, we discuss mutual relations between adjacent nonzero Bernoulli numbers by utilizing the expressions in (3.19). Consider the submatrix T_{n-2} of order $n-2$ of D_{n-1} obtained by deleting the $(n-2)$th row and the $(n-1)$th column from D_{n-1}, namely

$$T_{n-2} := \begin{bmatrix} \boldsymbol{p}_3 & & & \\ & \boldsymbol{p}_4 & O & \\ & \boldsymbol{p}_5 & & \\ & & \ddots & \\ & O & \boldsymbol{p}_{n-1} & \\ & & & \boldsymbol{p}_{n+1}^{**} \end{bmatrix} \quad (n \geq 2),$$

where $\boldsymbol{p}_{n+1}^{**}$ is the row vector defined in (3.10). Since both $\boldsymbol{p}_3^{**}$ and $\boldsymbol{p}_4^{**}$ are the empty vectors of length 0, assume that $T_0 = T_1 = [0]$ as convention.

Using the above matrices, we can state the following two kinds of mutual relations between adjacent nonzero Bernoulli numbers:

Corollary 3.5. *For $n \geq 2$ we have*

$$\begin{aligned} &\text{(i)} \quad B_{2n} + \frac{n(2n-1)}{30} B_{2(n-1)} = \frac{4 \det C_{n-2}}{(n-1)!}; \\ &\text{(ii)} \quad B_{2n} + \frac{n(n-1)(2n-1)}{6(2n+1)} B_{2(n-1)} = \frac{(-1)^n n \det T_{n-2}}{(2n+1) \cdot (n+1)!}. \end{aligned} \tag{3.20}$$

Proof. In order to obtain (3.20) (i), we have only to eliminate $\det D_{n-2}$ from two expressions in (3.19). For the proof of (3.20) (ii), consider the matrix D_{n-1} of order $n-1$ and expand its determinant along the last column

$$\left(O_{n-3}, \binom{n}{n-1}, \binom{n+1}{n-2}\right)^{\mathrm{T}} = (O_{n-3}, n, (n+1)n(n-1)/6)^{\mathrm{T}}.$$

Then we have

$$\det D_{n-1} = \frac{(n+1)n(n-1) \det D_{n-2}}{6} - n \det T_{n-2}, \tag{3.21}$$

and this leads to (3.20) (ii) from (3.19) (ii). □

By repeatedly making use of the expressions in (3.20), we can also derive two kinds of recurrence relations for 'any' number of consecutive nonzero Bernoulli numbers. On the occasion of practical use of them, it requires in advance the knowledge of some corresponding values of $\det C_n$ or $\det T_n$. Since $\det C_n$ can be obtained by using $\det D_n$, it is desirable to explore a method of how to efficiently calculate $\det D_n$ and $\det T_n$. As is observed empirically, they are growing so rapidly.

The following is a list of the first few values of $\det D_n$ and $\det T_n$ with $\det D_0 = 1$ and $\det T_0 = 0$ by convention:

n	0	1	2	3	4	5	6	7	8
$\det D_n$	1	1	4	36	600	16584	705600	43751232	3790108800
$\det T_n$	0	0	1	24	736	31872	1939896	162226560	18182807424

Here, we note that each $\det T_n$ can be calculated by using $\det D_n$ and $\det D_{n+1}$. Indeed, the relation (3.21) tells us that

$$\det T_n = \frac{(n+3)(n+1)}{6} \det D_n - \frac{1}{n+2} \det D_{n+1} \quad (n \geq 0).$$

A square matrix $H_n = (h_{i,j})_{1 \leq i,j \leq n}$ is called a lower *Hessenberg matrix* if $h_{i,j} = 0$ for all i, j with $j - i > 1$ and its determinant can be calculated based on a very effective recursive algorithm. As a matter of fact, letting $\det H_0 = 1$ by convention, it is easy to show by induction that $\det H_1 = h_{1,1}$ and for $n \geq 2$,

$$\det H_n = h_{n,n} \det H_{n-1} + \sum_{k=1}^{n-1} (-1)^{n-k} h_{n,k} \prod_{j=k}^{n-1} h_{j,j+1} \det H_{k-1}. \tag{3.22}$$

Since both D_n and T_n are just the lower Hessenberg-type matrices, (3.22) is available for computing their determinants recursively. However, it is needed to calculate successively the determinants of submatrices of D_n and T_n corresponding to $H_1, H_2, \ldots, H_{n-1}$. Note that D_k (resp. T_k) is not always the submatrix of D_n (resp. T_n) corresponding to H_k for each $k = 1, 2, \ldots, n-1$.

By way of experiment, we now attempt to utilize (3.22) for computing $\det T_n$. Denoting $T_n = (t_{i,j})_{1 \leq i,j \leq n}$ with the (i, j) entry $t_{i,j}$, we see from the definition of T_n that

$$\begin{aligned} &t_{k,1} = t_{k,2} = \cdots = t_{k,\lfloor k/2 \rfloor} = 0, \quad k = 2, 3, \ldots, n-1; \\ &t_{n,1} = t_{n,2} = \cdots = t_{n,n-\lfloor n/2 \rfloor} = 0. \end{aligned} \tag{3.23}$$

For each $k = 1, 2, \ldots, n-1$, let $T'_k := (t_{i,j})_{1 \leq i,j \leq k}$ be the submatrix of order k of T_n corresponding to H_k. Taking account of the conditions in (3.23) and using (3.22), we can derive the following recursive formulas for $\det T'_k$ and $\det T_n$.

Lemma 3.6. *For simplicity, denote $\alpha_k := \det T'_k$. Then, for an integer $n \geq 2$ we have, with $\alpha_0 = 1$ by convention,*

$$\begin{aligned} &\text{(i)} \quad \alpha_k = t_{k.k} \alpha_{k-1} + \sum_{l=\lfloor k/2 \rfloor+1}^{k-1} (-1)^{k-l} t_{k,l} \prod_{j=l}^{k-1} t_{j,j+1} \alpha_{l-1} \quad (1 \leq k \leq n-1); \\ &\text{(ii)} \quad \det T_n = t_{n,n} \alpha_{n-1} + \sum_{k=n+1-\lfloor n/2 \rfloor}^{n-1} (-1)^{n-k} t_{n,k} \prod_{j=k}^{n-1} t_{j,j+1} \alpha_{k-1}. \end{aligned} \tag{3.24}$$

Here, in particular, $t_{k,k} = \binom{k+2}{k-1}$ unless $k = n$ and $t_{n,n} = \binom{n+3}{n-2}$.

To illustrate the procedure of how to calculate $\det T_n$ based on Lemma 3.6, we present below a concrete example for $n = 2, 3, \ldots, 6$. By applying the recursive algorithm (3.24) (i), the following values of α_k $(1 \leq k \leq 5)$ can be easily obtained:

$$\begin{aligned} &\alpha_1 = \binom{3}{0} \alpha_0 = 1; \quad \alpha_2 = \binom{4}{1} \alpha_1 = 4; \quad \alpha_3 = \binom{5}{2} \alpha_2 - \binom{5}{0} \binom{4}{3} \alpha_1 = 36; \\ &\alpha_4 = \binom{6}{3} \alpha_3 - \binom{6}{1} \binom{5}{4} \alpha_2 = 600; \\ &\alpha_5 = \binom{7}{4} \alpha_4 - \binom{7}{2} \binom{6}{5} \alpha_3 + \binom{7}{0} \binom{5}{4} \binom{6}{5} \alpha_2 = 16584. \end{aligned}$$

By using these values, it is shown from (3.24) (ii) that

$$\det T_2 = \binom{5}{0}\alpha_1 = 1; \quad \det T_3 = \binom{6}{1}\alpha_2 = 24;$$
$$\det T_4 = \binom{7}{2}\alpha_3 - \binom{7}{0}\binom{5}{4}\alpha_2 = 736;$$
$$\det T_5 = \binom{8}{3}\alpha_4 - \binom{8}{1}\binom{6}{5}\alpha_3 = 31872;$$
$$\det T_6 = \binom{9}{4}\alpha_5 - \binom{9}{2}\binom{7}{6}\alpha_4 + \binom{9}{0}\binom{6}{5}\binom{7}{6}\alpha_3 = 1939896,$$

as indicated in the above table.

4. Basic Properties

In this section, in order to confirm the effectiveness of our determinantal expressions mentioned in Section 3, we recover some of basic properties of Bernoulli polynomials based on these expressions, not relying on the generating function $\mathcal{F}(t, x)$.

We begin by proving the following symmetric property:

(a) $B_n(1-x) = (-1)^n B_n(x) \quad (n \geq 0)$.

Proof. The case $n = 0$ is trivial, so assume $n \geq 1$. Since $(1-x)^2-(1-x) = x^2-x$ and $2(1-x)-1 = -(2x-1)$, we have $X_k(1-x) = X_k(x)$ and $Y_k(1-x) = -Y_k(x)$ for $k \geq 1$, which imply $\det \widetilde{P}_{n-1}(1-x) = \det \widetilde{P}_{n-1}(x)$ and $\det \widetilde{Q}_{n-1}(1-x) = -\det \widetilde{Q}_{n-1}(x)$, respectively. So that (a) follows immediately from (3.4) (i) and (ii). □

(b) $B_n(x+1) = (-1)^n B_n(-x) \quad (n \geq 0)$.

Proof. Since $(x+1)^2 - (x+1) = x^2 + x$ and $2(x+1) - 1 = 2x + 1$, we have for $k \geq 1$,

$$X_k(x+1) = \frac{1}{2}(x^2+x)^k + s_k = X_k(-x);$$
$$Y_k(x+1) = \frac{k}{2}(2x+1)(x^2+x)^{k-1} = -Y_k(-x),$$

which lead to

$$\det \widetilde{P}_{n-1}(x+1) = \det \widetilde{P}_{n-1}(-x); \quad \det \widetilde{Q}_{n-1}(x+1) = -\det \widetilde{Q}_{n-1}(-x),$$

respectively. Thus, make use of (3.4) (i) and (ii) in order to deduce (b). □

The following is the fundamental difference equation as mentioned in (1.7).

(c) $B_n(x+1) - B_n(x) = nx^{n-1} \quad (n \geq 0)$.

Proof. In view of (b), we see that (c) is equivalent to $(-1)^n B_n(-x) - B_n(x) = nx^{n-1}$. By a direct calculation we have

$$X_k(-x) - X_k(x) = \frac{1}{2}\left((x^2+x)^k - (x^2-x)^k\right) = \frac{1}{2}\sum_{i=0}^{k}\binom{k}{i}\left(1-(-1)^{k-i}\right)x^{k+i}$$
$$= \sum_{j=0}^{\lfloor (k-1)/2 \rfloor}\binom{k}{2j+1}x^{2k-1-2j} \quad (k \geq 2).$$

Here, we notice the fact that the ordered array of binomial coefficients is exactly the same as that of coefficients of $B_{2(k-i)}(x)/2(k-i)$, $i = 0, 1, \dots, \lfloor (k-1)/2 \rfloor$, on the left-hand side of (2.2) (i). Assuming that $n \geq 2$ is an even integer, consider the matrix $A_{n-1}(x)$ of order $n-1$, replacing the last column of $\widetilde{P}_{n-1}(x)$ by

$$(X_2(-x) - X_2(x), \dots, X_n(-x) - X_n(x))^{\mathrm{T}}.$$

Multiply the jth column of $\det A_{n-1}(x)$ by x^{2j+1} and subtract it from the last column for each $j = 1, 2, \dots, n-2$. Then, from the above fact we see that the last column of $\det A_{n-1}(x)$ becomes $(O_{n-2}, nx^{2n-1})^{\mathrm{T}}$. So that, expanding $\det A_{n-1}(x)$ along this column and using the last entry $\binom{k}{k-1} = k$ of $\boldsymbol{p}_k$ $(2 \leq k \leq n-1)$, we get

$$\det A_{n-1}(x) = nx^{2n-1}\det\begin{bmatrix} \boldsymbol{p}_2 & & & O \\ \boldsymbol{p}_3 & & & \\ & \boldsymbol{p}_4 & & \\ & \boldsymbol{p}_5 & & \\ O & & \ddots & \\ & & & \boldsymbol{p}_{n-1} \end{bmatrix} = nx^{2n-1}\prod_{k=2}^{n-1} k = n!x^{2n-1},$$

which yields from (3.4) (i) that

$$B_{2n}(-x) - B_{2n}(x) = \frac{2\det A_{n-1}(x)}{(n-1)!} = 2nx^{2n-1},$$

and this proves (c) in the case when n is even. The proof for an odd $n \geq 1$ can be given in almost the same way as done above by using $Y_k(x)$ instead of $X_k(x)$. □

Incidentally, by setting $x = 0$ in (c), we have

$$B_n(1) - B_n(0) = \begin{cases} 1 & \text{if } n = 1; \\ 0 & \text{if } n = 0 \text{ or } n \geq 2, \end{cases}$$

which tells us that $B_1(1) = 1/2$ and $B_n(1) = B_n$ otherwise.

(d) $B_n(x + 1/2) = (-1)^n B_n(-x + 1/2)$ $(n \geq 0)$.

Proof. Take $x - 1/2$ in place of x in (b). □

This property shows that $B_n(x+1/2)$ is an even or an odd function of x depending on the parity of n. We may prove (d) independently of (b) by using the facts that

$$X_k(x+1/2) = \frac{1}{2}(x^2 - 1/4)^k + s_k \text{ and } Y_k(x+1/2) = kx(x^2 - 1/4)^{k-1}.$$

The following two properties are the repetition of (1.3) (i) and (ii).

(e) $\dfrac{d}{dx}B_n(x) = nB_{n-1}(x) \quad (n \geq 1).$

Proof. Since $\frac{d}{dx}X_k(x) = Y_k(x)$ and $\frac{d}{dx}Y_k(x) = Z_k(x)$, from Theorem 3.2 we have

$$\frac{d}{dx}B_{2n}(x) = \frac{2}{(n-1)!}\sum_{k=2}^{n} c_k Y_k(x) = 2nB_{2n-1}(x);$$

$$\frac{d}{dx}B_{2n-1}(x) = \frac{1}{n!}\sum_{k=2}^{n} c_k Z_k(x) = (2n-1)B_{2(n-1)}(x),$$

and thus, (e) follows. □

(f) $\displaystyle\int_0^x B_n(t)dt = \frac{B_{n+1}(x) - B_{n+1}}{n+1} \quad (n \geq 0).$

Proof. Since $\int_0^x Y_k(t)dt = X_k(x) - X_k(0)$ and $\int_0^x Z_k(t)dt = Y_k(x) - Y_k(0)$ for $k \geq 2$, it follows from Theorem 3.2 that

$$\begin{aligned}\int_0^x B_{2n-1}(t)dt &= \frac{1}{n!}\sum_{k=2}^{n} c_k \int_0^x Y_k(t)dt = \frac{1}{n!}\sum_{k=2}^{n} c_k\{X_k(x) - X_k(0)\} \\ &= \frac{B_{2n}(x) - B_{2n}(0)}{2n};\end{aligned}$$

$$\begin{aligned}\int_0^x B_{2(n-1)}(t)dt &= \frac{1}{(2n-1)\cdot n!}\sum_{k=2}^{n} c_k \int_0^x Z_k(t)dt \\ &= \frac{1}{(2n-1)\cdot n!}\sum_{k=2}^{n} c_k\{Y_k(x) - Y_k(0)\} = \frac{B_{2n-1}(x) - B_{2n-1}(0)}{2n-1},\end{aligned}$$

which imply (f) by summarizing the both cases. □

(g) (1) $\displaystyle\int_0^1 B_n(x)dx = 0 \quad (n \geq 1)$; (2) $\displaystyle\int_0^{1/2} B_n(x)dx = 0 \quad (n \geq 2 \text{ an even}).$

Proof. Since $\int_0^1 Y_k(x)dx = [X_k(x)]_0^1 = 0$ for $k \geq 1$ and $\int_0^1 Z_k(x)dx = [Y_k(x)]_0^1 = 0$ for $k \geq 2$, we obtain (1) immediately from (3.5) (ii) and (iii). Further, since $\int_0^{1/2} Z_k(x)dx = [Y_k(x)]_0^{1/2} = 0$ for $k \geq 2$, we can deduce (2) from (3.5) (iii). □

(h) (Trivial zeros) $B_1(1/2) = 0$, $B_{2n+1}(0) = B_{2n+1}(1/2) = B_{2n+1}(1) = 0$ $(n \geq 1)$.

Proof. Since $(2x-1) \mid 2Y_1(x)$ and $(2x-1)(x^2-x) \mid 2Y_k(x)$ for $k \geq 2$, the results obviously hold true from (3.5) (ii). □

It should be noted here that Nörlund [15, p. 22] has shown that these trivial zeros are all the zeros of $B_{2n+1}(x)$, $n \geq 1$, in the interval $[0,1]$. Furthermore, Inkeri [13] has proved that these are the only rational zeros of any Bernoulli polynomial.

Although all the above trivial zeros of $B_{2n+1}(x)$ are single ones, from (3.5) (ii) one can see that the improved polynomials

$$\mathcal{B}_{2n+1}(x) := (n+1)!B_{2n+1}(x) + (-1)^n(2x-1)(x^2-x)\det D_{n-1},\ n = 1,2,3,\ldots,$$

have always the zeros $0, 1/2$ and 1 whose multiplicities are $2, 1$ and 2, respectively, because of $c_2 = (-1)^{n+1}\det D_{n-1}$ and $(2x-1)x^2(x-1)^2 \mid Y_k(x)$ for all $k \geq 3$. For example, if we take $n = 2$ and 3, then

$$\mathcal{B}_5(x) = 3!B_5(x) + (2x-1)(x^2-x)\det\left[\,\tbinom{3}{0}\,\right] = 3(2x-1)x^2(x-1)^2;$$

$$\begin{aligned}\mathcal{B}_7(x) &= 4!B_7(x) - (2x-1)(x^2-x)\det\begin{bmatrix}\binom{3}{0} & \binom{3}{2}\\ 0 & \binom{4}{1}\end{bmatrix}\\ &= 12(2x-1)x^2(x-1)^2(x+1)(x-2).\end{aligned}$$

It is possible to prove even more properties of Bernoulli polynomials by applying our determinantal expressions, but let us move to the next subject.

5. An Encounter With the Golden Ratio

In this section we will talk about some special topics related to the golden ratio referring to the discussion in [4].

Although the zeros of Bernoulli polynomials have been studied quite extensively from various points of view, it is very curious that the concretely known odd-index Bernoulli polynomial that has non-trivial zeros is only $B_{11}(x)$. As a matter of fact, denoting $\phi := (1+\sqrt{5})/2$ (the golden ratio), it can be shown that

Theorem 5.1. *We have $B_{11}(\phi) = B_{11}(\widehat{\phi}) = 0$, where $\widehat{\phi}$ is the conjugate of ϕ.*

Proof. Since $B_n(x) \in \mathbb{Q}[x]$, it suffices to prove only $B_{11}(\phi) = 0$. From (3.4) (ii) we have $B_{11}(\phi) = \det\widetilde{Q}_5(\phi)/5!$. In addition, since $\phi^2 - \phi = 1$, it is clear that

$$Y_k(\phi) = \frac{k}{2}(2\phi-1)(\phi^2-\phi)^{k-1} = \frac{2\phi-1}{2}k\ \ (k \geq 1).$$

Thus, by a direct calculation we can confirm that

$$\det \widetilde{Q}_5(\phi) = \det \begin{bmatrix} \binom{2}{1} & 0 & 0 & 0 & Y_2(\phi) \\ \binom{3}{0} & \binom{3}{2} & 0 & 0 & Y_3(\phi) \\ 0 & \binom{4}{1} & \binom{4}{3} & 0 & Y_4(\phi) \\ 0 & \binom{5}{0} & \binom{5}{2} & \binom{5}{4} & Y_5(\phi) \\ 0 & 0 & \binom{6}{1} & \binom{6}{3} & Y_6(\phi) \end{bmatrix} = \frac{2\phi - 1}{2} \det \begin{bmatrix} 2 & 0 & 0 & 0 & 2 \\ 1 & 3 & 0 & 0 & 3 \\ 0 & 4 & 4 & 0 & 4 \\ 0 & 1 & 10 & 5 & 5 \\ 0 & 0 & 6 & 20 & 6 \end{bmatrix} = 0,$$

which completes the proof. □

Of course we may prove the above fact by directly substituting $x = \phi$ into (1.2) with $n = 11$, but it requires in advance the knowledge of B_n for $n = 0, 1, \ldots, 11$.

Since $X_k(\phi) = 1/2 + s_k$, we see from (3.5) (i) that all the values of even-index Bernoulli polynomials at $x = \phi$ are rational numbers, i.e., $B_{2n}(\phi) \in \mathbb{Q}$ for all $n \geq 0$. More generally, if θ_b is a real or complex root of the quadratic equation $x^2 - x = b$ for any given $b \in \mathbb{Q}$, then the same is true for the values at $x = \theta_b$, i.e., $B_{2n}(\theta_b) \in \mathbb{Q}$, because $B_0(\theta_b) = 1$ and $X_k(\theta_b) \in \mathbb{Q}$ for $k \geq 1$. Needless to say, there is no such guarantee for odd-index Bernoulli polynomials. However, we can state the following fact by applying (3.4) (ii).

($*$) *There exist infinitely many $b \in \mathbb{Q}$ such that $B_{2n-1}(\theta_b) \in \mathbb{Q}$ for all $n \geq 1$.*

Indeed, if we put $b = r(r+1)$ for an integer $r \geq 1$, then $\sqrt{4b+1} = 2r+1$; thus, we have for $k \geq 1$,

$$\frac{2}{k} Y_k(\theta_b) = (2\theta_b - 1)(\theta_b^2 - \theta_b)^{k-1} = \sqrt{4b+1} \cdot b^{k-1} = (2r+1)(r^2+r)^{k-1} \in \mathbb{Z},$$

which implies that $\det \widetilde{Q}_{n-1}(\theta_b) = n! B_{2n-1}(\theta_b) \in \mathbb{Z}$ for all $n \geq 1$ from (3.4) (ii). Since one can give infinitely many such the integers b, the above ($*$) holds true.

We cannot say for certain at this time, but the possibility of that the above θ_b with $b \neq 0, 1$ satisfies either one of the equations $\det \widetilde{P}_{n-1}(x) = 0, \det \widetilde{Q}_{n-1}(x) = 0$ and $\det \widetilde{R}_{n-1}(x) = 0$ for some $n \geq 2$ cannot be denied, as well as in the case of $\det \widetilde{Q}_5(\phi) = 0$ for $b = 1$ and $n = 6$. If it happens once, then we are able to find a new non-trivial zero of the corresponding Bernoulli polynomial. Just to be sure, it may be worth mentioning that for $k \geq 2$,

$$X_k(\theta_b) = \frac{1}{2} b^k + s_k; \; Y_k(\theta_b) = \frac{k}{2} \sqrt{1+4b} \cdot b^{k-1}; \; Z_k(\theta_b) = \binom{2k}{2} b^{k-1} + \binom{k}{2} b^{k-2}.$$

Next, we discuss a certain infinite series constructed by Bernoulli numbers that has a closed-form expression in terms of the golden ratio. The series we will deal with is not directly related to our determinantal expressions, but it concerns with the special case of (1.5) for $k = m$; that is, the identity (2.2) (i) with $x = 0$.

We prepare in advance the following lemma.

Lemma 5.2. *It follows that*

$$\sum_{k=0}^{\infty} \frac{(-1)^k}{2k+1} \binom{2k}{k}^{-1} = \frac{4 \log \phi}{\sqrt{5}}. \tag{5.1}$$

Proof. For any numbers $x, y \in \mathbb{C}$ satisfying $\mathrm{Re}(x) > 0$ and $\mathrm{Re}(y) > 0$, denote by $B(x, y)$ the beta function defined by

$$B(x,y) := \int_0^1 t^{x-1}(1-t)^{y-1} dt = \frac{\Gamma(x)\Gamma(y)}{\Gamma(x+y)},$$

where Γ is the Gamma function. Setting here $x = y = k+1$ for an integer $k \geq 0$, we have

$$B(k+1, k+1) = \frac{\Gamma(k+1)\Gamma(k+1)}{\Gamma(2k+2)} = \frac{(k!)^2}{(2k+1)!} = \frac{1}{2k+1}\binom{2k}{k}^{-1}.$$

Therefore, the alternating sum taken over $k = 0, 1, 2, \ldots$ can be given by

$$\begin{aligned}\sum_{k=0}^{\infty}(-1)^k B(k+1,k+1) &= \int_0^1 \sum_{k=0}^{\infty}(-1)^k t^k (1-t)^k dt \\ &= \int_0^1 \frac{1}{1+t-t^2} dt = \frac{4\log\phi}{\sqrt{5}},\end{aligned}$$

which proves (5.1). □

Note that (5.1) is not new. Indeed, Sprugnoli [20] has derived a number of closed formulas for infinite sums involving the reciprocals of central binomial coefficients; we can find (5.1) in [20, Theorem 3.5]). See also [4].

For given integers m, a with $m \geq a \geq 1$ let us denote

$$\gamma_{m,a} := \sum_{i=0}^{a-1} \binom{m+i}{2i+1}.$$

In particular, when $a = m$, we simply write as $\gamma_m := \gamma_{m,m}$, i.e., in plain terms,

$$\gamma_m = \binom{m}{1} + \binom{m+1}{3} + \cdots + \binom{2m-1}{2m-1} \quad (m \geq 1).$$

Using such binomial sums, we will prove the following formula:

Theorem 5.3. *We have*

$$\sum_{m=1}^{\infty} \gamma_m \frac{B_{2m}}{2m} = \frac{1}{2} - \frac{2\log\phi}{\sqrt{5}} \approx 0.069591059\cdots. \tag{5.2}$$

Proof. Setting $x = 0$ in (2.2) (i), we get

$$\sum_{i=0}^{\lfloor (k-1)/2 \rfloor} \binom{k}{2i+1} \frac{B_{2(k-i)}}{2(k-i)} = s_k \quad (k \geq 1), \tag{5.3}$$

where s_k is the constant term of $X_k(x)$ defined in (3.7). For an integer $a \geq 1$, take $k = 1, 2, \ldots, a$ in (5.3) and add up all of them. Then, gathering the terms involving the same Bernoulli numbers in one place, the sum of the left-hand side of (5.3) can be written accurately as

$$\begin{aligned} \sum_{k=1}^{a} \sum_{i=0}^{\lfloor (k-1)/2 \rfloor} \binom{k}{2i+1} \frac{B_{2(k-i)}}{2(k-i)} &= \sum_{j=1}^{\lfloor (a+1)/2 \rfloor} \gamma_j \frac{B_{2j}}{2j} \\ &+ \sum_{j=\lfloor (a+1)/2 \rfloor + 1}^{a} \gamma_{j,a+1-j} \frac{B_{2j}}{2j}. \end{aligned} \tag{5.4}$$

Here we notice that as a increases the terms in the second sum on the right-hand side of (5.4) are embedded successively into the first sum, because the coefficient of $B_{2m}/2m$ grows taking a step according to $0, \gamma_{m,1}, \gamma_{m.2}, \ldots, \gamma_{m,m-1}$, and eventually, it arrives at

$$\gamma_{m,m-1} + \binom{2m-1}{2m-1} = \gamma_m.$$

From the above, we see that the limiting sum of the left-hand side of (5.4) as $a \to \infty$ can be expressed as follows:

$$\lim_{a \to \infty} \sum_{k=1}^{a} \sum_{i=0}^{\lfloor (k-1)/2 \rfloor} \binom{k}{2i+1} \frac{B_{2(k-i)}}{2(k-i)} = \sum_{j=1}^{\infty} \gamma_j \frac{B_{2j}}{2j}. \tag{5.5}$$

On the other hand, from (5.1) the limiting sum of the right-hand side of (5.3) as $k \to \infty$ is given by

$$\lim_{a \to \infty} \sum_{k=1}^{a} s_k = \sum_{k=1}^{\infty} s_k = -s_0 + \sum_{k=0}^{\infty} s_k = \frac{1}{2} - \frac{2 \log \phi}{\sqrt{5}}. \tag{5.6}$$

So that, by equating (5.4) and (5.6) we can derive (5.2). □

6. Additional Remarks

In this final section, as additional remarks, we wish to make a brief summary on determinantal expressions for Euler polynomials that have an intimate connection with those for Bernoulli polynomials discussed in Section 3.

The Euler polynomials $E_n(x)$, $n = 0, 1, 2, \ldots$, are defined by means of the generating function

$$\mathcal{E}(t,x) := \frac{2e^{xt}}{e^t+1} = \sum_{n=0}^{\infty} \frac{E_n(x)t^n}{n!} \quad (|t| < \pi).$$

A large number of recurrence relations for Euler polynomials have been developed over the years in parallel with those for Bernoulli polynomials (see, e.g., [19, 14, 12]). Among them, the most basic linear one is

$$\sum_{i=0}^{n} \binom{n}{i} E_i(x) + E_n(x) = E_n(x+1) + E_n(x) = 2x^n \quad (n \geq 0), \tag{6.1}$$

which can be easily shown from the relation $\mathcal{E}(t, x+1) + \mathcal{E}(t,x) = 2e^{xt}$.

Many kinds of determinantal expressions for Euler polynomials can be derived by making use of various recurrence relations. As an easy example, dividing (6.1) by $2 \cdot n!$, we have

$$\frac{1}{n!}E_n(x) + \frac{1}{2}\sum_{i=0}^{n-1} \frac{1}{i!(n-i)!}E_i(x) = \frac{x^n}{n!}.$$

Similar to (1.8), according to Cramer's rule this identity yields that, by using ordinary determinantal operations,

$$\frac{2^n}{n!}E_n(x) = \det \begin{bmatrix} 2 & 0 & 0 & \cdots & 0 & 1 \\ \frac{1}{1!} & 2 & 0 & \cdots & 0 & \frac{x}{1!} \\ \frac{1}{2!} & \frac{1}{1!} & 2 & \cdots & 0 & \frac{x^2}{2!} \\ \vdots & \vdots & \vdots & \ddots & \vdots & \vdots \\ \frac{1}{(n-1)!} & \frac{1}{(n-2)!} & \frac{1}{(n-3)!} & \cdots & 2 & \frac{x^{n-1}}{(n-1)!} \\ \frac{1}{n!} & \frac{1}{(n-1)!} & \frac{1}{(n-2)!} & \cdots & \frac{1}{1!} & \frac{x^n}{n!} \end{bmatrix} \quad (n \geq 0).$$

We now pick out the following shortened recurrence relation instead of (6.1), in which some of the preceding Euler polynomials are completely excluded:

$$\sum_{i=0}^{k} \binom{k}{i} E_{m+i}(x) + \sum_{j=0}^{m} (-1)^{m-j} \binom{m}{j} E_{k+i}(x) = 2x^k(x-1)^m, \tag{6.2}$$

which is valid for arbitrary integers $k, m \geq 0$ (see [1, 2] for an elementary proof). In particular, setting $m = k$ in (6.2) and dividing by 2, we get immediately

$$\sum_{i=0}^{\lfloor k/2 \rfloor} \binom{k}{2i} E_{2(k-i)}(x) = (x^2 - x)^k = 2(X_k(x) - s_k), \tag{6.3}$$

which is the recurrence relation for even-index Euler polynomials.

Since $E_0(x) = 1$ and $\frac{d}{dx}E_n(x) = nE_{n-1}(x)$ for $n \geq 1$, one sees that the Euler polynomials form an Appell sequence. Based on this fact, taking a derivative of both sides of (6.3) with respect to x, we can obtain the following recurrence relation for odd-index Euler polynomials:

$$\sum_{i=0}^{\lfloor k/2 \rfloor} \binom{k}{2i} 2(k-i)E_{2(k-i)-1}(x) = k(2x-1)(x^2-x)^{k-1} = 2Y_k(x). \tag{6.4}$$

It can be seen that (6.3) and (6.4) are just corresponding to (2.2) (i) and (ii), respectively, in the Bernoulli polynomial case. So that, by using these shortened recurrence relations we are able to establish new determinantal expressions for Euler polynomials in almost the same way as done in Section 3.

Furthermore, from the functional identity $t\mathcal{E}(t,x) = 2\{\mathcal{F}(t,x) - \mathcal{F}(2t,x/2)\}$ we obtain a mutual relation between Euler and Bernoulli polynomials such that

$$(n+1)E_n(x) = 2\left(B_{n+1}(x) - 2^{n+1}B_{n+1}(x/2)\right) \quad (n \geq 0).$$

By way of this identities, one can observe some interesting relationships between determinantal expressions for Euler and Bernoulli polynomials.

References

[1] T. Agoh, Recurrences for Bernoulli and Euler polynomials and numbers, *Expo. Math.* **18** (2000), 197–214.

[2] T. Agoh, Shortened recurrence relations for generalized Bernoulli polynomials, *J. Number Theory* **176** (2017), 149–173.

[3] T. Agoh and K. Dilcher, Reciprocity relations for Bernoulli numbers, *Amer. Math. Monthly* **115** (2008), 237–244.

[4] T. Amdeberhan and H. Cohen, Bernoulli sum meets golden number, *MathOverflow*, version of 2017-06-15, https://mathoverflow.net/questions/272228.

[5] R. Booth and H. D. Nguyen, Bernoulli polynomials and Pascal's square, *Fibonacci Quart.* **46/47** (2008/2009), 38–47.

[6] P. Costabile, F. Dell'Accio and M. I. Gualtieri, A new approach to Bernoulli polynomials, *Rend. Mat. Appl.* **26** (2006), 1–12.

[7] F. A. Costabile and E. Longo, A determinantal approach to Appell polynomials. *J. Comput. Appl. Math.* **234** (2010), 1528–1542.

[8] K. Dilcher, *Bernoulli and Euler polynomials*, NIST Digital Library of Mathematical Functions, Chap. 24, Cambridge Univ. Press, Cambridge-New York, 2010, http://dlmf.nist.gov.

[9] K. Dilcher, Multiple zeros of Bernoulli polynomials, *Acta Arith.* **134** (2008), 149–155, https://eudml.org/doc/278788.

[10] A. von Ettingshausen, *Vorlesungen über die Höhere Mathematik*, Vol.1, C. Gerold, Vienna, 1827.

[11] M. B. Gelfand, A note on a certain relation among Bernoulli numbers (Russian),*Bashkir. Gos. Univ. Uchen. Zap. Ser. Mat.* **31** (1968), 215–216.

[12] E. R. Hansen, *A Table of Series and Products*, Prentice-Hall, Inc., Englewood Cliffs, NJ, 1975.

[13] K. Inkeri, The real roots of Bernoulli polynomials, *Ann. Univ. Turku.* Ser. AI **37** (1959), 20pp.

[14] N. Nielsen, *Traité élémentaire des nombres de Bernoulli*, Gauthier-Villars, Paris, 1923.

[15] N. E. Nörlund, *Vorlesungen über Differenzenrechnung*, Springer, Berlin, 1924.

[16] F. Qi and R. J. Chapman, Two closed forms for the Bernoulli polynomials, *J. Number Theory* **159** (2016), 89–100.

[17] F. Qi and B.-N. Guo, Some determinantal expressions and recurrence relations of the Bernoulli polynomials, *Mathematics*, **65** (20016), 1–11.

[18] L. Saalschütz, Verkürtzte Recursionsformeln für die Bernoullischen Zahlen, *Z. Math. Phys.* **37** (1892), 374–378.

[19] L. Saalschütz, *Vorlesungen über die Bernoullischen Zahlen, ihren Zusammenhang mit den Secanten-Coefficienten und ihre wichtigeren Anwendungen*, Springer-Verlag, Berlin, 1893.

[20] R. Sprugnoli, Sums of reciprocals of the central binomial coefficients, *Integers* **6** (2006), #A 27, 18pp.

[21] M. A. Stern, Beiträge zur Theorie der Bernoullischen und Eulerschen Zahlen, *Abh. Gesellsch. Wiss. Göttingen Math.* **23** (1878), 1–44.

4

POLYCHROMATIC COLORINGS ON THE INTEGERS

Maria Axenovich
Karlsruhe Institute of Technology, Karlsruhe, Germany
maria.aksenovich@kit.edu

John Goldwasser
West Virginia University, Morgantown, West Virginia
jgoldwas@math.wvu.edu

Bernard Lidický [1]
Iowa State University, Ames, Iowa
lidicky@iastate,edu

Ryan R. Martin[2]
Iowa State University, Ames, Iowa
rymartin@iastate.edu

David Offner [3]
Westminster College, New Wilmington, Pennsylvania
offnerde@westminster.edu

John Talbot
University College London, London, UK
j.talbot@ucl.ac.uk

Michael Young
Iowa State University, Ames, Iowa
myoung@iastate.edu

Abstract

We show that for any set $S \subseteq \mathbb{Z}$ with $|S| = 4$ there exists a 3-coloring of $\mathbb{Z}$ in which every translate of S receives all three colors. This implies that S has a codensity of at most $1/3$, proving a conjecture of Newman. We also consider related questions in $\mathbb{Z}^d$, $d \geq 2$.

[1]Supported in part by NSF grant DMS–1600390
[2]Supported in part by Simons Foundation Collaboration Grant (#353292, to R.R. Martin)
[3]Supported in part by Westminster College McCandless Research Award

1. Introduction

Throughout the paper, let G denote an arbitrary abelian group. Given $S, T \subseteq G$, $n \in G$, define $S+T = \{s+t : s \in S, t \in T\}$ and $n+S = \{n\}+S$. Any set of the form $n+S$ is called a *translate* of S. Given a subset S of G, a coloring of the elements of G is *S-polychromatic* if every translate of S contains an element of each color. Define the *polychromatic number* of S, denoted $p_G(S)$, to be the largest number of colors allowing an S-polychromatic coloring of the elements of G. We just write $p(S)$ when the choice of G is clear from context.

We begin with some elementary observations that will be used repeatedly. First, if S' is a subset of S, then any S'-polychromatic coloring is also an S-polychromatic coloring, and thus $p(S') \leq p(S)$. Also, if $S' = n+S$ is a translate of S, then S' and S have the same set of translates, so $p(S') = p(S)$.

We are primarily concerned with the setting where $G = \mathbb{Z}$ and S is finite. If $S \subseteq \mathbb{Z}$ has cardinality 1 or 2, $p(S) = |S|$. For $|S| = 3$, $p(S)$ can be 2 or 3. For example, if $S = \{0,1,5\}$ then every translate of S contains three elements which are each in different congruence classes (mod 3). Thus a 3-coloring of the integers where each congruence class (mod 3) is colored with a different color is S-polychromatic, and $p(\{0,1,5\}) = 3$. However $p(\{0,1,3\}) = 2$. To see that $p(\{0,1,3\}) \neq 3$, let χ be a 3-coloring of $\mathbb{Z}$ with $\chi(0)$, $\chi(1)$, and $\chi(3)$ all different. Some element $s \in \{0,1,3\}$ has $\chi(s) = \chi(2)$, and there is a translate of $\{0,1,3\}$ that contains both s and 2, so the coloring is not polychromatic. Our main result concerns the polychromatic numbers of sets with cardinality 4.

Theorem 1. *If $S \subseteq \mathbb{Z}$ and $|S| = 4$, then $p(S) \geq 3$.*

The proof of Theorem 1 is given in Section 2. For larger sets S, Alon, Kříž, and Nešetřil [2] proved that $p(S) \geq \frac{(1+o(1))|S|}{3\ln|S|}$, while there exists some set S where $p(S) \leq \frac{(1+o(1))|S|}{\ln|S|}$. Subsequently, Harris and Srinivasan [6] established a tight asymptotic lower bound on polychromatic numbers.

Theorem 2 ([2], [6]). *For a finite set $S \subseteq \mathbb{Z}$, $p(S) \geq \frac{(1+o(1))|S|}{\ln|S|}$. Moreover, there exists some set S where $p(S) \leq \frac{(1+o(1))|S|}{\ln|S|}$.*

One motivation for studying polychromatic numbers is that they provide bounds for Turán type problems (see for example [1], [10], [11]). Call $T \subseteq G$ a *blocking set* for S if $G \backslash T$ contains no translate of S, i.e., if for all $n \in G$, $n+S \nsubseteq G \backslash T$. A Turán type problem asks for the smallest blocking set for a given set S. In the case where S is finite and $G = \mathbb{Z}$, any blocking set is countably infinite, so we ask how small the density of a blocking set can be. Following the notation of Newman [8], (he worked in the setting of the natural numbers, but the definitions are equivalent), define for any set $T \subseteq \mathbb{Z}$ its *upper density* $\overline{d}(T)$ and *lower density* $\underline{d}(T)$ as

$$\overline{d}(T) = \limsup_{n\to\infty} \frac{|T \cap [-n,n]|}{2n+1} \qquad \text{and} \qquad \underline{d}(T) = \liminf_{n\to\infty} \frac{|T \cap [-n,n]|}{2n+1}.$$

If $\overline{d}(T) = \underline{d}(T)$, we call this quantity the *density* of T and denote it by $d(T)$. Define $\alpha(S)$ to be a measure of how small the density of a blocking set for S can be. Let

$$\alpha(S) = \inf\{d(T) : \ T \text{ is a blocking set for } S \text{ and } d(T) \text{ exists}\}.$$

In Section 3, we describe the relationship between polychromatic colorings and blocking sets, and prove Lemma 1.

Lemma 1. *For any finite set $S \subseteq \mathbb{Z}$, $\alpha(S) \leq 1/p(S)$.*

One of the main consequences of Theorem 1 concerns covering densities of sets of integers. Given a set $S \subseteq G$, we say $T \subseteq G$ is a *complement set* for S if $S + T = G$. We say S *tiles G by translation* if it has a complement set T such that if $s_1, s_2 \in S$, $t_1, t_2 \in T$, then $s_1 + t_1 = s_2 + t_2$ implies $s_1 = s_2$ and $t_1 = t_2$. We call such a complement set T a *tiling complement set* for S. Note that the set S tiles G by translation if all the translates $S + t$, $t \in T$, are disjoint and every $n \in G$ is an element of some translate $S+t$. In this paper we only consider tilings by translation, so if S tiles G by translation with tiling complement set T we will simply say S tiles G and write $G = S \oplus T$.

Again, our primary interest will be the case where $G = \mathbb{Z}$ and S is finite. For example, if $S = \{0, 1, 5\}$, then S tiles $\mathbb{Z}$ with complement set $T = \{3n : n \in \mathbb{Z}\}$. However $S = \{0, 1, 3\}$ does not tile $\mathbb{Z}$. Newman [9] proved necessary and sufficient conditions for a finite set S to tile $\mathbb{Z}$ if $|S|$ is a power of a prime.

Theorem 3 (Newman [9]). *Let $S = \{s_1, \ldots, s_k\}$ be distinct integers with $|S| = p^\alpha$ where p is prime and α is a positive integer. For $1 \leq i < j \leq k$ let $p^{e_{ij}}$ be the highest power of p that divides $s_i - s_j$. Then S tiles $\mathbb{Z}$ if and only if $|\{e_{ij} : 1 \leq i < j \leq k\}| \leq \alpha$.*

Later Coven and Meyerowitz [5] gave necessary and sufficient conditions for S to tile $\mathbb{Z}$ when $|S| = p_1^{\alpha_1} p_2^{\alpha_2}$, where p_1 and p_2 are primes. The general question is still open. Kolountzakis and Matolcsi [7] and Amiot [3] have published recent work motivated by what are called rhythmic tilings in music.

If a finite set S tiles $\mathbb{Z}$, it has a complement set of density $1/|S|$. Following Newman [8], we define the *codensity* of a set S, denoted $c(S)$, as a measure of how small the density of a complement set can be. Let

$$c(S) = \inf\{d(T) : S + T = \mathbb{Z} \text{ and } d(T) \text{ exists}\}.$$

We are interested in the largest codensities for sets of a given cardinality. Define

$$c_k = \sup_{\{S : |S| = k\}} c(S).$$

An example of a complement set for $\{0,1,3\}$ is $\{t \in \mathbb{Z} : t \equiv 0 \text{ or } 1 \pmod 5\}$, so $c(\{0,1,3\}) \leq 2/5$. The following theorem and conjecture on c_4 are due to Newman.

Theorem 4 (Newman [8]).

- $c(\{0,1,3\}) = 2/5$.
- $c_3 = 2/5$.
- $c(\{0,1,2,4\}) = 1/3$.

Conjecture 1 (Newman [8]). $c_4 = 1/3$.

Conjecture 1 is stated and attributed to Newman by Weinstein [16], who proved that $c_4 < .339934$. Based on a computer search, Bollobás, Janson, and Riordan [4] confirmed Newman's conjecture for sets with diameter at most 22, where the *diameter* of a nonempty finite set of integers is defined to be the difference between the largest and smallest elements in the set. They also conjectured that $c_5 = 3/11$ and $c_6 = 1/4$ (See Remark 5.6 and Question 5.7 in [4]. Note they use different notation).

In Section 3 we prove the following lemma relating blocking sets and complement sets.

Lemma 2. *For any finite set $S \subseteq \mathbb{Z}$, $c(S) = \alpha(S)$.*

Theorem 1, along with Lemmas 1 and 2, suffice to resolve Conjecture 1.

Theorem 5. $c_4 = 1/3$.

Proof. Theorem 4 implies $c(\{0,1,2,4\}) = 1/3$, so it remains to show that for any other set S with cardinality four, $c(S) \leq 1/3$. Let $S \subseteq \mathbb{Z}$ have four elements. Then Theorem 1 implies that $p(S) \geq 3$, and by Lemmas 1 and 2,

$$c(S) = \alpha(S) \leq 1/p(S) \leq 1/3.$$

□

In Subsection 3.1 we consider the relationship between polychromatic colorings and tilings. The main result is Theorem 6, which states that a set S tiles an abelian group G by translation if and only if $p(S) = |S|$.

Finally, in Section 4 we turn our attention to polychromatic numbers and tilings for finite sets in $\mathbb{Z}^d$. We begin by proving in Theorem 7 that the bound of Theorem 2 applies to subsets of $\mathbb{Z}^d$. We then show that if a set of points in $\mathbb{Z}^d$ is collinear, determining its polychromatic number is equivalent to determining the polychromatic number of a specific projection of this set into $\mathbb{Z}$. Theorem 6 implies that a set S tiles $\mathbb{Z}^d$ if and only if $p_{\mathbb{Z}_d}(S) = |S|$, so we use this to restate some well-known results on tilings of $\mathbb{Z}^d$ by finite sets in the language of polychromatic colorings. We conclude by applying these results to determine polychromatic numbers of sets with cardinality 3 and 4 in $\mathbb{Z}^d$.

2. Sets of Cardinality Four

In this section we prove that every set of four integers has polychromatic number at least 3. We begin by stating some general lemmas that reduce the problem of finding an S-polychromatic coloring of $\mathbb{Z}$ to finding an S-polychromatic coloring of $\mathbb{Z}_m = \{0, 1, \ldots, m-1\}$ for a specific choice of m.

Lemma 3. *If G and H are abelian groups and $\phi : G \to H$ is a homomorphism, then for all $S \subseteq G$,*

$$p_G(S) \geq p_H(\phi(S)).$$

Proof. Let $S \subseteq G$ and let χ' be a $\phi(S)$-polychromatic coloring of H with $p_H(\phi(S))$ colors. Define the coloring χ on G such that $\chi(g) = \chi'(\phi(g))$. Consider a translate $g + S$ of the set S. Since

$$\chi(g+S) = \chi'(\phi(g+S)) = \chi'(\phi(g) + \phi(S)),$$

and $\phi(g) + \phi(S) \subseteq H$ is a translate of $\phi(S)$, $\chi(g+S)$ contains all $p_H(\phi(S))$ colors, and χ is S-polychromatic. □

Corollary 1. *If G and H are abelian groups and $\phi : G \to H$ is an isomorphism, then for all $S \subseteq G$,*

$$p_G(S) = p_H(\phi(S)).$$

Lemma 4. *If H is a subgroup of an abelian group G, and $S \subseteq H$, then $p_H(S) = p_G(S)$.*

Proof. Since H is a subset of G, $p_H(S) \geq p_G(S)$. To prove the other inequality, suppose χ' is an S-polychromatic coloring of H with $p_H(S)$ colors. Let $V \subseteq G$ be a set containing exactly one element from each coset of H. For every $g \in G$, there is a unique $h \in H$ and $v \in V$ such that $g = h + v$. Define a coloring χ of G such that $\chi(g) = \chi'(h)$, i.e., the summand v is ignored. We show that χ is S-polychromatic. Given $g \in G$ with $g = h + v$ for some $h \in H$, $v \in V$, consider the translate $g + S$, and note

$$\chi(g+S) = \chi(h+v+S) = \chi'(h+S).$$

Since $h + S \subseteq H$ is a translate of S, $\chi(g+S)$ contains all $p_H(\phi(S))$ colors, and χ is S-polychromatic. □

Lemma 5. *Suppose $a, b, c, k \in \mathbb{Z}$ with $0 < a < b < c$ and $k \geq 1$. Let $S = \{0, ka, kb, kc\}$, $S_1 = \{0, a, b, c\}$, and $S_2 = \{0, b-a, b, 2b-a\}$. Then*

(i) $p_{\mathbb{Z}}(S) = p_{\mathbb{Z}}(S_1)$.

(ii) *If $q \in \mathbb{N}$, then* $p_{\mathbb{Z}}(S_1) \geq p_{\mathbb{Z}_q}(S_1)$.

(iii) If $m = c - a + b$, then $p_{\mathbb{Z}_m}(S_1) = p_{\mathbb{Z}_m}(S_2)$.

(iv) If $q \in \mathbb{N}$, with $\gcd(k, q) = 1$, then $p_{\mathbb{Z}_q}(S) = p_{\mathbb{Z}_q}(S_1)$.

Proof.

(i) Define $\phi : \mathbb{Z} \to k\mathbb{Z}$ such that $\phi(n) = kn$. Then ϕ is an isomorphism where $\phi(S_1) = S$, so Corollary 1 implies $p_{k\mathbb{Z}}(S) = p_{\mathbb{Z}}(S_1)$. Since $k\mathbb{Z}$ is a subgroup of $\mathbb{Z}$ and $S \subseteq k\mathbb{Z}$, Lemma 4 implies $p_{k\mathbb{Z}}(S) = p_{\mathbb{Z}}(S)$. By combining these equations, we conclude $p_{\mathbb{Z}}(S) = p_{k\mathbb{Z}}(S) = p_{\mathbb{Z}}(S_1)$.

(ii) This part follows from Lemma 3 using the homomorphism $\phi : \mathbb{Z} \to \mathbb{Z}_q$ where $\phi(n) = n \pmod q$.

(iii) In $\mathbb{Z}_m$, with addition $\pmod m$, $S_2 = S_1 + (b - a)$. Thus in $\mathbb{Z}_m$, S_1 and S_2 are translates of each other and have the same polychromatic number.

(iv) Define $\phi : \mathbb{Z}_q \to \mathbb{Z}_q$ so that $\phi(n) = kn$. Since $\gcd(k, q) = 1$, ϕ is an isomorphism. Since $\phi(S_1) = S$, Corollary 10 implies $p_{\mathbb{Z}_q}(S) = p_{\mathbb{Z}_q}(S_1)$.

□

We now prove Theorem 1.

Proof. Let $S \subseteq \mathbb{Z}$ have cardinality four. Since all translates of S have the same polychromatic number, we may assume that 0 is the smallest element of S, and by Lemma 5, Part (i), it suffices to prove the theorem in the case that $S = \{0, a, b, c\}$ with $0 < a < b < c$ and $\gcd(a, b, c) = 1$.

It is possible, though tedious, to prove the entire theorem by hand. Thus in the interest of simplifying the exposition, we verified using a computer search that for every S with diameter at most 288 there exists an S-polychromatic 3-coloring of $\mathbb{Z}_q$ for some q depending on S. The code for this search has been included as an ancillary file with the preprint of this paper at arxiv.org/abs/1704.00042. By Lemma 5, Part (ii), this gives a periodic S-polychromatic 3-coloring of $\mathbb{Z}$. Hence we suppose that $c \geq 289$.

For the remainder of the proof, let $m = c - a + b$. By Lemma 5, Parts (ii) and (iii), it suffices to show that we can 3-color $\mathbb{Z}_m = \{0, 1, \ldots, m-1\}$ so that the translates of $\{0, b-a, b, 2b-a\}$ are polychromatic. So for the remainder of the proof we assume $S = \{0, b-a, b, 2b-a\}$ and seek an S-polychromatic 3-coloring of $\mathbb{Z}_m$. The key observation regarding S is that it contains two repeated differences: $b - a$ and b.

Define $d_1 = \gcd(b, m)$ and $d_2 = \gcd(b - a, m)$. Since $1 = \gcd(a, b, c) = \gcd(b - a, b, c - a + b) = \gcd(b - a, b, m)$, we know $\gcd(d_1, d_2) = 1$. We distinguish two main cases. In the first case, which we call "single cycle," we assume $\min\{d_1, d_2\} = 1$ and give a coloring of $\mathbb{Z}_m$. In the second case, which we call "multiple cycle," we assume $\min\{d_1, d_2\} > 1$ and partition $\mathbb{Z}_m$ into multiple cycles of length m/d_i for one of the choices of i. We then give a rule for coloring each cycle.

`Main Case 1 (Single cycle):` Suppose $\min\{d_1, d_2\} = 1$. Without loss of generality, assume $d_1 = 1$ (if not, then simply switch all occurences of b and $b - a$ in the argument below). Let $2 \le g \le m - 2$ satisfy $gb \equiv b - a \pmod{m}$, so that $S = \{0, bg, b, b(g+1)\}$. Applying Lemma 5, Part (iv), with $q = m$ and $k = b$, we can instead work with $S = \{0, g, 1, g+1\} = \{0, 1, g, g+1\}$.

We may assume that $g \le m/2$, as otherwise we could work with the translate $(m - g) + S = \{0, 1, m - g, m - g + 1\}$. Let s be the smallest multiple of 3 such that $g > \lceil m/s \rceil$. We consider four subcases: The first two are (1a) $g = 2$, 3, or 4 and (1b) $5 \le g < 2\lfloor m/s \rfloor$. In the remaining subcases (1c) and (1d), $2\lfloor m/s \rfloor \le g \le \lceil m/(s-3) \rceil$. For $m > 8$, if $2\lfloor m/s \rfloor \le g \le m/2$ then $s > 3$, and for $m > 44$, if $2\lfloor m/s \rfloor \le g \le \lceil m/(s-3) \rceil$ then $s < 9$. Since $m > c \ge 289 > 44$, we can assume $s = 6$, so $2\lfloor m/6 \rfloor \le g \le \lceil m/3 \rceil$. This implies $m = 3g + k$ where $-2 \le k \le 5$ and there are two further subcases to consider, depending on the residue class of m modulo 6: (1c) $m = 3g - 2$, $3g - 1$, $3g + 1$, $3g + 2$, $3g + 4$, or $3g + 5$, and (1d) $m = 3g$ or $3g + 3$.

`Subcase (1a):` Suppose $g = 2$, 3, or 4. Then $S = \{0, 1, 2, 3\}$, $\{0, 1, 3, 4\}$, or $\{0, 1, 4, 5\}$, respectively. In Subcase (1c) we will construct S-polychromatic 3-colorings of $\mathbb{Z}_m$ for each of these sets.

`Subcase (1b):` Suppose $5 \le g < 2\lfloor m/s \rfloor$. Then split $\mathbb{Z}_m$ into s intervals as equally as possible (i.e., of lengths $\lfloor m/s \rfloor$ and $\lceil m/s \rceil$) and color these intervals $010101\ldots$, followed by $121212\ldots$, then $202020\ldots$, repeating $s/3$ times. Since $\lceil m/s \rceil < g < 2\lfloor m/s \rfloor$, any translate of S' where the pairs $\{0, 1\}$ and $\{g, g+1\}$ lie in different intervals gets all three colors. If one of the pairs $\{0, 1\}$ or $\{g, g+1\}$ straddles two consecutive intervals, this pair may get only the single color common to these two intervals, but then the other pair lies fully inside a third interval which is colored with the remaining two colors.

`Subcase (1c):` Suppose $m = 3g - 2$, $3g - 1$, $3g + 1$, $3g + 2$, $3g + 4$, or $3g + 5$. In this case we know that $m \not\equiv 0 \pmod{3}$ so we can apply Lemma 5, Part (iv), with $q = m$ and $k = 3$, and instead work with one of the sets in $\mathcal{S} = \{\{0, 2, 3, 5\}, \{0, 1, 3, 4\}, \{0, 1, 2, 3\}, \{0, 3, 4, 7\}, \{0, 3, 5, 8\}\}$. For example, if $m = 3g - 2$, then multiplying by 3, S is transformed into $\{0, 3, 3g, 3g+3\} \equiv \{0, 2, 3, 5\}$, while if $m = 3g + 4$, then multiplying by 3, S is transformed into $\{0, 3, 3g, 3g + 3\} \equiv \{0, 3, -4, -1\}$, which is a translate of $\{0, 3, 4, 7\}$.

Thus we have reduced the problem to finding an S-polychromatic 3-coloring of $\mathbb{Z}_m$ for each of the sets $S \in \mathcal{S}$. For each $S \in \mathcal{S}$, in Table 1 we list one interval of length r and one of length $r + 1$ obtained by adding an initial 0 to the other interval. We also include an interval for $\{0, 1, 4, 5\}$ to cover Subcase (1a). Each of the intervals has the property that concatenating the intervals of length r and $r+1$ in any way results in an S-polychromatic coloring for the corresponding set. One can

S	r	period r	period $r+1$
$\{0,2,3,5\}$	6	001122	0001122
$\{0,1,3,4\}$	6	001212	0001212
$\{0,1,2,3\}$	3	012	0012
$\{0,3,4,7\}$	9	000111222	0000111222
$\{0,3,5,8\}$	9	000111222	0000111222
$\{0,1,4,5\}$	7	0001212	00001212

Table 1: One interval of a periodic coloring for sets in Subcases (1a) and (1c).

check this by hand, using the fact that, in each case, a translate of $S \in \mathcal{S}$ intesects at most two consecutive intervals. Hence if m can be expressed as a positive integer combination of r and $r+1$, $m = hr + k(r+1)$, we can obtain an S-polychromatic coloring with period m. For $r = 3, 6, 7, 9$, by the 2-coin Frobenius problem, m can be expressed as a positive integer combination of r and $r+1$ for any m greater than $r^2 - r - 1 \leq 71 < 289$.

`Subcase (1d):` Suppose $m = 3g$ or $3g+3$. If $g \not\equiv 0 \pmod 3$ then simply color $\mathbb{Z}_m$ with the pattern $0120120\ldots012$. If $g \equiv 0 \pmod 3$ and $m = 3g$, color $\mathbb{Z}_m$ in 3 equal intervals, each of length g: $012012\ldots012$ followed by $120120\ldots120$ followed by $201201\ldots201$. Finally, if $g \equiv 0 \pmod 3$ and $m = 3g+3$ we color $\mathbb{Z}_m$ in 3 equal intervals, each of length $g+1$: $012012\ldots0120$ followed by $201201\ldots2012$ followed by $120120\ldots1201$.

`Main case 2 (Multiple cycles):` Suppose $\min\{d_1, d_2\} > 1$. Since d_1 and d_2 are relatively prime, at most one of them can be a multiple of 3. Choose the smallest of these numbers that is not a multiple of 3, and as in the single cycle case, without loss of generality assume it is d_1.

Let $e_1 = m/d_1$ and $e_2 = m/d_2$. For $0 \leq i < d_1$, let

$$C_i = \{(b-a)i + bj \pmod m : 0 \leq j < e_1\}.$$

Since

$$\mathbb{Z}_m = \{(b-a)i + bj \pmod m : 0 \leq i < d_1, 0 \leq j < e_1\},$$

the C_i's form a partition of $\mathbb{Z}_m$ into d_1 cycles, each with e_1 elements.

Let $c_{i,j}$ denote the jth element of C_i, i.e., $c_{i,j} = i(b-a) + jb \pmod m$. Note that any translate of S contains two consecutive elements of two consecutive cycles, i.e., any translate of S has the form $\{c_{i,j}, c_{i,j+1}, c_{i+1,j}, c_{i+1,j+1}\}$, where the first entry in the subscript is taken $\bmod d_1$ and the second entry is taken $\bmod e_1$. We describe an S-polychromatic 3-coloring for each of four subcases: (2a) e_1 is even, (2b) d_1 is even and e_1 is odd, (2c) d_1 and e_1 are both odd, with $e_1 \leq 17$, and (2d) d_1 and e_1 are both odd, with $e_1 \geq 19$.

`Subcase (2a):` Suppose e_1 is even. For $i = 0, \ldots, \lfloor d_1/2 \rfloor - 1$, color each C_{2i} by

01010...01 and each C_{2i+1} by 02020...02. Finally, if d_1 is odd, color C_{d_1-1} by 1212...12.

`Subcase (2b)`: Suppose d_1 is even and e_1 is odd. For $i = 0, \dots, d_1/2 - 1$, color each C_{2i} by 01010...011 and each C_{2i+1} by 22020...02.

`Subcase (2c)`: Suppose d_1 and e_1 are both odd, with $e_1 \leq 17$. Since $e_1 e_2 \geq m > c \geq 289$, one of e_1 and e_2 is larger than 17, so $e_2 > e_1$ and hence $d_1 > d_2$. Since d_1 is the smaller of d_1 and d_2 that is not a multiple of 3, d_2 must be a multiple of 3, and thus so is e_1.

We color each C_i with one of three patterns: 012012...012, 120120...120, or 201201...201. Such a coloring is S-polychromatic so long as for all i, C_i and C_{i+1} are colored with different patterns. For $0 \leq i \leq (d_1 - 3)/2$, color C_{2i} with the first pattern and color C_{2i+1} with the second pattern. Finally, color C_{d_1-1} with the third pattern.

`Subcase (2d)`: Suppose d_1 and e_1 are both odd, with $e_1 \geq 19$. Since d_1 is not divisible by 3 and $\min\{d_1, d_2\} > 1$, $d_1 \geq 5$. Let $e_1 = u + v + w$ be a sum of odd integers u, v, w with $u \geq v \geq w \geq u - 2$. Color C_0 in intervals of size u, v, w, using the patterns 0101...010 then 1212...121 and then 2020...202. For each $i \geq 1$, color C_i by taking a "counterclockwise rotation" of length r_i of the coloring of C_{i-1}, so that the color of $c_{i,j+r}$ is the same as the color of $c_{i-1,j}$. For $1 \leq i \leq d_1 - 1$, if $u \leq r_i \leq v + w = e_1 - u$, then each translate of S meeting C_{i-1} and C_i receives all 3 colors.

It remains to show that there are choices of $r_1, \dots, r_{d_1-1}$ with $u \leq r_i \leq v + w = e_1 - u$ so that of the translates of S meeting C_{d_1-1} and C_0 receive all three colors. The coloring of C_0 is a "clockwise rotation" of length $R = -r_1 - r_2 - \cdots - r_{d_1-1}$ of the coloring of C_{d_1-1}, i.e., the color of $c_{0,j-R}$ is the same as the color of $c_{d_1-1,j}$. Since for each i, $u \leq r_i \leq v + w = e_1 - u$, it suffices to show that there is a multiple of e_1 in the interval $[d_1 u, d_1(e_1 - u)]$, ensuring there are choices for the r_i's such that R is congruent to a number between u and $e_1 - u \pmod{e_1}$. This certainly holds if $d_1(e_1 - 2u) \geq e_1 - 1$ which, since $d_1 \geq 5$, holds if $4e_1 \geq 10u - 1$. This inequality is true for $e_1 \geq 19$.

This completes the multiple cycles case and the proof. □

3. Colorings, Blocking Sets, Coverings, and Tilings

In this section we prove the results necessary to resolve Newman's conjecture. The key insight in proving Lemma 1 is that the elements of a given color in an S-polychromatic coloring form a blocking set for S. While it is possible for $\alpha(S)$ to be equal to $1/p(S)$ (e.g. if $|S| = 2$ then $\alpha(S) = 1/2 = 1/p(S)$), in general these two quantities are not equal. For example, $p(\{0,1,3\}) = 2$, but by Lemma 2 and Theorem 4, $\alpha(\{0,1,3\}) = 2/5 < 1/2$.

We now prove Lemma 1.

Proof. Let χ be an S-polychromatic coloring of $\mathbb{Z}$ with $p(S)$ colors. Suppose $d \in \mathbb{Z}$ is greater than the diameter of S and let $I_j = \{n \in \mathbb{Z} : jd \leq n < (j+1)d\}$. By the pigeonhole principle, for some $0 \leq j_1 < j_2 \leq (p(S))^d$ the coloring of the intervals I_{j_1} and I_{j_2} are identical, i.e., for $0 \leq k < d$, $\chi(j_1 d + k) = \chi(j_2 d + k)$. Let $m = (j_2 - j_1)d$. For any $n \in \mathbb{Z}$, denote by r the remainder when n is divided by m, so $0 \leq r < m$. Let χ' be the coloring of $\mathbb{Z}$ where $\chi'(n) = \chi(j_1 d + r)$. Note that χ' uses $p(S)$ colors and is periodic with period m, i.e., for all $n \in \mathbb{Z}$, $\chi(n) = \chi(n+m)$. Furthermore, the coloring under χ' of any d consecutive integers is identical to the coloring under χ of some d consecutive integers, so χ' is S-polychromatic. Let $T_i = \{n \in \mathbb{Z} : \chi'(n) = i\}$. Since any periodic set has a defined density, $d(T_i)$ is defined for each i, and $\sum_{i=1}^{p(S)} d(T_i) = 1$. Since χ' is S-polychromatic, for each i, each translate of S contains an element of T_i, i.e., T_i is also a blocking set for S. Thus for some i, T_i is a blocking set for S with density at most $1/p(S)$, which implies that $\alpha(S) \leq 1/p(S)$. □

For any subset T of an abelian group G, let $-T$ denote the set $\{-t : t \in T\}$. Lemma 6 is well-known (see e.g. [14]) but for completeness we present a proof.

Lemma 6. *Let G be an abelian group, and $S \subseteq G$. Then $T \subseteq G$ is a complement set for S if and only if $-T$ is a blocking set for S.*

Proof. Suppose T is a complement set for S. For any $n \in G$, we have $-n \in S+T$, so there must be some $t \in T, s \in S$ such that $t + s = -n$. This implies $t = -n - s$, so $-n - s \in T$, and $n + s \in -T$. Thus for every n, some element of $n + S$ is in $-T$, and $-T$ is a blocking set for S.

Conversely, suppose $-T$ is a blocking set for S. For the sake of contradiction, assume T is not a complement set for S, i.e., there is some $-n \in G$ such that $-n \notin S + T$. This implies that for all $s \in S$, $-n - s \notin T$, which means for all $s \in S$, $n + s \notin -T$. Thus $n + S \subseteq G \setminus -T$, and so $-T$ is not a blocking set for S, a contradiction. □

We now prove Lemma 2.

Proof. Lemma 6 implies that T is a complement set for S if and only if $-T$ is a blocking set for S. If they exist, the densities of T and $-T$ are the same. □

3.1. Polychromatic Colorings and Tilings

We now describe some relationships between polychromatic colorings and tilings.

Theorem 6. *Let G be any abelian group. A finite set $S \subseteq G$ tiles G by translation if and only if $p(S) = |S|$. Moreover, if χ is an S-polychromatic coloring of G with*

$|S|$ colors and T is the set of elements of G colored by χ with any given color, then $S \oplus T = G$.

Proof. Let $S = \{s_1, s_2, \ldots, s_k\}$, and suppose S tiles G with complement set $T \subseteq G$. For each $n \in G$, define a coloring χ on G so that $\chi(n) = i$ if $n = s_i + t$ for some $t \in T$. By the definition of tiling, this coloring is well-defined. For the sake of contradiction, assume χ is not S-polychromatic. Then for some l where $1 \leq l \leq k$, there exists $n \in G$ and $s_i, s_j \in S$ with $i \neq j$ such that $\chi(n + s_i) = \chi(n + s_j) = l$. Then there exist $t_1, t_2 \in T$, $t_1 \neq t_2$, such that $n + s_i = t_1 + s_l$ and $n + s_j = t_2 + s_l$. Subtracting these equations, we find that $s_i - s_j = t_1 - t_2$. Thus $t_2 + s_i = t_1 + s_j$, which is a contradiction.

Conversely, let $S = \{s_1, s_2, \ldots, s_k\}$, suppose $p(S) = |S|$, and let χ be an S-polychromatic coloring of G with $|S|$ colors. Then for all $n \in G$, if $i \neq j$ then $\chi(n + s_i) \neq \chi(n + s_j)$. Let $T \subseteq G$ be the set of elements colored with a given color. We show that $S \oplus T = G$. First assume for the sake of contradiction that two translates of S share an element, i.e., there exist $s_i, s_j \in S$, $i \neq j$, $t_1, t_2 \in T$, $t_1 \neq t_2$, such that $s_i + t_1 = s_j + t_2$. Let $n = t_1 - s_j = t_2 - s_i$, so $t_1 = n + s_j$ and $t_2 = n + s_i$. Since $\chi(t_1) = \chi(t_2)$ we get $\chi(n + s_j) = \chi(n + s_i)$, so two elements of $n + S$ are colored identically, which is a contradiction.

It remains to show that $S + T = G$. Suppose there is some $n \in G$ such that $n \notin S + T$. Then for all i, $n - s_i \notin T$, which implies that the $|S|$ elements of $n - S$ are colored with at most $|S| - 1$ colors, i.e., two are colored identically. Suppose $\chi(n - s_i) = \chi(n - s_j)$, where $i \neq j$. Let $m = n - s_j - s_i$. Then $m + S$ contains both $m + s_i = n - s_j$ and $m + s_j = n - s_i$. Since these integers are colored identically, $m + S$ is a translate of S that does not contain all colors, which is a contradiction. □

Sets of integers with cardinality $n = 3$ or 4 always have polychromatic number n or $n - 1$, and a corollary of Theorem 6 is that they have polychromatic number $n - 1$ if and only if they do not tile $\mathbb{Z}$. According to Remark 5.6 in [4], $c(\{0, 1, 3, 4, 8\}) = 3/11 > 1/4$. Thus by Lemma 1, $\{0, 1, 3, 4, 8\}$ is an example of a set with cardinality 5 and polychromatic number 3. The results of [2] and [6] imply that for sets S with large cardinality n the cardinality and polychromatic number of S can differ by a factor of $1/\ln n$.

We now state some other corollaries of Theorem 6.

Corollary 2. *If a finite set S tiles an abelian group G by translation, then any S-polychromatic coloring of G with $|S|$ colors is also a $(-S)$-polychromatic coloring.*

Proof. Suppose S tiles G. By Theorem 6, there exists an S-polychromatic coloring χ of G with $|S|$ colors. Let $T \subseteq G$ be the set of all elements of a given color. Again by Theorem 6, $S + T = G$. Therefore by Lemma 6, $-T$ is a blocking set for S, i.e., for all $n \in G$, $n + S \nsubseteq G \setminus (-T)$. This implies that for all $n \in G$, $-n - S \nsubseteq G \setminus T$, i.e., T is a blocking set for $-S$. Since T is a blocking set for $-S$

for every color choice, every translate of $-S$ contains every color, i.e., the coloring χ is $(-S)$-polychromatic. $\square$

Define $t(S)$ to be the cardinality of the largest subset of S that tiles G.

Corollary 3. *For any finite subset S of an abelian group G, $p(S) \geq t(S)$.*

If $S \subseteq \mathbb{Z}$, $|S| \leq 3$, then $p(S) = t(S)$. But these parameters can be different for sets of integers with at least four elements. For example, $S = \{0, 1, 3, 7\}$ is an example of a set where $t(S) = 2$, but $p(S) = 3$.

Question 1. For sets S of a given cardinality, how large can the gap between $t(S)$ and $p(S)$ be?

4. Polychromatic Colorings in $\mathbb{Z}^d$

In this section we consider polychromatic numbers in the case where $G = \mathbb{Z}^d$, $d \geq 2$. We will frequently "project" a set $S \subseteq \mathbb{Z}^d$ to another set $S' \subseteq \mathbb{Z}^{d-1}$ as follows. Let $d \geq 2$, and $\boldsymbol{w} = (w_1, \ldots, w_{d-1}, 1) \in \mathbb{Z}^d$. Define $f_{\boldsymbol{w}} : \mathbb{Z}^d \to \mathbb{Z}^{d-1}$ so that if $\boldsymbol{s} = (v_1, \ldots, v_d) \in \mathbb{Z}^d$,

$$f_{\boldsymbol{w}}(\boldsymbol{s}) = (v_1, \ldots, v_{d-1}) - v_d(w_1, \ldots, w_{d-1}).$$

We call $f_{\boldsymbol{w}}(\boldsymbol{s})$ the *projection* of $\boldsymbol{s}$ along $\boldsymbol{w}$. Given a set $S \subseteq \mathbb{Z}^d$, we call the set $f_{\boldsymbol{w}}(S) \subseteq \mathbb{Z}^{d-1}$ the *projection* of S along $\boldsymbol{w}$.

For example, if $\boldsymbol{s} = (2, 7, 4)$ and $\boldsymbol{w} = (3, 1, 1)$, the projection of $\boldsymbol{s}$ along $\boldsymbol{w}$ is $f_{\boldsymbol{w}}(\boldsymbol{s}) = (2, 7) - 4(3, 1) = (-10, 3)$. As another example, note that if $\boldsymbol{s} = (v_1, \ldots, v_d) \in \mathbb{Z}^d$, the vector $\boldsymbol{s}' = (v_1, \ldots, v_{d-1}) \in \mathbb{Z}^{d-1}$ is the projection of $\boldsymbol{s}$ along $\boldsymbol{w} = (0, \ldots, 0, 1)$.

Lemma 7. *Let $d \geq 2$, and $\boldsymbol{w} = (w_1, \ldots, w_{d-1}, 1) \in \mathbb{Z}^d$. Let $S \subseteq \mathbb{Z}^d$, and suppose $S' \subseteq \mathbb{Z}^{d-1}$ is the projection of S along $\boldsymbol{w}$. Then $p_{\mathbb{Z}^d}(S) \geq p_{\mathbb{Z}^{d-1}}(S')$.*

Proof. Since $f_{\boldsymbol{w}} : \mathbb{Z}^d \to \mathbb{Z}^{d-1}$ is a homomorphism, the result follows from Lemma 3. $\square$

Proposition 1. *Let $d \geq 2$. For any $S \subseteq \mathbb{Z}^d$, there is a projection $S' \subseteq \mathbb{Z}^{d-1}$ where $|S| = |S'|$.*

Proof. Let $S = \{\boldsymbol{s}_1, \ldots, \boldsymbol{s}_k\} \subseteq \mathbb{Z}^d$ and suppose $\boldsymbol{w} = (w_1, \ldots, w_{d-1}, 1) \in \mathbb{Z}^d$. For $1 \leq i \leq k$ let $\boldsymbol{s}_i' = f_{\boldsymbol{w}}(\boldsymbol{s}_i)$. For $1 \leq i \leq k$, let s_{id} denote the last coordinate of $\boldsymbol{s}_i$ and note that if $i \neq j$, $\boldsymbol{s}_i' = \boldsymbol{s}_j'$ if and only if

$$\boldsymbol{w} = \frac{1}{s_{id} - s_{jd}}(\boldsymbol{s}_i - \boldsymbol{s}_j).$$

In other words $s'_i = s'_j$ if and only if $\boldsymbol{w}$ is parallel to $s_i - s_j$. Since the number of differences $s_i - s_j$ is finite, we can choose $\boldsymbol{w}$ so that it is not parallel to any of these. For this choice of $\boldsymbol{w}$, for all $1 \leq i \neq j \leq k$, $s'_i \neq s'_j$ Then $S' = \{s'_1, \ldots, s'_k\}$, is a projection S with $|S'| = |S|$. □

Theorem 7. *Fix $d \geq 2$. For a finite set $S \subseteq \mathbb{Z}^d$, $p(S) \geq \frac{(1+o(1))|S|}{\ln |S|}$.*

Proof. Given $S \subseteq \mathbb{Z}^d$, Proposition 1 implies we can project $d-1$ times to ultimately obtain a set $S' \subseteq \mathbb{Z}$, with $|S'| = |S|$. Theorem 2, along with repeated application of Lemma 7, implies $p_{\mathbb{Z}^d}(S) \geq p_{\mathbb{Z}}(S') \geq \frac{(1+o(1))|S'|}{\ln |S'|} = \frac{(1+o(1))|S|}{\ln |S|}$. □

We will be interested in the case where $S = \{s_0, s_1, s_2, \ldots, s_k\} \subseteq \mathbb{Z}^d$ contains a set of collinear points. In the general case, we can assume for each i that $s_i = (b_1 + l_i a_1, b_2 + l_i a_2, \ldots, b_d + l_i a_d)$, where $0 = l_0 < l_1 < l_2 < \cdots < l_k$, $a_i, b_i \in \mathbb{Z}$, and $\gcd(a_1, a_2, \ldots, a_d) = 1$. However since translation does not affect the polychromatic number, we will restrict our attention to the case where $a_1 > 0$ and for all i, $b_i = 0$.

Theorem 8. *Let $d \geq 2$. Let $S = \{s_0, s_1, s_2, \ldots, s_k\}$ be a set of $k+1$ collinear points in $\mathbb{Z}^d$ where for each i, $s_i = (l_i a_1, l_i a_2, \ldots, l_i a_d)$, where $0 = l_0 < l_1 < l_2 < \cdots < l_k$, $a_i \in \mathbb{Z}$, $a_1 > 0$, and $\gcd(a_1, a_2, \ldots, a_d) = 1$. Let $S' = \{0, l_1, l_2, \ldots, l_k\} \subseteq \mathbb{Z}$. Then $p_{\mathbb{Z}^d}(S) = p_{\mathbb{Z}}(S')$.*

Proof. Let $S'' = \{0, l_1 a_1, l_2 a_1, \ldots, l_k a_1\} \subseteq \mathbb{Z}$. By Lemma 5, Part (i), $p_{\mathbb{Z}}(S') = p_{\mathbb{Z}}(S'')$. Since S'' can be obtained from S by a sequence of $d-1$ projections, Lemma 7 implies $p_{\mathbb{Z}^d}(S) \geq p_{\mathbb{Z}}(S'') = p_{\mathbb{Z}}(S')$. For the other direction, let $\boldsymbol{a} = (a_1, \ldots, a_d) \in \mathbb{Z}^d$ and note that the function $\phi : \mathbb{Z} \to \mathbb{Z}^d$ where $\phi(n) = n\boldsymbol{a}$ is a homomorphism where $\phi(S') = S$. Thus by Lemma 3, $p_{\mathbb{Z}}(S') \geq p_{\mathbb{Z}^d}(S)$ □

Now we return to the subject of tilings. Lemma 8 and Theorems 9, 10, and 11 are well-known in the field of discrete geometry (see e.g. Section III of [14]) as simple examples of "splitting" groups. We restate them here using the language of polychromatic colorings.

Lemma 8. *If a set $S \subseteq G$ tiles a nontrivial subgroup H of G, then S tiles G.*

Proof. Suppose $S \subseteq G$ tiles a nontrivial subgroup H of G. Theorem 6 implies $p_H(S) = |S|$, so by Lemma 4, $p_G(S) = |S|$. By Theorem 6, S tiles G. □

For any $d \geq 1$, let $\mathbf{0}$ denote the element $(0, 0, \ldots, 0) \in \mathbb{Z}^d$ and let e_i denote the element $(0, \ldots, 0, 1, 0, \ldots, 0) \in \mathbb{Z}^d$ with all 0's except for a 1 in the ith position. For $s = (v_1, \ldots, v_d) \in \mathbb{Z}^d$, let $-s = (-v_1, \ldots, -v_d)$. Define the d-semicross $SC_d = \{\mathbf{0}, e_1, \ldots, e_d\}$ and the d-cross $C_d = \{\mathbf{0}, e_1, -e_1, e_2, -e_2, \ldots, e_d, -e_d\}$. Theorem 6 implies that any finite set $S \subseteq G$ with $p(S) = |S|$ tiles G, and we use this insight to show that these sets tile $\mathbb{Z}^d$.

Theorem 9. *For all $d \geq 1$, the d-semicross $SC_d = \{\mathbf{0}, \mathbf{e}_1, \ldots, \mathbf{e}_d\}$ tiles $\mathbb{Z}^d$.*

Proof. Consider the coloring $\chi : \mathbb{Z}^d \to [d+1]$ where $\chi(v_1, \ldots, v_d) = v_1 + 2v_2 + 3v_3 + \cdots + dv_d \pmod{d+1}$. On any translate $\boldsymbol{n} + SC_d \subseteq \mathbb{Z}^d$, the colors $\chi(\boldsymbol{n} + \mathbf{0}), \chi(\boldsymbol{n} + \mathbf{e}_1), \chi(\boldsymbol{n} + \mathbf{e}_2), \ldots, \chi(\boldsymbol{n} + \mathbf{e}_d)$ are $\chi(\boldsymbol{n}), \chi(\boldsymbol{n}) + 1, \chi(\boldsymbol{n}) + 2, \ldots, \chi(\boldsymbol{n}) + d \pmod{d+1}$. They are all different, so χ is SC_d-polychromatic with $|SC_d| = d+1$ colors. By Theorem 6, SC_d tiles $\mathbb{Z}^d$. □

Theorem 10. *For all $d \geq 1$, the d-cross $C_d = \{\mathbf{0}, \mathbf{e}_1, -\mathbf{e}_1, \mathbf{e}_2, -\mathbf{e}_2, \ldots, \mathbf{e}_d, -\mathbf{e}_d\}$ tiles $\mathbb{Z}^d$.*

Proof. The $(2d+1)$-coloring $\chi : \mathbb{Z}^d \to [2d+1]$ where $\chi(v_1, \ldots, v_d) = v_1 + 2v_2 + 3v_3 + \cdots + dv_d \pmod{2d+1}$ is C_d-polychromatic: On any translate $\boldsymbol{n} + C_d \subseteq \mathbb{Z}^d$, the colors $\chi(\boldsymbol{n}+\mathbf{0}), \chi(\boldsymbol{n}+\mathbf{e}_1), \chi(\boldsymbol{n}-\mathbf{e}_1), \chi(\boldsymbol{n}+\mathbf{e}_2), \chi(\boldsymbol{n}-\mathbf{e}_2), \ldots, \chi(\boldsymbol{n}+\mathbf{e}_d), \chi(\boldsymbol{n}-\mathbf{e}_d)$ are $\chi(\boldsymbol{n}), \chi(\boldsymbol{n})+1, \chi(\boldsymbol{n})-1, \chi(\boldsymbol{n})+2, \chi(\boldsymbol{n})-2, \ldots, \chi(\boldsymbol{n})+d, \chi(\boldsymbol{n})-d \pmod{2d+1}$. □

Theorem 11. *Let $d \geq 2$. Let $S \subseteq \mathbb{Z}^d$ be a set that contains $\mathbf{0}$ and $j \leq d$ other elements $\boldsymbol{s}_1, \ldots, \boldsymbol{s}_j$, where no nontrivial integer linear combination of $\{\boldsymbol{s}_1, \ldots, \boldsymbol{s}_j\}$ is $\mathbf{0}$. Then S tiles $\mathbb{Z}^d$.*

Proof. Let $H \subseteq \mathbb{Z}^d$ be the set of all integer linear combinations of $\{\boldsymbol{s}_1, \ldots, \boldsymbol{s}_j\}$. By Theorem 9, there is a set $T \subseteq \mathbb{Z}^j$ such that $\{\mathbf{0}, \mathbf{e}_1, \ldots, \mathbf{e}_j\} \oplus T = \mathbb{Z}^j$. Let $M : \mathbb{Z}^j \to \mathbb{Z}^d$ be the unique linear transformation which maps $\mathbf{e}_i$ to $\boldsymbol{s}_i$ for each $i \leq j$. Then $\{\mathbf{0}, \boldsymbol{s}_1, \ldots, \boldsymbol{s}_j\}$ tiles H with complement set $\{M(t) : t \in T\}$. Since H is a subgroup of $\mathbb{Z}^d$, by Lemma 8, S tiles $\mathbb{Z}^d$. □

We can now determine the polychromatic number of any set S of cardinality 3 or 4 in $\mathbb{Z}^d$, $d \geq 2$. Since translation does not affect polychromatic numbers, in all cases we may assume $\mathbf{0} \in S$.

Theorem 12. *Let $d \geq 2$ and suppose $S \subseteq \mathbb{Z}^d$ has cardinality 3, with $\mathbf{0} \in S$. Then $p_{\mathbb{Z}^d}(S) = 3$ if the three points are in general position or if they are collinear and there exists $S' \subseteq \mathbb{Z}$ with $p_{\mathbb{Z}}(S') = 3$ such that S' is the image of S after $d-1$ projections. Otherwise $p_{\mathbb{Z}^d}(S) = 2$.*

Proof. Theorem 11 implies that if $d \geq 2$ and $S \subseteq \mathbb{Z}^d$ consists of three points in general position, then S tiles $\mathbb{Z}^d$, and thus $p(S) = 3$. If $S \subseteq \mathbb{Z}^d$ has three collinear points, then Theorem 8 implies the problem is equivalent to finding the polychromatic number of a set of three integers, which is either 2 or 3 and can be determined using Theorem 3. □

Theorem 13. *Let $d \geq 2$ and suppose $S \subseteq \mathbb{Z}^d$ has cardinality 4, with $\mathbf{0} \in S$. Then*

1. *If all points of S are collinear, $p_{\mathbb{Z}^d}(S)$ is 3 or 4.*
2. *If exactly three points of S are collinear, $p_{\mathbb{Z}^d}(S) = 4$.*

3. *If $d \geq 3$ and S has four points in general position, $p_{\mathbb{Z}^d}(S) = 4$.*

4. *If $d = 2$ and S has four points in general position, $p_{\mathbb{Z}^2}(S)$ is 3 or 4.*

Proof. For $d \geq 2$ and a set $S \subseteq \mathbb{Z}^d$ with $|S| = 4$, Proposition 1 implies that there is a set $S' \subseteq \mathbb{Z}$ where $|S'| = 4$ and S' can be obtained by $d-1$ projections of S. Thus Theorem 1 and Lemma 7 imply that $p(S) \geq 3$. Determining whether $p(S)$ is 3 or 4 is equivalent to determining whether S tiles $\mathbb{Z}^d$. As with the $|S| = 3$ case, we can examine cases depending on how many points of S are collinear.

If the four points of S are in general position, then if none is a nontrivial integer linear combination of the others, $p(S) = 4$ by Theorem 11. Otherwise, we can assume $S \subseteq \mathbb{Z}^2$. In this case, $p(S)$ can be 3, e.g., if $S = \{(0,0), (1,0), (0,1), (1,2)\} \subseteq \mathbb{Z}^2$. It can also be 4, for example if $S = \{(0,0), (1,0), (0,1), (1,1)\} \subseteq \mathbb{Z}^2$. Szegedy [15] gave an algorithm to determine if a set of cardinality 4 tiles $\mathbb{Z}^2$.

If the four points of S are all collinear, then $p(S)$ is determined by applying Theorems 8 and 3.

If exactly three of the four points are collinear, then without loss of generality assume $S = \{\mathbf{0}, (a, 0, \ldots, 0), (b, 0, \ldots 0), \boldsymbol{s}\}$, with $0 < a < b$ and $\boldsymbol{s} = (s_1, \ldots, s_d)$. Then S is in the subgroup $\{x \in \mathbb{Z}^d : x_i \in s_i\mathbb{Z} \text{ for } 2 \leq i \leq d\}$. By Lemmas 3 and 4 we may assume that $s_i = 1$ for some $2 \leq i \leq d$, and thus by a sequence of projections, S can be projected to the set $\{0, a, b, c\}$, for any $c \in \mathbb{Z}$. By Lemma 7 it suffices to show that there exists $c \in \mathbb{Z}$ such that $p_{\mathbb{Z}}(\{0, a, b, c\}) = 4$. Without loss of generality, we may assume that a and b have different parity, and in this case Theorem 3 implies that $S = \{0, a, b, a+b\}$ has polychromatic number 4. □

The fact that $p_{\mathbb{Z}^d}(S) = 4$ if S contains exactly three collinear points implies that for any set S of three integers, there is a 4-coloring of $\mathbb{Z}$ so that every translate of S gets three different colors. Here is an explicit example of one such coloring. Without loss of generality we need only consider sets of the following form: Let $S = \{0, a, b\} \subseteq \mathbb{Z}$ where a and b are positive with a even and b odd (note that we do not specify which is larger). Define the *alternating block 4-coloring relative to* S as follows: Given any $m \in \mathbb{Z}$, let q_m and r_m be the unique integers such that $m = 2aq_m + r_m$, where $-a \leq r_m < a$. Let $X(m) = 0$ if $r_m \geq 0$, $X(m) = 1$ otherwise. Let $Y(m) = 0$ if m is even, $Y(m) = 1$ otherwise. Define χ, the alternating block 4-coloring relative to S, so that $\chi(m) = (X(m), Y(m))$.

Theorem 14. *Let $S = \{0, a, b\} \subseteq \mathbb{Z}$ with $a, b > 0$, a even, and b odd. If the integers are colored with the alternating block 4-coloring relative to S then every translate of S has elements of three different colors.*

Proof. For any translate $n + S = \{n, n+a, n+b\}$ of S, $X(n) \neq X(n+a)$, while $Y(n) = Y(n+a) \neq Y(n+b)$. Thus χ has the property that any translate of S contains elements with three different colors. □

Given a set of three integers, the alternating block 4-coloring shows that there is a 4-coloring of the integers so that every translate gets three different colors. If $S \subset \mathbb{Z}$, $|S| = 4$, is there a 5-coloring of $\mathbb{Z}$ so that every translate of S has 4 colors? More generally, we ask the following question.

Question 2. Let $d \geq 1$. Given $k, n \in \mathbb{Z}$ with $k \leq n$, let $p(n, k)$ denote the minimum r so that any $S \subseteq \mathbb{Z}$ with $|S| = n$ has an r-coloring where every translate of S gets at least k colors. What is an asymptotic upper bound on $p(n, k(n))$ for natural choices of $k(n)$?

Acknowledgement. This research was conducted at a workshop made possible by the Alliance for Building Faculty Diversity in the Mathematical Sciences (DMS 0946431), held at the Institute for Computational and Experimental Research in Mathematics.

The authors wish to thank the anonymous referee for generalizing some results and greatly improving the exposition.

References

[1] N. Alon, A. Krech, and T. Szabó, Turán's theorem in the hypercube. *SIAM J. Discrete Math.* **21** (2007), 66–72.

[2] N. Alon, I. Kříž, and J. Nešetřil, How to color shift hypergraphs, *Studia Sci. Math. Hungar.* **30** (1995), 1–11.

[3] E. Amiot, *Music Through Fourier Space*, Springer, Cham, 2016.

[4] B. Bollobás, S. Janson, and O. Riordan, On covering by translates of a set. *Random Structures Algorithms* **38** (2011), 33–67.

[5] E. M. Coven and A. Meyerowitz, Tiling the integers with translates of one finite set, *J. Algebra* **212** (1999), 161–174.

[6] D. Harris and A. Srinivasan, A note on near-optimal coloring of shift hypergraphs, *Random Structures Algorithms* **48** (2016), 53–56.

[7] M. Kolountzakis and M. Matolcsi, Algorithms for translational tiling, *J. Math. Music* **3** (2009), 85–97.

[8] D. J. Newman, Complements of finite sets of integers, *Michigan Math. J.* **14** (1967), 481–486.

[9] D. J. Newman, Tesselation of integers, *J. Number Theory* **9** (1977), 107–111.

[10] D. Offner, Some Turán type results on the hypercube, *Discrete Math.* **309** (2009), 2905–2912.

[11] L. Özkahya and B. Stanton, On a covering problem in the hypercube, *Graphs Combin.* **31** (2015), 235–242.

[12] W. Schmidt and D. Tuller, Covering and packing in $\mathbb{Z}^n$ and R^n I, *Monatsh. Math.* **153** (2008), 265–281.

[13] W. Schmidt and D. Tuller, Covering and packing in $\mathbb{Z}^n$ and R^n II, *Monatsh. Math.* **160** (2010), 195–210.

[14] S. Stein, Tiling, packing, and covering by clusters, *Rocky Mountain J. Math.* **16** (1986), 277–321.

[15] M. Szegedy, Algorithms to tile the infinite grid with finite clusters, *Proceedings of the 39th Annual Symposium on the Foundations of Computer Science* (1998), 137–145.

[16] G. Weinstein, Some covering and packing results in number theory, *J. Number Theory* **8** (1976), 193–205.

5

SOME EXPLICIT FORMULAS FOR EULER-GENOCCHI POLYNOMIALS

H. Belbachir
Department of Mathematics, USTHB, RECITS Laboratory, Algiers, Algeria
hbelbachir@usthb.dz

S. Hadj-Brahim
Department of Mathematics, USTHB, RECITS Laboratory, Algiers, Algeria
bhadj@usthb.com

Abstract

As an extension to both Euler and Genocchi polynomials, we study the Euler-Genocchi family of polynomials introduced recently via its generating function. We give some combinatorial properties and expressions in terms of alternate power sums and related sums for the Stirling numbers of the second kind. Moreover, we give a formula for the Jacobi-Stirling numbers in terms of the Euler-Genocchi family of polynomials using the determinantal approach.

1. Introduction

The falling factorial is defined as

$$(n)_r := \begin{cases} 1, & r = 0; \\ n(n-1)\cdots(n-r+1), & r \in \mathbb{N}. \end{cases}$$

The Bernoulli polynomials $B_n(x)$, Euler polynomials $E_n(x)$ and Genocchi polynomials $G_n(x)$ admit, respectively, the following generating functions (see [17]):

$$\begin{aligned} \frac{t}{e^t-1}e^{xt} &= \sum_{n\geq 0} B_n(x)\frac{t^n}{n!} \qquad (|t| < 2\pi), \\ \frac{2}{e^t+1}e^{xt} &= \sum_{n\geq 0} E_n(x)\frac{t^n}{n!} \qquad (|t| < \pi), \\ \frac{2t}{e^t+1}e^{xt} &= \sum_{n\geq 0} G_n(x)\frac{t^n}{n!} \qquad (|t| < \pi). \end{aligned} \tag{1}$$

The Bernoulli numbers B_n, the Euler numbers E_n and the Genocchi numbers G_n are given by $B_n := B_n(0)$, $E_n := E_n(0)$ and $G_n := G_n(0)$, (see [9, 13]). Several approaches have been developed to deal with Appell polynomials. Together with the classical generating function methods, we mention the umbral calculus approach (see, for instance, Roman [17], Tempesta [19], and the references therein), and the determinantal approach introduced by Costabile and Longo [8]. Recently, Merca [14] expressed a convolution alterning formula for the Jacobi-Stirling numbers of the first kind $J^s_\gamma(n,k)$ and the second kind $J^S_\gamma(n,k)$ in terms of the Bernoulli polynomials. For other properties, see [2, 3, 11, 12].

In Section 2, we study some new properties of the Euler-Genocchi polynomials introduced in [6], such as the expression of the power of a variable, the Raabe-like formula, the linear recurrence and the difference equations. In Section 3, we evaluate the alternate power sums in terms of the Euler-Genocchi family of polynomials. In Section 4, related sums of the Stirling numbers of second kind with the Euler-Genocchi family of polynomials are given. In Section 5, we present a determinantal approach for the Euler-Genocchi polynomials. Finally, an extension in a determinantal representation for the Jacobi-Stirling numbers in terms of both Bernoulli and Euler-Genocchi polynomials is studied.

2. The Euler-Genocchi Polynomials

For $r \in \mathbb{N} \cup \{0\}$, Belbachir et al. [6] introduced the class of Euler-Genocchi polynomials $\{A_n^{(r)}(x)\}_{n\geq 0}$ defined by the generating function

$$\frac{2t^r}{e^t+1}e^{xt} = \sum_{n=0}^{\infty} A_n^{(r)}(x)\frac{t^n}{n!} \qquad (|t| < \pi), \tag{2}$$

with $A_j^{(r)}(x) = 0$ for $j < r$. We denote by $A_n^{(r)} := A_n^{(r)}(0)$, the Euler-Genocchi numbers of order r. The Euler and Genocchi polynomials are $E_n(x) = A_n^{(0)}(x)$ and $G_n(x) = A_n^{(1)}(x)$ respectively.

In [5], the author established the ordinary generating function for this class of polynomials. From (2), it follows that

$$\sum_{n=0}^{\infty} A_n^{(r)}(x+1)\frac{t^n}{n!} = \frac{2t^r}{e^t+1}e^{xt}e^t = \left(\sum_{n=0}^{\infty} A_n^{(r)}(x)\frac{t^n}{n!}\right)\left(\sum_{m=0}^{\infty}\frac{t^m}{m!}\right),$$

and using Cauchy's rule, we get the formula

$$A_n^{(r)}(x+1) = \sum_{k=0}^{n}\binom{n}{k}A_k^{(r)}(x). \tag{3}$$

On the other hand, we have

$$\begin{aligned}\sum_{n=0}^{\infty}\left(A_n^{(r)}(x+1)+A_n^{(r)}(x)\right)\frac{t^n}{n!} &= \frac{2t^r}{e^t+1}e^{(x+1)t}+\frac{2t^r}{e^t+1}e^{xt}\\ &= 2t^re^{xt}\\ &= \sum_{n=0}^{\infty}2x^n\frac{t^{n+r}}{n!}\\ &= \sum_{n=0}^{\infty}2(n)_rx^{n-r}\frac{t^n}{n!}.\end{aligned}$$

Equating the two series and using (3), we obtain the following theorem.

Theorem 2.1. *Let n and r be two non-negative integers such that $r \leq n$. The following formula holds:*

$$x^n = \frac{1}{2(n+r)_r}\left[A_{n+r}^{(r)}(x)+\sum_{k=0}^{n}\binom{n+r}{k+r}A_{k+r}^{(r)}(x)\right].$$

Corollary 2.2. [15, 18] *We have,*

$$x^n = \frac{1}{2}\left[E_n(x)+\sum_{k=0}^{n}\binom{n}{k}E_k(x)\right] \tag{4}$$

and

$$x^n = \frac{1}{2(n+1)}\left[G_{n+1}(x)+\sum_{k=0}^{n}\binom{n+1}{k+1}G_{k+1}(x)\right]. \tag{5}$$

Remark 2.3. Formulas (4) and (5) has been proved by Pintér and Srivastava [18, 15]. The corresponding formula for Bernoulli polynomials is (see for instance [18, Eq (27)])

$$x^n = \frac{1}{n+1}\sum_{k=0}^{n}\binom{n+1}{k}B_k(x).$$

Classically, a multiplication theorem for both Euler and Bernoulli polynomials has been proved by Raabe [16] (see also Bayad and Komatsu [4]). For odd m, we have

$$\sum_{k=0}^{m-1}B_n\left(\frac{x+k}{m}\right) = \frac{1}{m^{n-1}}B_n(x)$$

and

$$\sum_{k=0}^{m-1}(-1)^k E_n\left(\frac{x+k}{m}\right) = \frac{1}{m^n}E_n(x).$$

The corresponding formula for Euler-Genocchi polynomials is given by the following theorem.

Theorem 2.4. *Let r and m be non-negative integers with m odd, and $\{A_n^{(r)}(x)\}_{n\geq 0}$ be the class of polynomials given by (2). Then*

$$\sum_{k=0}^{m-1}(-1)^k A_n^{(r)}\left(\frac{x+k}{m}\right) = \frac{1}{m^{n-r}}A_n^{(r)}(x).$$

Proof. From (2), it follows that

$$\begin{aligned}
\sum_{n\geq 0}\sum_{k=0}^{m-1}(-1)^k A_n^{(r)}\left(\frac{x+k}{m}\right)\frac{t^n}{n!} &= \sum_{k=0}^{m-1}(-1)^k \sum_{n\geq 0} A_n^{(r)}\left(\frac{x+k}{m}\right)\frac{t^n}{n!} \\
&= \sum_{k=0}^{m-1}(-1)^k \frac{2t^r}{e^t+1}\exp\left((x+k)t/m\right) \\
&= \frac{2t^r}{exp(t)+1}\exp(xt/m)\frac{[1-(-\exp(t/m))^m]}{1+\exp(t/m)} \\
&= \frac{2t^r}{\exp(t/m)+1}\exp(xt/m) \\
&= m^r \sum_{n\geq 0} A_n^{(r)}(x)\frac{(t/m)^n}{n!} \\
&= \sum_{n\geq 0}\frac{1}{m^{n-r}}A_n^{(r)}(x)\frac{t^n}{n!}.
\end{aligned}$$

Equating the coefficients of t^n on both sides, we get the result. □

Theorem 2.5. *The Euler-Genocchi polynomials satisfy the following linear homogeneous recurrence relation*

$$\left(1-\frac{r}{n+1}\right)A_{n+1}^{(r)}(x) + (2-x)A_n^{(r)}(x) = \sum_{k=0}^{n}\binom{n}{k}\frac{A_{k+r}^{(r)}}{(k+r)_r}A_{n-k}^{(r)}(x). \qquad (6)$$

Proof. Differentiating both sides of (2) with respect to t yields

$$\begin{aligned}
\frac{d}{dt}\sum_{n\geq 0} A_n^{(r)}(x)\frac{t^n}{n!} &= \frac{2xt^r}{e^t+1}e^{xt} + \frac{2rt^{r-1}(e^t+1)-2t^re^t}{(e^t+1)^2}e^{xt}\\
&= r\sum_{n\geq 0} A_n^{(r)}(x)\frac{t^{n-1}}{n!} + (x-2)\sum_{n\geq 0} A_n^{(r)}(x)\frac{t^n}{n!} + \frac{1}{e^t+1}\sum_{n\geq 0} A_n^{(r)}(x)\frac{t^n}{n!}\\
&= r\sum_{n=0}^{\infty}\frac{1}{n+1}A_{n+1}^{(r)}(x)\frac{t^n}{n!} + (x-2)\sum_{n=0}^{\infty} A_n^{(r)}(x)\frac{t^n}{n!}\\
&+ \left(\sum_{n\geq 0} A_n^{(r)}\frac{t^{n-r}}{n!}\right)\left(\sum_{n\geq 0} A_n^{(r)}(x)\frac{t^n}{n!}\right),
\end{aligned}$$

$$\begin{aligned}
\sum_{n\geq 0} A_{n+1}^{(r)}(x)\frac{t^n}{n!} &= r\sum_{n\geq 0}\frac{1}{n+1}A_{n+1}^{(r)}(x)\frac{t^n}{n!} + (x-2)\sum_{n\geq 0} A_n^{(r)}(x)\frac{t^n}{n!}\\
&+ \sum_{n\geq 0}\sum_{k=0}^{n}\binom{n}{k}\frac{1}{(k+r)_r}A_{k+r}^{(r)}A_{n-k}^{(r)}(x)\frac{t^n}{n!}.
\end{aligned}$$

Then, using simple manipulations and equating the coefficients of t^k on both sides leads to the result. □

As $A_n^{(r)}(x)$ belongs to the family of Appell polynomials, we have

$$D_x^k A_n^{(r)}(x) = (n)_k A_{n-k}^{(r)}(x), \quad \text{where} \quad D_x^k := \frac{d^k}{dx^k}.$$

The recurrence relation (6) can be written as

$$A_{n+1}^{(r)}(x) = \frac{n+1}{n-r+1}\left[(x-2) + \sum_{k=0}^{n}\frac{A_{k+r}^{(r)}}{k!(k+r)_r}D_x^k\right]A_n^{(r)}(x).$$

Introducing the shift operator

$$E := \frac{n+1}{n-r+1}\left[(x-2) + \sum_{k=0}^{n}\frac{A_{k+r}^{(r)}}{k!(k+r)_r}D_x^k\right]$$

and the derivative operator

$$\widehat{E} := \frac{1}{n+1}D_x$$

and applying both operators $\widehat{E}$ and E to $A_n^{(r)}(x)$, we obtain

$$\widehat{E}EA_n^{(r)}(x) = A_n^{(r)}(x). \tag{7}$$

From (6) and (7), we get the following theorem:

Theorem 2.6. *The Euler-Genocchi polynomials $A_n^{(r)}(x)$ satisfy the differential equation*

$$\frac{A_{r+n-1}^{(r)}}{(n-1)!(r+n-1)_r}y^{(n)}+\cdots+\frac{A_{r+1}^{(r)}}{(r+1)_r}y^{(2)}+\left(\frac{A_r^{(r)}}{(r)_r}+(x-2)\right)y'+(r-n)y=0.$$

3. The Power Sum and the Alternate Power Sum

The power sum $S_k(n)$ and the alternate power sum $T_k(n)$ are defined respectively by:

$$S_k(n):=\sum_{i=0}^{n} i^k \qquad \text{and} \qquad T_k(n):=\sum_{i=0}^{n}(-1)^i i^k.$$

The exponential generating function of $(S_k(n))_k$ and $(T_k(n))_k$ are given by

$$\sum_{k=0}^{\infty} S_k(n)\frac{t^k}{k!}=\frac{1-e^{(n+1)t}}{1-e^t} \qquad \text{and} \qquad \sum_{k=0}^{\infty} T_k(n)\frac{t^k}{k!}=\frac{1-(-1)^{n+1}e^{(n+1)t}}{1+e^t}. \tag{8}$$

The power sum $S_k(n)$ and the alternate power sum $T_k(n)$ are related to Bernoulli and Euler polynomials in integer variable, (see [1, Eq. (23.1.4)]) via:

$$S_k(n)=\frac{B_{k+1}(n+1)-B_{k+1}}{k+1} \qquad \text{and} \qquad T_k(n)=\frac{(-1)^nE_k(n+1)+E_k}{2}. \tag{9}$$

For the alternate sums, we have an extension to the Euler-Genocchi polynomials in integer variables.

Theorem 3.1. *Let n and k be nonnegative integers. The following formula holds:*

$$T_k(n)=\frac{(-1)^nA_{k+r}^{(r)}(n+1)+A_{k+r}^{(r)}}{2(k+r)_r}. \tag{10}$$

Proof. From (8), we have

$$\sum_{k=0}^{\infty} T_k(n)\frac{t^k}{k!}=\frac{1-(-1)^{n+1}e^{(n+1)t}}{e^t+1} = \frac{2}{2(e^t+1)}-\frac{2(-1)^{n+1}e^{(n+1)t}}{2(e^t+1)}.$$

Multiplying both sides by t^r, we get

$$\sum_{k=0}^{\infty} T_k(n)\frac{t^{k+r}}{k!}=\frac{2t^r}{2(e^t+1)}-\frac{2t^r(-1)^{n+1}e^{(n+1)t}}{2(e^t+1)}.$$

Then, using (2) and a direct computation gives

$$\sum_{k=0}^{\infty}(k)_r T_{k-r}(n)\frac{t^k}{k!} = \frac{1}{2}\sum_{k=0}^{\infty} A_k^{(r)}\frac{t^k}{k!} - \frac{(-1)^{n+1}}{2}\sum_{k=0}^{\infty} A_k^{(r)}(n+1)\frac{t^k}{k!}.$$

Then equating the coefficients of t^k on both sides leads to the result. □

Remark 3.2. For $r = 0$, Expression (10) gives the alternate sum in terms of Euler polynomials (9). For $r = 1$, we express the alternate sum in terms of Genocchi polynomials:

$$T_k(n) = \frac{(-1)^n G_{k+1}(n+1) + G_{k+1}}{2(k+1)}.$$

4. Related sums with Stirling numbers of the second kind

Given nonnegative integers n and k ($k \leq n$), it is known that the Stirling numbers of the second kind $S(n,k)$ satisfy [7, p. 206]

$$S(n,k) = \frac{1}{k!}\sum_{j=0}^{k}(-1)^{k-j}\binom{k}{j} j^n.$$

In the following theorem, we give a link between the Euler-Genocchi numbers of order r, Bernoulli numbers, and Stirling numbers.

Theorem 4.1. *We have*

$$\sum_{k=0}^{n}\binom{n}{k} A_{n-k}^{(r)} B_k = 2^{n-r}(n)_r \sum_{k=0}^{n-r}\frac{(-1)^k}{k+1} k! S(n-r,k). \tag{11}$$

Proof. Let t be a real number with $|t| < \pi$, and set $\mathbb{T}(t) := \dfrac{2t^r}{e^t+1} \times \dfrac{t}{e^t-1}$. Taking into account the right-hand side of (1) and (2) for $x = 0$, a direct computation gives

$$\mathbb{T}(t) = \sum_{n\geq 0}\left\{\sum_{k=0}^{n}\binom{n}{k} A_{n-k}^{(r)} B_k\right\}\frac{t^n}{n!}.$$

On the other hand, we have

$$\begin{aligned}\mathbb{T}(t) = \frac{2t^{r+1}}{e^{2t}-1} = \frac{t^r \ln\left((e^{2t}-1)+1\right)}{e^{2t}-1} &= t^r \sum_{k\geq 0} \frac{(-1)^k}{k+1}(e^{2t}-1)^k \\ &= t^r \sum_{k\geq 0} \frac{(-1)^k}{k+1} \sum_{j=0}^{k} (-1)^{k-j} \binom{k}{j} e^{2jt} \\ &= t^r \sum_{k\geq 0} \frac{(-1)^k}{k+1} \sum_{j=0}^{k} (-1)^{k-j} \binom{k}{j} \sum_{n\geq 0} 2^n j^n \frac{t^n}{n!} \\ &= \sum_{n\geq 0} 2^n \frac{t^{n+r}}{n!} \sum_{k\geq 0} \frac{(-1)^k}{k+1} \sum_{j=0}^{k} (-1)^{k-j} \binom{k}{j} j^n \\ &= \sum_{n\geq 0} 2^n \frac{t^{n+r}}{n!} \sum_{k\geq 0} \frac{(-1)^k}{k+1} k! S(n,k) \\ &= \sum_{n\geq 0} 2^{n-r} (n)_r \frac{t^n}{n!} \sum_{k\geq 0} \frac{(-1)^k}{k+1} k! S(n-r,k).\end{aligned}$$

Also, $S(n-r,k) = 0$ when $k > n-r$. Hence

$$\mathbb{T}(t) = \sum_{n=0}^{\infty} 2^{n-r} (n)_r \sum_{k=0}^{n-r} \frac{(-1)^k}{k+1} k! S(n-r,k) \frac{t^n}{n!}.$$

Comparing the preceding two expansions of $\mathbb{T}(t)$, we get Equation (11). □

In particular, we get the following corollary

Corollary 4.2. *For $r = 0$ and $r = 1$ we have*

$$\sum_{k=0}^{n} \binom{n}{k} E_{n-k} B_k = 2^n \sum_{k=0}^{n} \frac{(-1)^k}{k+1} k! S(n,k)$$

and

$$\sum_{k=0}^{n} \binom{n}{k} G_{n-k} B_k = 2^{n-1} n \sum_{k=0}^{n-1} \frac{(-1)^k}{k+1} k! S(n-1,k).$$

5. A Determinantal Approach for Euler-Genocchi Polynomials

Belbachir et al. [6] proposed a determinantal form to express x^n in terms of both Euler-Genocchi and Bernoulli polynomials. We give an alternate version to Theorem 3.1. in [6].

Theorem 5.1. *For every integer $n \geq r$ $(r \geq 1)$, the following formula holds:*

$$x^n = \sum_{k=0}^{n+r} \Lambda_{n,k}^{(r)} \times \nabla_{n-(k-r),k}^{(r-1)}(x+1, x), \tag{12}$$

where $\nabla_{n,k}^{(r)}(x,y) = \begin{vmatrix} B_n(x) & A_k^{(r)}(y) \\ B_n(y) & A_k^{(r)}(x) \end{vmatrix}$ *and* $\Lambda_{n,k}^{(r)} = \frac{1}{2^{n+1}(n+r)_r}\binom{n+r}{k}$.

Proposition 5.2. *Let n and k be nonnegative integers and $r \geq 1$. Then the following identity holds:*

$$\sum_{s=0}^{m-1} \nabla_{n,k}^{(r)}(s+1, s) = \nabla_{n,k}^{(r)}(m, 0).$$

Proof. We have

$$\begin{aligned}
\sum_{s=0}^{m-1} \nabla_{n,k}^{(r)}(s+1, s) &= \sum_{s=0}^{m-1} \begin{vmatrix} B_n(s+1) & A_k^{(r)}(s) \\ B_n(s) & A_k^{(r)}(s+1) \end{vmatrix} \\
&= \sum_{s=0}^{m-1} \left[B_n(s+1)A_k^{(r)}(s+1) - B_n(s)A_k^{(r)}(s)\right] \\
&= \left[B_n(m)A_k^{(r)}(m) - B_n(0)A_k^{(r)}(0)\right] \\
&= \begin{vmatrix} B_n(m) & A_k^{(r)}(0) \\ B_n(0) & A_k^{(r)}(m) \end{vmatrix}.
\end{aligned}$$

□

Theorem 5.3. *For all integers n, m and $r \geq 0$, we have*

$$S_n(m) = \sum_{k=0}^{n+r} \Lambda_{n,k}^{(r)} \times \nabla_{n-(k-r),k}^{(r-1)}(m+1, 0). \tag{13}$$

Proof. Letting $x = s$ in (12), we get:

$$s^n = \sum_{k=0}^{n+r} \Lambda_{n,k}^{(r)} \times \nabla_{n-(k-r),k}^{(r-1)}(s+1, s).$$

We conclude by summing up the s^n and using Proposition 5.2. □

6. Jacobi-Stirling Numbers in a Determinantal Form Representation

For n and k nonnegative integers and γ a positive real number, the Jacobi-Stirling numbers of the first kind $J_\gamma^s(n,k)$ and the second kind $J_\gamma^S(n,k)$ are given respectively by

$$E(t) := \prod_{k=1}^{n}(1+k(k+2\gamma-1)t) = \sum_{k=0}^{\infty} J_\gamma^s(n+1,n+1-k)\,t^k \tag{14}$$

and

$$H(t) := \prod_{k=1}^{n}(1-k(k+2\gamma-1)t)^{-1} = \sum_{k=0}^{\infty} J_\gamma^S(n+k,n)\,t^k. \tag{15}$$

Merca [14] expresses a convolution alternating formula with Jacobi-Stirling numbers of both kinds in terms of Bernoulli polynomials in an integer variable as follows:

Theorem 6.1. *We have*

$$\sum_{j=1}^{k}(-1)^{j-1} j J_\gamma^s(n+1,n+1-j)\,J_\gamma^S(n+k-j,n) = \sum_{j=0}^{k} \frac{B_{k+j+1}(n+1)-B_{k+j+1}}{k+j+1} \times \binom{k}{j}(2\gamma-1)^{k-j}.$$

In the following Theorem, we propose another version to express the convolution alternating formula with Jacobi-Stirling numbers of both kinds in determinantal form.

Theorem 6.2. *We have*

$$\sum_{j=1}^{k}(-1)^{j-1} j J_\gamma^s(n+1,n+1-j)\,J_\gamma^S(n+k-j,n) = \sum_{j=0}^{k}\binom{k}{j}(2\gamma-1)^{k-j}\sum_{t=0}^{k+j+r}\Lambda_{k+j,t}^{(r)} \times \nabla_{k+j-(t-r),t}^{(r-1)}(n,0). \tag{16}$$

Proof. It is easy to verify that

$$\frac{d}{dt}\ln H(t) = E^{'}(-t)H(t).$$

We replace t by $-t$ in the power series (14) and we calculate the derivative. We get:

$$E^{'}(-t) = \sum_{k=0}^{\infty} k J_\gamma^s(n+1,n+1-k)\,(-t)^{k-1}.$$

On the one hand, we have

$$\frac{d}{dt}\ln H(t) = \sum_{k=1}^{n}\frac{k(k+2\gamma-1)}{1-k(k+2\gamma-1)t} = \sum_{k=1}^{n}\sum_{j=1}^{\infty} k^j(k+2\gamma-1)^j t^{j-1}.$$

Interchanging the order of the resulting double sums, we find

$$\sum_{k=1}^{\infty}\sum_{j=1}^{n} j^k(j+2\gamma-1)^k t^{k-1} = \left(\sum_{k=1}^{\infty} kJ_{\gamma}^{s}\left(n+1,n+1-k\right)(-t)^{k-1}\right)\left(\sum_{k=0}^{\infty} J_{\gamma}^{S}\left(n+k,n\right)t^k\right).$$

Taking into account the series product and equating coefficients of t^{k-1} on both sides, we find the following identity:

$$\sum_{j=1}^{n} j^k(j+2\gamma-1)^k = \sum_{j=1}^{k}(-1)^{j-1} jJ_{\gamma}^{s}\left(n+1,n+1-j\right)J_{\gamma}^{S}\left(n+k-j,n\right). \quad (17)$$

On the other hand, taking into account the left hand side of (17), we get

$$\begin{aligned}\sum_{j=1}^{n} j^k(j+2\gamma-1)^k &= \sum_{j=1}^{n} j^k \sum_{i=0}^{k}\binom{k}{i} j^i(2\gamma-1)^{k-i}\\ &= \sum_{j=0}^{n-1}\sum_{i=0}^{k}\binom{k}{i} j^{k+i}(2\gamma-1)^{k-i}\\ &= \sum_{j=0}^{k}\binom{k}{j}(2\gamma-1)^{k-j}\sum_{i=0}^{n-1} i^{k+j}.\end{aligned}$$

Now, upon replacing the factor i^{k+j} in Formula (13) by its expression, we obtain

$$\sum_{j=1}^{n} j^k(j+2\gamma-1)^k = \sum_{j=0}^{k}\binom{k}{j}(2\gamma-1)^{k-j}\sum_{t=0}^{k+j+r}\Lambda_{k+j,t}^{(r)}\times\nabla_{k+j-(t-r),t}^{(r-1)}(n,0). \quad (18)$$

Comparing the two expansions (17) and (18), we get the result. □

For $\gamma = 1$, (16) gives Legendre-Stirling numbers of both kinds introduced by Everitt et al. [10] in 2002. Some of their properties may be found in [14, 2, 3]. We have the following property.

Corollary 6.3. *Let $J_1^s(n,k)$ and $J_1^S(n,k)$ be the Legendre-Stirling numbers of the first and second kind respectively, we have*

$$\sum_{j=1}^{k}(-1)^{j-1} jJ_1^{s}\left(n+1,n+1-j\right)J_1^{S}\left(n+k-j,n\right) = \sum_{j=0}^{k}\binom{k}{j}\sum_{t=0}^{k+j+r}\Lambda_{k+j,t}^{(r)}\nabla_{k+j-(t-r),t}^{(r-1)}(n,0).$$

Letting $\gamma = 1/2$ in Formula (17), and using the same approach as in the proof of Theorem 6.2, allows us to express the Chebyshev-Stirling numbers (see [12]) in terms of a determinantal representation in the following corollary.

Corollary 6.4. *Let $J^{s}_{1/2}(n,k)$ and $J^{S}_{1/2}(n,k)$ be Chebyshev-Stirling numbers of the first and second kind respectively. We have*

$$\sum_{j=1}^{k}(-1)^{j-1}jJ^{s}_{1/2}(n+1,n+1-j)\,J^{S}_{1/2}(n+k-j,n)=\sum_{t=0}^{2k+r}\Lambda^{(r)}_{2k,t}\times\nabla^{(r-1)}_{2k-(t-r),t}(n+1,1).$$

For $r=1$ and $r=2$, we have the determinantal representation for Jacobi-Stirling numbers in terms of Bernoulli-Euler numbers and Bernoulli-Genocchi numbers respectively. Hence, we get the following corollary.

Corollary 6.5. *We have*

$$\sum_{j=1}^{k}(-1)^{j-1}jJ^{s}_{\gamma}(n+1,n+1-j)\,J^{S}_{\gamma}(n+k-j,n)=\sum_{j=0}^{k}\binom{k}{j}(2\gamma-1)^{k-j}$$
$$\times\sum_{t=0}^{k+j+1}\frac{1}{2^{k+j+1}(k+j+1)}\binom{k+j+1}{t}\begin{vmatrix}B_{k+j-(t-1)}(n) & E_t\\ B_t & E_{k+j-(t-1)}(n)\end{vmatrix} \tag{19}$$

and

$$\sum_{j=1}^{k}(-1)^{j-1}jJ^{s}_{\gamma}(n+1,n+1-j)\,J^{S}_{\gamma}(n+k-j,n)=\sum_{j=0}^{k}\binom{k}{j}(2\gamma-1)^{k-j}$$
$$\times\sum_{t=0}^{k+j+2}\frac{1}{2^{k+j+1}(k+j+1)(k+j+2)}\binom{k+j+2}{t}\begin{vmatrix}B_{k+j-(t-2)}(n) & G_t\\ B_t & G_{k+j-(t-2)}(n)\end{vmatrix}.$$

The determinantal representation (19) of Jacobi-Stirling numbers involves the Bernoulli and Euler polynomials. However, these formulas can be expressed in terms of the Bernoulli polynomials, using some well known closed relations between Bernoulli and Euler polynomials. Indeed, it is well-known that (see [9, 18]):

$$E_n(x)=\frac{2}{n+1}\left[B_{n+1}(x)-2^{n+1}B_{n+1}\left(\frac{x}{2}\right)\right]. \tag{20}$$

Combining (20) and (19), we obtain

$$\nabla^{(0)}_{m,k}(n,0)=\frac{2}{k+1}\begin{vmatrix}B_m(n) & \left(1-2^{k+1}\right)B_{k+1}\\ B_m & B_{k+1}(n)-2^{k+1}B_{k+1}(\frac{n}{2})\end{vmatrix}.$$

For all integers s, k and a, we let

$$\mathbb{B}_{s,k}(a):=\begin{vmatrix}B_s(a) & \left(1-2^{k}\right)B_k\\ B_s & B_k(a)-2^kB_k(\frac{a}{2})\end{vmatrix}.$$

Expression (19) may then be written in terms of Bernoulli polynomials:

$$\sum_{j=1}^{k}(-1)^{j-1}jJ_{\gamma}^{s}(n+1,n+1-j)\,J_{\gamma}^{S}(n+k-j,n)$$
$$=\sum_{j=0}^{k}\binom{k}{j}(2\gamma-1)^{k-j}\sum_{t=0}^{k+j+1}\frac{2}{t+2}\Lambda_{k+j,t}^{(0)}\times\mathbb{B}_{k+j-(t-1),t+1}(n).$$

References

[1] M. Abramowitz and I. A. Stegun, *Handbook of Mathematical Functions with Formulas, Graphs, and Mathematical Tables*, Reprint of the 1972 edition, Dover Publications, Inc., New York, 1992.

[2] G. E. Andrews and L. L. Littlejohn, A combinatorial interpretation of the Legendre-Stirling numbers, *Proc. Amer. Math. Soc.* **137** (2009), 2581–2590.

[3] G. E. Andrews, W. Gawronski and L. L. Littlejohn, The Legendre-Stirling numbers, *Discrete Math.* **311** (2011), 1255–1272.

[4] A. Bayad and T. Komatsu, New characterization of Appell polynomials, *Integral Transforms Spec. Funct.* **28** (2017), 212–222.

[5] H. Belbachir, Ordinary generating function for a class of Appell polynomials and Stirling polynomials, Submitted.

[6] H. Belbachir, S. Hadj-Brahim and M. Rachidi, Another determinental approach for a family of Appell polynomials, *Filomat* **12** (2018), 4155–4164.

[7] L. Comtet, *Advanced Combinatorics, The Art of Finite and Infinite Expansions*, revised and enlarged edition, D. Reidel Publishing Co., Dordrecht and Boston, 1974.

[8] F. A. Costabile and E. Longo, A determinantal approach to Appell polynomials, *J. Comput. Appl. Math.* **234** (2010), 1528–1542.

[9] A. Erdélyi, W. Magnus, F. Oberhettinger and F. G. Tricomi, *Higher Transcendental Functions*, Volume I, McGraw-Hill, New York, (1953).

[10] W. N. Everitt, L. L. Littlejohn and R. Wellman, Legendre polynomials, Legendre-Stirling numbers, and the left-definite analysis of the Legendre differential expression, *J. Comput. Appl. Math.* **148** (2002), 213–238.

[11] W. N. Everitt, K. H. Kwon, L. L. Littlejohn, R. Wellman and G. J. Yoon, Jacobi-Stirling numbers, Jacobi polynomials and the left-definite analysis of the classical Jacobi differential expression, *J. Comput. Appl. Math.* **208** (2007), 29–56.

[12] W. Gawronski, L. L. Littlejohn and T. Neuschel, Asymptotics of Stirling and Chebyshev-Stirling numbers of the second kind, *Stud. Appl. Math.* **133** (2014), 1–17.

[13] T. Kim, Some identities for the Bernoulli, the Euler and the Genocchi numbers and polynomials, *Adv. Stud. Contemp. Math.* (Kyungshang) **20** (2010), 23–28.

[14] M. Merca, A connection between Jacobi-Stirling numbers and Bernoulli polynomials, *J. Number Theory* **151** (2015), 223–229.

[15] Á. Pintér and H. M. Srivastava, Addition theorems for the Appell polynomials and the associated classes of polynomial expansions, *Aequationes Math.* **85** (2013), 483–495.

[16] J. L. Raabe, Zurückführung einiger Summen and bestimmten Integrale auf die Jacob Bernoullische Function, *Journal für die reine and angrew. Math.* **42** (1851) 348–376, especially p. 356.

[17] S. Roman, *The Umbral Calculus*, Academic Press, New York, 1984.

[18] H. M. Srivastava and A. Pinter, Remarks on some relationships between the Bernoulli and Euler polynomials, *Appl. Math. Lett.* **17** (2004), 375–380.

[19] P. Tempesta, On Appell sequences of polynomials of Bernoulli and Euler type, *J. Math. Anal. Appl.* **341** (2008), 1295–1310.

6

ON GENERALIZED DERANGEMENTS AND SOME ORTHOGONAL POLYNOMIALS

István Mező[1]
School of Mathematics and Statistics, Nanjing University of Information Science & Technology, Nanjing, P.R. China
istvanmezo81@gmail.com

José L. Ramírez [2]
Departamento de Matemáticas, Universidad Nacional de Colombia, Bogotá, Colombia
jlramirezr@unal.edu.co

Chen-Ying Wang[3]
School of Mathematics and Statistics, Nanjing University of Information Science & Technology, Nanjing, P.R. China
wang.chenying@163.com

Abstract

We make use of orthogonal polynomials to give new identities for a recently defined class of combinatorial numbers, the r-derangements. In the second part, we carry out the complex analysis of the derangement function. In the third part of the paper we present a combinatorial generalization of the r-derangement numbers.

1. Introduction

The derangement number $D(n)$ denotes the number of fixed point-free permutations (FPF for short) on n letters. In a recent paper [21], the authors defined a new class of derangements: an *r-derangement* is an FPF-permutation such that the first r elements are restricted to be in different cycles. For example,

$$(172)(35)(468)$$

[1]The research of István Mező was supported by the National Natural Science Foundation of China. Grant no. 11501299.

[2]The research of José L. Ramírez was partially supported by Universidad Nacional de Colombia, Project No. 37805.

[3]The research of Chen-Ying Wang was supported by the National Natural Science Foundation of China (Grant no. 11601239).

is not a 2-derangement (1 and 2 share a common cycle), while

$$(197)(2845)(36)$$

is a 2-derangement and even a 3-derangement (but not a 4-derangement). The number of r-derangements on $n+r$ letters is denoted by $D_r(n)$.

First we are going to study the r-derangement numbers $D_r(n)$ by their relation to some orthogonal polynomials. Then we define the complex function $D(z)$, extending $D(n)$ to complex numbers, and study its properties. In the final part of the paper we look at a possible generalization of the r-derangement numbers. This generalization, among other things, leads to the generalization of the recurrence

$$D(n) = nD(n-1) + (-1)^n$$

for the r-derangements (see (17)).

Our motivation comes from the sources [1, 7, 8, 9]. In these papers, M. Ismail et al. applied the theory of orthogonal polynomials to prove identities and develop asymptotic approximations with respect to certain generalized derangement numbers. Those tools can also be applied for the r-derangement numbers, which were defined later.

2. Identities for $D_r(n)$ Related to the Charlier Polynomials

The r-derangements are intimately related to the Charlier polynomials [6], and thus we recall some of the properties of these polynomials.

The Charlier polynomials $C_n^{(a)}(y)$ can be defined by the following equivalent relations:

- (Generating function) $\sum_{n\geq 0} C_n^{(a)}(y)\frac{x^n}{n!} = e^{-ax}(1+x)^y$,
- (Explicit formula) $C_n^{(a)}(y) = \sum_{k=0}^{n} \binom{n}{k}\binom{y}{k} k!(-a)^{n-k}$,
- (Recursion) $C_{n+1}^{(a)}(y) = (y-n-a)C_n^{(a)}(y) - anC_{n-1}^{(a)}(y)$,
- (Hypergeometric representation) $C_n^{(a)}(y) = {}_2F_0\left[\begin{matrix}-n,\ -y\end{matrix}; -\frac{1}{a}\right]$,
- (Forward shift) $C_n^{(a)}(y+1) - C_n^{(a)}(y) = nC_{n-1}^{(a)}(y)$ (with appropriate initial conditions).

We apply the Charlier polynomials now to deduce some simple closed form and recursive identities with respect to $D_r(n)$. Instead of giving the detailed symbolic deduction we give only the combinatorial proof wherever we have found one.

It is known [21, Theorem 3] that the exponential generating function of the r-derangement numbers is

$$\sum_{n\geq 0} D_r(n)\frac{x^n}{n!} = \frac{x^r e^{-x}}{(1-x)^{r+1}}. \tag{1}$$

The exponential generating function of the Charlier polynomials immediately gives the relation

$$D_r(n) = (-1)^{n-r}\, n^{\underline{r}}\, C_{n-r}^{(-1)}(-(r+1)). \tag{2}$$

Here $a^{\underline{b}} = a(a-1)\cdots(a-b+1)$ is the falling factorial symbol $((a)^{\underline{0}} = 1)$. We will also use the symbol $a^{\overline{b}} = a(a+1)\cdots(a+b-1)$ for the raising factorial, also known as Pochhammer symbol.

A consequence of this relation and the above explicit formula for the Charlier polynomials results in the next statement that we will prove by combinatorial arguments.

Theorem 1. *For any $r, n \geq 0$ we have*

$$D_r(n) = n^{\underline{r}} \sum_{k=0}^{n-r} \binom{n-r}{k} (-1)^k (r+1)^{\overline{n-r-k}}. \tag{3}$$

Proof. The number of those permutations in which the first r elements are in disjoint cycles and are not fixed points is, as it is easy to see, equal to

$$n^{\underline{r}}(r+1)^{\overline{n-r}}. \tag{4}$$

Indeed, to guarantee that the elements $1, 2, \ldots, r$ are not fixed points, we choose an element from n to go to the cycle of 1, then we choose another element from $n-1$ to go to the block of 2, and so on. There are $n^{\underline{r}}$ such selections. Denote these r elements by A. The remaining $n-r$ elements go, one by one, to one of the cycles of $1, 2, \ldots, r$ or they form a new permutation σ. In order to assure that σ has no fixed points, we subtract from (4) (the $k = 0$ term in the theorem's statement) those permutations which have one fixed point among the elements $\{r+1, r+2, \ldots, r+n\} \setminus A$. This is the $k = 1$ term. Then we continue by the inclusion-exclusion principle. □

The basic recursion we found for $D_r(n)$ in [21] is

$$D_r(n) = rD_{r-1}(n-1) + (n-1)D_r(n-2) + (n+r-1)D_r(n-1) \quad (n > 2, r > 0). \tag{5}$$

By using the above described recursion for the Charlier polynomials, it is possible to find a recursion for $D_r(n)$ which is simpler than (5). This is the content of the following theorem.

Theorem 2. *We have for $n \geq 2$ and $r \geq 0$*

$$(n-r)D_r(n) = n(n-1)D_r(n-1) + n(n-1)D_r(n-2).$$

Proof. The left-hand side can be interpreted as follows: we choose an r-derangement σ on $n+r$ elements, then we paint (with red, say) one of the n elements which is not the direct successor of the $1, 2, \ldots, r$ elements in the cycle decomposition of σ. For example, in the permutation

$$(159)(264)(387)$$

we can choose $9, 4, 7$ but we are not permitted to choose $5, 6, 8$. There are $n+r-r-r = n-r$ elements to paint, and altogether we have $(n-r)D_r(n)$ choices – this is counted by the left-hand side. We now show that the same objects are counted by the right-hand side.

1) We can paint initially one of the elements, say i in the set $\{r+1, \ldots, n+r\}$ (n options), then we construct an r-derangement on the other $n+r-1$ elements in $D_r(n-1)$ ways. Now we insert i among the $n+r-1$ elements or at the end of the permutation, but i cannot go to the right-hand side of any of the $1, \ldots, r$ elements. We have $n+r-1-r = n-1$ places for i. Thus we have $n(n-1)D_r(n-1)$ options. This is the first term on the right-hand side.

2) It is possible to choose an r-derangement on $n+r-2$ elements in $D_r(n-2)$ ways if i and the other element form a 2-cycle. In order to be sure that i does not go to one of the $1, \ldots, r$ elements, we must choose from the set $\{r+1, \ldots, n+r\} \setminus \{i\}$ ($n-1$ possibilities). As the painted element i can be chosen in n ways, we have $n(n-1)D_r(n-2)$ possibilities, this is what the second term on the right-hand side counts.

It can be seen that we have covered all the possibilities during counting our r-derangements with i painted. □

Note that this theorem, in particular, gives the well known recursion for the classical derangements:

$$D(n) = (n-1)D(n-1) + (n-1)D(n-2).$$

Two results of the Charlier polynomials we have not used yet: the hypergeometric representation and the forward shift. These, together with (2), results in the following two statements.

Theorem 3. *For all $n \geq r \geq 0$ we have the following hypergeometric function representation for the r-derangement numbers:*

$$D_r(n) = (-1)^{n-r} n^{\underline{r}} \cdot {}_2F_0\left[\begin{matrix} -(n-r),\ r+1 \\ - \end{matrix} ; 1\right].$$

The forward difference results in the following recursion, which goes one step back in both n and r.

Theorem 4. *For all $n \geq r \geq 0$*

$$D_r(n) = n(D_r(n-1) + D_{r-1}(n-1)).$$

Proof. We give a combinatorial proof. To do this, we recall the standard form of a permutation: we shift the elements in each of the individual cycles so that the minimal elements (the cycle leaders) are the first, then we order the cycles in decreasing order with respect to the cycle leaders.

Then we have two options for the construction of an r-derangement on $n+r$ letters:

1) the first element 1 is in a cycle of length two. Then we choose an element from n to put it in this cycle by the side of 1, and the rest of the permutation is constructed $D_{r-1}(n-1)$ ways, resulting in the second term of the right-hand side.

2) the first element 1 is in a cycle of length greater than two. Then we construct an arbitrary permutation on $n-1+r$ elements. In any of these 1 will have at least one companion. We insert the remaining element (which can be one of the n) so 1 will have at least two companions in its cycle. This gives the first term on the right-hand side. □

To close this section we give another closed form representation for the r-derangement sequence. The Laguerre polynomials can be defined by the generating function

$$\sum_{n=0}^{\infty} L_n^{(\alpha)}(x)t^n = (1-t)^{-\alpha-1}\exp\left(\frac{-xt}{1-t}\right).$$

Putting $x = 1-t$ and multiplying by t^r, we can transform this generating function into the exponential generating function (1) of the r-derangement numbers. By comparison of the coefficients, and recalling that

$$L_n^{(\alpha)}(x) = \sum_{i=0}^{n}(-1)^i\binom{n+\alpha}{n-i}\frac{x^i}{i!},$$

we arrive at the following statement.

Theorem 5. *For any $n, r \geq 0$*

$$D_r(n) = n!\sum_{k=0}^{n}\frac{(r+1)_{k-r}}{(k-r)!}\sum_{\ell=n-k}^{k-r}\binom{\ell}{n-k}\frac{(r-k)^{\overline{\ell}}}{\ell!(r+1)^{\overline{\ell}}}(-1)^{n-k}.$$

3. An Integral Formula and Some of Its Corollaries

We now give an integral representation formula for $D_r(n+r)$ which is a corollary of (3).

Corollary 1. *For any $n, r \geq 0$ we have*

$$D_r(n+r) = (-1)^n (n+r)^{\underline{r}} \int_0^\infty e^{-t} {}_2F_1\left[{-n,\ r+1 \atop 1}; t\right] dt.$$

If $0 \leq n \leq r$, we additionally have

$$D_r(n+r) = (-1)^n (n+r)^{\underline{r}} \frac{(-r)^{\overline{n}}}{n!} \int_0^\infty e^{-t} {}_2F_1\left[{-n,\ r+1 \atop -n+r+1}; 1-t\right] dt. \tag{6}$$

Here ${}_2F_1\left[{a,\ b \atop c}; t\right]$ is the Gaussian hypergeometric function.

Proof. It is not hard to see that (3) is equivalent to

$$D_r(n+r) = (-1)^n (n+r)^{\underline{r}} \sum_{k=0}^{n} \binom{n}{k} \binom{-(r+1)}{k} k!.$$

Substituting the standard integral $k! = \int_0^\infty e^{-t} t^k dt$, we have

$$D_r(n+r) = (-1)^n (n+r)^{\underline{r}} \int_0^\infty e^{-t} \left(\sum_{k=0}^{n} \binom{n}{k} \binom{-(r+1)}{k} t^k \right) dt.$$

The inner sum can be shown to be equal to

$$\sum_{k=0}^{\infty} \binom{n}{k} \binom{-(r+1)}{k} t^k = {}_2F_1\left[{-n,\ r+1 \atop 1}; t\right],$$

and the result comes.

To see the validity if (6), we recall the known transformation formula

$${}_2F_1\left[{-n,\ b \atop c}; t\right] = \frac{(c-b)^{\overline{n}}}{c^{\overline{n}}} {}_2F_1\left[{-n,\ b \atop -n+b+c-1}; 1-t\right].$$

This was proven combinatorially by Labelle and Yeh [15]. Setting $b = r+1$ and $c = 1$ we get (6) when $n \leq r$ (when $n > r$ the rising factorial $(-r)^{\overline{n}}$ is obviously zero, while the left-hand side is not). □

Note that when $r = 0$, the hypergeometric function encodes the binomial expansion:

$${}_2F_1\left[{-n,\ 1 \atop 1}; t\right] = {}_1F_0\left[{-n \atop }; t\right] = (1-t)^n,$$

and we get a well known fact [1, p. 855]:

$$D_n = \int_0^\infty e^{-t}(t-1)^n \mathrm{d}t. \tag{7}$$

We can deduce a summation formula from (7):

$$\sum_{n=0}^\infty D_n a_n = \int_0^\infty e^{-t} \sum_{n=0}^\infty a_n (t-1)^n \mathrm{d}t.$$

To make this summation work, a_n must be appropriate in the sense that the sum on the left-hand side converges, as well as the integral on the right-hand side so that the integral and summation can be interchanged.

For example, taking $a_n = 1/(n!)^2$, we get

$$\sum_{n=0}^\infty \frac{D_n}{n!^2} = \int_0^\infty e^{-t} I_0(2\sqrt{t-1})\mathrm{d}t,$$

where I_0 is the modified Bessel function of the first kind [3, 22]. More generally, for $m \geq 0$

$$\sum_{n=0}^\infty \frac{D_n}{n!(n+m)!} = \int_0^\infty \frac{e^{-t}}{(x-1)^{m/2}} I_m(2\sqrt{t-1})\mathrm{d}t.$$

4. The Derangement Function $D(z)$

A quick look at (7) reveals how the sequence $D(n)$ can be extended to complex arguments. Let us define the function

$$D(z) = \int_0^\infty e^{-t}(t-1)^z \mathrm{d}t \quad (\Re(z) > -1),$$

and outside of the region of validity of the integral by analytic extension via the recursion

$$D(z+1) = zD(z) + zD(z-1). \tag{8}$$

We must be careful with this definition, because at the negative integers $D(z)$ is not defined (see the below Theorem for the full statement).

Note that the $D(z)$ function is not completely new, by the integral transformation one can see that

$$D(z) = \int_0^\infty e^{-t}(t-1)^z \mathrm{d}t = \frac{1}{e}\int_{-1}^\infty e^{-t} t^z \mathrm{d}t = \frac{1}{e}\Gamma(z+1,-1),$$

where the last function is the complementary incomplete Gamma function

$$\Gamma(a,z) = \int_z^\infty e^{-t} t^{a-1} \mathrm{d}t. \tag{9}$$

The idea of studying this particular incomplete Gamma (or derangement) function came from the study of the related left factorial function

$$K(n) = \sum_{i=0}^{n-1} i! \quad (n > 0),$$

done by Kurepa [13, 14]. Kurepa extended $K(n)$ to the complex plane, and studied the analyticity of the $K(z)$ function. In the more recent paper [10], the main properties of this function are listed. Moreover, in [17] the alternating sum

$$A(n) = \sum_{i=1}^{n} (-1)^{n-i} i!,$$

as well as the function

$$A_1(z) = \sum_{n=0}^{\infty} (-1)^n \Gamma(z+1-n)$$

are studied.

Our main observations with respect to the function $D(z)$ are contained in the following two theorems.

Theorem 6. *The function $D(z)$ is an analytic function on the domain $z \in \mathbb{C} \setminus \{-1, -2, \dots\}$, and can be holomorphically extended to the whole complex plane $\mathbb{C}$ to a function $\overline{D}$ such that*

$$\overline{D}(z) = \begin{cases} D(z), & \text{if } z \neq -1, -2, \dots; \\ \overline{D}(k), & \text{if } z = k,\ k = -1, -2, \dots, \end{cases}$$

where

$$\overline{D}(-1) = -\frac{\mathrm{Ei}(1)}{e} - \frac{\pi}{e} i, \quad \overline{D}(-2) = -1 + \frac{\mathrm{Ei}(1)}{e} + \frac{\pi}{e} i,$$

and

$$\overline{D}(k) = \frac{1}{k+1} \overline{D}(k+2) - \overline{D}(k+1) \quad (k = -3, -4, \dots).$$

Moreover, $z = 1$ is a zero of $D(z)$ of multiplicity one. More concretely,

$$\lim_{z \to 1} \frac{D(z)}{z-1} = 1 - \frac{\mathrm{Ei}(1)}{e} - \frac{\pi}{e} i.$$

Here and above

$$\mathrm{Ei}(1) = -\int_{-1}^{\infty} \frac{e^{-t}}{t} \mathrm{d}t \approx 1.8951178$$

is the exponential integral function at $z = 1$.

Proof. That the integral is convergent for $\Re(z) > -1$ is easy to see. Now let $z = -1 + is$ with some $s \in \mathbb{R}$. Then, by (8),

$$D(-1+is) = \frac{1}{is}D(is+1) - D(is).$$

The right-hand side is regular and defined by the integral representation, except the only point when $s = 0$. At this point the function $D(z)$ has a removable singularity, and the limit $\lim_{z\to-1+} D(z)$ will be denoted by $\overline{D}(-1)$. To prove this statement we calculate the following limit:

$$\begin{aligned}\overline{D}(-1) := \lim_{z\to-1+} D(z) &= \lim_{z\to-1+}\left(\frac{D(z+2)}{z+1} - D(z+1)\right)\\ &= \lim_{z\to0+}\left(\frac{D(z+1)}{z} - D(z)\right) = \lim_{z\to0+}\frac{D(z+1)}{z} - 1.\end{aligned} \tag{10}$$

The limit can be determined by making the expansion of $(t-1)^{z+1}$ around the point $z = 0$ behind the integral defining $D(z+1)$:

$$\begin{aligned}D(z+1) &= \int_0^\infty e^{-t}((t-1) + (t-1)\log(t-1)z + O(z^2))\mathrm{d}t\\ &= \int_0^\infty e^{-t}(t-1)\mathrm{d}t + z\int_0^\infty e^{-t}(t-1)\log(t-1)\mathrm{d}t + O(z^2).\end{aligned}$$

The first integral is zero, thus

$$\lim_{z\to0+}\frac{D(z+1)}{z} = \int_0^\infty e^{-t}(t-1)\log(t-1)\mathrm{d}t + O(z) = 1 - \frac{\mathrm{Ei}(1)}{e} - \frac{\pi}{e}i.$$

Substituting this into (10) we see that

$$\overline{D}(-1) = 1 - \frac{\mathrm{Ei}(1)}{e} - \frac{\pi}{e}i - 1 = -\frac{\mathrm{Ei}(1)}{e} - \frac{\pi}{e}i.$$

We define the holomorphic extension of $D(z)$ at this point by the value $\overline{D}(-1)$. In any vertical line the function is determined by the function values taken on the line shifted to the right by one and two. Thus $D(z)$ is well defined on all the complex plane except the points $z = -1, -2, \dots$. Here it has removable singularities, where the limits

$$\overline{D}(-k) := \lim_{z=-k} D(z)$$

exist. We have just seen that $\overline{D}(-1) = -\frac{\mathrm{Ei}(1)}{e} - \frac{\pi}{e}i$, and we can also see that

$$\overline{D}(-2) = -D(0) - \overline{D}(-1) = -1 + \frac{\mathrm{Ei}(1)}{e} + \frac{\pi}{e}i.$$

In general, by the recursion (8),

$$\overline{D}(k) = \frac{1}{k+1}\overline{D}(k+2) - \overline{D}(k+1) \quad (k = -3, -4, \dots).$$

Now we turn to the only zero we know of the $D(z)$ function, $D(1) = 0$, as there are no fixed point free partitions on one element. Now we prove that $z = 1$ is a zero of the complex $D(z)$ function of multiplicity one by showing that

$$\lim_{z \to 1} \frac{D(z)}{z-1} \neq 0.$$

Indeed,

$$\lim_{z\to 1}\frac{D(z)}{z-1} = \lim_{z\to 1}\frac{1}{z-1}\int_0^\infty e^{-t}(t-1)^z \mathrm{d}t = \lim_{z\to 1}\frac{1}{z-1}\int_0^\infty e^{-t}(t-1)\mathrm{d}t$$
$$+ \lim_{z\to 1}\frac{1}{z-1}\int_0^\infty e^{-t}(t-1)\log(t-1)(z-1)\mathrm{d}t + \lim_{z\to 1}\frac{1}{z-1}O(z-1)^2.$$

The integral $\int_0^\infty e^{-t}(t-1)\mathrm{d}t$ vanishes as well as the $\lim_{z\to 1}\frac{1}{z-1}O(z-1)^2$ term. Thus

$$\lim_{z\to 1}\frac{D(z)}{z-1} = \lim_{z\to 1}\frac{1}{z-1}\int_0^\infty e^{-t}(t-1)\log(t-1)(z-1)\mathrm{d}t$$
$$= \int_0^\infty e^{-t}(t-1)\log(t-1)\mathrm{d}t = 1 - \frac{\mathrm{Ei}(1)}{e} - \frac{\pi}{e}i \neq 0,$$

as we stated. □

The Taylor coefficients of $D(z)$ around the origin can be expressed by the derivatives of the incomplete Gamma function. The exact statement is in the following theorem.

Theorem 7. *Around the origin we have*

$$D(z) = \sum_{n=0}^{\infty} D^{(n)}(0)\frac{z^n}{n!},$$

where $D^{(0)}(0) = 1$, *and*

$$D^{(n)}(0) = (i\pi)^n + \frac{n}{e}\Gamma^{(n-1)}(0,-1) \quad (n \geq 1).$$

The expression $\Gamma^{(n-1)}(0,-1)$ *is meant to be the* $(n-1)$*-th derivative of* $\Gamma(z,-1)$ *at the point* $z = 0$.

Proof. It is obvious that $D^{(0)}(0) = D(0) = 1$. Moreover, by the definition of $D(z)$ we have that

$$D^{(n)}(0) = \int_0^\infty e^{-t} \log^n(t-1)\mathrm{d}t.$$

By partial integration, then by a change of the integration variable,

$$\int_0^\infty e^{-t} \log^n(t-1)\mathrm{d}t = [-e^{-t}\log^n(t-1)]_{t=0}^{t=\infty} + n\int_0^\infty e^{-t}\frac{\log^{n-1}(t-1)}{t-1}\mathrm{d}t$$

$$= (i\pi)^n + \frac{n}{e}\int_{-1}^\infty \frac{e^{-t}}{t}\log^{n-1} t\mathrm{d}t.$$

We only need to prove that

$$\int_{-1}^\infty \frac{e^{-t}}{t}\log^{n-1} t\mathrm{d}t = \Gamma^{(n-1)}(0,-1). \tag{11}$$

This, however, is obvious if we consider the integral as a sequence of the parameter n, and look for its generating function:

$$\sum_{n=1}^\infty \left(\int_{-1}^\infty \frac{e^{-t}}{t}\log^n t\mathrm{d}t\right)\frac{z^n}{n!} = \int_{-1}^\infty \frac{e^{-t}}{t}\sum_{n=1}^\infty \log^n t\frac{z^n}{n!}\mathrm{d}t = \int_{-1}^\infty \frac{e^{-t}}{t}t^z\mathrm{d}t = \Gamma(z,-1),$$

by (9). By the standard theory of generating functions we get (11). □

We have seen that $D(z)$ has a zero at $z = 1$. Does the function $D(z)$ have any other zeros? This question might be interesting to study.

5. A Generalization of the r-derangement Numbers

We introduce now a generalization which will permit us to deduce new identities for $D_r(n)$. Among others, we will be able to find the r-version of the famous recursion of Euler:

$$D(n) = nD(n-1) + (-1)^n. \tag{12}$$

For this result, see (17). A combinatorial proof for (12) can be found in [2].

We consider the restriction that every cycle contains at least m elements (cf. [4, 18, 20]). The number of permutations on n letters with the above restriction is denoted by $D_{\geq m}(n)$. It is clear that $D_{\geq 2}(n) = D(n)$. In this section we introduce a subclass of the FPF r-permutations stemming out from the above restriction. (We will always assume that $m \geq 2$.)

Definition 1. An (r,m)-*permutation* on $n+r$ letters is a permutation such that in its cycle decomposition the first r elements are in distinct cycles, with the restriction that every cycle contains at least m elements. The elements $\{1, 2, \ldots, r\}$ will be called *special elements*. A cycle is called *special* if it contains a special element.

For example,

$$(164)(239)(587\,10)$$

is a $(2,3)$-permutation.

Let us denote by $D_{r,\geq m}(n)$ the number of (r,m)-permutations on $n+r$ letters. We shall call the number $D_{r,\geq m}(n)$ *generalized r-derangement numbers.* In particular, $D_{0,\geq m}(n) = D_{\geq m}(n)$ and $D_{r,\geq 2}(n) = D_r(n)$. A basic recursion for these numbers is the following.

Theorem 8. *For any $r > 0$ and $n > m$, we have that*

$$D_{r,\geq m}(n) = r(m-1)(n-1)_{m-2}D_{r-1,\geq m}(n-m+1) \\ + (n-1)^{\underline{m-1}}D_{r,\geq m}(n-m) + (n+r-1)D_{r,\geq m}(n-1). \quad (13)$$

Proof. For any (r,m)-permutation on $n+r$ letters we can do the following construction

- the element $n+r$ is in a cycle of length m. There are two such cases: the element $n+r$ is either in a special cycle or not. In the first case, the element $n+r$ can be placed into one of the r cycles, and we can choose the remaining $m-2$ elements from $n-1$. This is done in $r(m-1)!\binom{n-1}{m-2} = r(m-1)(n-1)^{\underline{m-2}}$ ways. Note that we multiply by the factorial $(m-1)!$ because the order of the letters counts. The remaining elements can be choose in $D_{r-1,\geq m}(n-m+1)$ options. Thus, we have $r(m-1)(n-1)^{\underline{m-2}}D_{r-1,\geq m}(n-m+1)$ ways. For the second case the argument is similar. So, we have the second term in (13).
- The element $n+r$ is in an existing cycle of longer than m, after constructing a (r,m)-permutation on $n+r-1$ letters. For this we have $(n+r-1)D_{r,\geq m}(n-1)$ possibilities.

We therefore obtain the relation (13). □

By setting $m = 2$ we recover (5). Theorem 9 can be obtained by a slight change in the proof of Theorem 8.

Theorem 9. *For any $r > 0$ and $n > m$ we have that*

$$D_{r,\geq m}(n) = r\sum_{j=m-2}^{n-1}(j+1)(n-1)^{\underline{j}}D_{r-1,\geq m}(n-j-1) \\ + \sum_{j=m-1}^{n-1}(n-1)^{\underline{j}}D_{r,\geq m}(n-j-1).$$

We now give an additional recurrence by analyzing the first special element.

Theorem 10. *For any $r > 0$ and $n \geq m$ we have that*

$$D_{r,\geq m}(n) = \sum_{j=m-1}^{n-(r-1)(m-1)} n^{\underline{j}} D_{r-1,\geq m}(n-j).$$

Proof. For any (r,m)-permutation on $n+r$ letters we have that the first special element can be in a cycle of length j, for $j = m-1, m, \ldots, n-(r-1)(m-1)$. First we have to choose j non-special elements from that of n. This is done in $j!\binom{n}{j} = n^{\underline{j}}$ ways. The remaining cycles can be constructed in $D_{r-1,\geq m}(n-j)$ ways. Altogether, we have $(n)_j D_{r-1,\geq m}(n-j)$ possibilities. Finally, we sum over j. □

5.1. The Generating Function of $D_{r,\geq m}(n)$

In this section we give the exponential generating function for the sequence $(D_{r,\geq m}(n))_n$ and we derive several combinatorial identities.

To make our formulas more comprehensible, we introduce the notation

$$S_m(x) = \sum_{k=1}^{m} \frac{x^k}{k}.$$

Theorem 11. *For $r \geq 0$, the exponential generating function of the sequence of generalized r-derangement numbers is*

$$D_{r,m}(x) := \sum_{n=0}^{\infty} D_{r,\geq m}(n)\frac{x^n}{n!} = \left(\frac{x^{m-1}}{1-x}\right)^r \exp\left(\ln\left(\frac{1}{1-x}\right) - S_{m-1}(x)\right).$$

Proof. Each $(r,m)-$permutation may be uniquely decomposed into cycles as

$$(1\cdots)(2\cdots)\cdots(r\cdots)P,$$

where P is a $(0,m)$-permutation, and the first r cycles are of length at least m. Briggs and Remmel [4] (see also [23]) gave the exponential generating function for the number of $(0,m)$-permutations:

$$\exp\left(\ln\left(\frac{1}{1-x}\right) - S_{m-1}(x)\right).$$

Therefore, from the symbolic method [5] we obtain the following expression

$$D_{r,m}(x) = \left(\frac{x^m}{1-x}\right)^r D_{0,m}(x) = \left(\frac{x^{m-1}}{1-x}\right)^r \exp\left(\ln\left(\frac{1}{1-x}\right) - S_{m-1}(x)\right).$$

□

Theorem 12. *Let $r \geq 0$ and $s \in \{1, \ldots, r\}$. Then for each $n \geq s$ we have*

$$D_{r,\geq m}(n) = \sum_{j=s}^{n-(m-2)s} \binom{j-1}{s-1} \frac{n!}{(n-(m-2)s-j)!} D_{r-s,\geq m}(n-(m-2)s-j). \tag{14}$$

In particular, if $r = s$ then

$$D_{r,\geq m}(n) = \sum_{j=r}^{n-(m-2)r} \binom{j-1}{r-1} \frac{n!}{(n-(m-2)r-j)!} D_{\geq m}(n-(m-2)r-j). \tag{15}$$

Proof. From Theorem 11 we have

$$D_{r,m}(x) = \sum_{n=0}^{\infty} D_{r,\geq m}(n)\frac{x^n}{n!} = \left(\frac{x^{m-1}}{1-x}\right)^r \exp\left(\ln\left(\frac{1}{1-x}\right) - S_{m-1}(x)\right)$$

$$= \left(\frac{x^{m-1}}{1-x}\right)^s D_{r-s,m}(x) = [x^{(m-2)s}\left(\sum_{n=s}^{\infty}\binom{n-1}{s-1}x^n\right)\left(\sum_{n=0}^{\infty} D_{r-s,\geq m}(n)\frac{x^n}{n!}\right)$$

$$= x^{(m-2)s}\sum_{n=s}^{\infty}\left(\sum_{j=s}^{\infty}\binom{j-1}{s-1}\frac{D_{r-s,\geq m}(n-j)}{(n-j)!}\right)x^n.$$

The comparison of the coefficients of x^n yields the identity (14). $\square$

Theorem 13. *For any $r > 0$ and $n > m$ we have that*

$$D_{\geq m}(n) = n! \sum_{s=0}^{n} \frac{1}{s!} \sum_{(\ell_1,\ldots,\ell_{m-1})\in\Omega(s,m-1)} (-1)^{\ell_1+\ell_2+\cdots+\ell_{m-1}}$$

$$\times \binom{s}{\ell_1, 2\ell_2, \ldots, (m-1)\ell_{m-1}} \prod_{j=1}^{m-1} \frac{(j\ell_j)!}{j^{\ell_j}(\ell_j!)},$$

where

$$\Omega(n, m-1) = \left\{(\ell_1, \ell_2, \ldots, \ell_{m-1}) \in \mathbb{Z}_{n\geq 0}^{m-1} : \ell_1 + 2\ell_2 + \cdots + (m-1)\ell_{m-1} = n\right\}.$$

Proof. The generating function (with $r = 0$) in Theorem 11 can be written as

$$\sum_{n=0}^{\infty} D_{\geq m}(n)\frac{x^n}{n!} = \frac{1}{1-x}\exp\left(-x-\frac{x^2}{2}-\cdots-\frac{x^{m-1}}{m-1}\right)$$
$$= \frac{1}{1-x}\prod_{j=1}^{m-1}\sum_{\ell_j=0}^{\infty}\frac{(-1)^{\ell_j}x^{j\ell_j}}{j^{\ell_j}(\ell_j)!}$$
$$= \frac{1}{1-x}\sum_{n=0}^{\infty}\left(\sum_{(\ell_1,\ldots,\ell_{m-1})\in\Omega(n,m-1)}(-1)^{\ell_1+\ell_2+\cdots+\ell_{m-1}}\right.$$
$$\left.\binom{n}{\ell_1, 2\ell_2, \ldots, (m-1)\ell_{m-1}}\prod_{j=1}^{m-1}\frac{(j\ell_j)!}{j^{\ell_j}(\ell_j!)}\right)\frac{x^n}{n!}$$
$$= \sum_{n=0}^{\infty}\left(n!\sum_{s=0}^{n}\frac{1}{s!}\sum_{(\ell_1,\ldots,\ell_{m-1})\in\Omega(s,m-1)}(-1)^{\ell_1+\ell_2+\cdots+\ell_{m-1}}\right.$$
$$\left.\binom{n}{\ell_1, 2\ell_2, \ldots, (m-1)\ell_{m-1}}\prod_{j=1}^{m-1}\frac{(j\ell_j)!}{j^{\ell_j}(\ell_j!)}\right)\frac{x^n}{n!}.$$

The result now comes by comparing the coefficients of $x^n/n!$. □

In particular, if $m = 2$, we recover a basic identity for the derangement numbers:

$$D(n) = n!\left(1-\frac{1}{1!}+\frac{1}{2!}+\cdots+\frac{(-1)^n}{n!}\right). \tag{16}$$

If $m = 3$, we obtain a relation with the Hermite polynomials $H_n(x)$. Remember that the Hermite polynomials can be defined (cf. [19] or [12, Section F]) by the following generating function:

$$\sum_{n=0}^{\infty} H_n(x)\frac{z^n}{n!} = e^{2xz-z^2}.$$

Therefore we have the expression

$$\frac{D_{\geq 3}(n)}{n!} = \sum_{\ell=0}^{n}\frac{(-1)^{\ell}H_{\ell}(\sqrt{2}/2)}{(\sqrt{2})^{\ell}\ell!}.$$

Theorem 14. *The generalized r-derangement numbers have the below representa-*

tion:

$$D_{r,\geq m}(n) = \sum_{s=r}^{n-r(m-2)} \binom{s}{r} \frac{n!}{(n-r(m-2)-s)!} \sum_{(\ell_1,\dots,\ell_{m-1})\in\Omega(n-r(m-2)-s,m-1)} (-1)^{\ell_1+\ell_2+\cdots+\ell_{m-1}} \binom{n-r(m-2)-s}{\ell_1, 2\ell_2, \dots, (m-1)\ell_{m-1}} \prod_{j=1}^{m-1} \frac{(j\ell_j)!}{j^{\ell_j}(\ell_j!)}.$$

Proof. The proof is a corollary of (15) and Theorem 13. □

If $m = 2$, we have a closed formula for the r-derangement numbers (see also Theorem 4 in [21])

$$D_r(n) = \sum_{s=r}^{n} \binom{s}{r} \frac{n!}{(n-s)!} (-1)^{n-s}, \quad n \geq r.$$

In particular, if $r = 0$ and $m = 2$ we obtain (16) again.

Theorem 15. *The generalized r-derangement numbers satisfy the relation*

$$D_{r,\geq m}(n) = \sum_{j=1}^{r+1} n^{\underline{j}} \binom{r+1}{j} (-1)^{j+1} D_{r,\geq m}(n-j) + n^{\underline{r(m-1)}} \sum_{(\ell_1,\dots,\ell_{m-1})\in\Omega(n-r(m-1),m-1)} (-1)^{\ell_1+\ell_2+\cdots+\ell_{m-1}} \times \binom{n-r(m-1)}{\ell_1, 2\ell_2, \dots, (m-1)\ell_{m-1}} \prod_{j=1}^{m-1} \frac{(j\ell_j)!}{j^{\ell_j}(\ell_j!)}.$$

Proof. The proof follows from the equality

$$(1-x)^{r+1} \sum_{n=0}^{\infty} D_{r,\geq m}(n) \frac{x^n}{n!} = x^{(m-1)r} \exp\left(-S_{m-1}(x)\right).$$

□

If $m = 2$ we have the relation for the r-derangement numbers

$$D_r(n) = \sum_{j=1}^{r+1} n^{\underline{j}} \binom{r+1}{j} (-1)^{j+1} D_r(n-j) + n^{\underline{r}} (-1)^{n-r}. \tag{17}$$

Remarkably, if $r = 0$ and $m = 2$ we recover the Euler recursion (12), again.

5.2. A Connection With the Associated Lah Numbers

The Lah numbers $L(n,k)$ count the number of set partitions of a set with n elements into k non-empty ordered lists. This sequence satisfies the following recurrence

$$L(n,k) = L(n-1,k-1) + (n+k-1)L(n-1,k)$$

with the initial values $L(0,0) = 1$, $L(n,0) = 0$ if $n > 0$, and $L(n,1) = n!$.

They can be computed by the following explicit formula

$$L(n,k) = \frac{n!}{k!}\binom{n-1}{k-1},$$

or by the generating function

$$\sum_{n\geq k} L(n,k)\frac{x^n}{n!} = \frac{1}{k!}\left(\frac{x}{1-x}\right)^k.$$

The associated Lah numbers $L_{\geq m}(n,k)$ are defined as the number of set partitions of a set with n elements into k non-empty ordered lists such that every list contains at least m elements. It is not difficult to show that the exponential generating function of this sequence is given by

$$\sum_{n\geq k} L_{\geq m}(n,k)\frac{x^n}{n!} = \frac{1}{k!}\left(\frac{x^m}{1-x}\right)^k. \tag{18}$$

Theorem 16. *For $r \geq 0$ and $n \geq 1$, such that $n \geq r$, we have*

$$(r+1)!L_{\geq m-1}(n,r-1) = \sum_{s=0}^{n-m+1}\binom{n}{s}(n-s)^{\underline{m-1}}D_{r,\geq m}(s)$$
$$\sum_{(\ell_1,\ldots,\ell_{m-1})\in\Omega(n-m-s+1,m-1)}\binom{n-m-s+1}{\ell_1,2\ell_2,\ldots,(m-1)\ell_{m-1}}\prod_{j=1}^{m-1}\frac{(j\ell_j)!}{j_j^{\ell}(\ell_j!)}.$$

Proof. Let us first introduce the shorthand notation

$$B(n,m-1) := \sum_{(\ell_1,\ldots,\ell_{m-1})\in\Omega(n,m-1)}\binom{n}{\ell_1,2\ell_2,\ldots,(m-1)\ell_{m-1}}\prod_{j=1}^{m-1}\frac{(j\ell_j)!}{j^{\ell_j}(\ell_j!)}.$$

From the generating functions given in (18) and Theorem 11 we have

$$(r+1)!\sum_{n=0}^{\infty} L_{\geq m}(n,r+1)\frac{x^n}{n!} = \left(\frac{x^m}{1-x}\right)^{r+1} =$$

$$x^{m-1}\exp\left(S_{m-1}(x)\right)\left(\sum_{n=0}^{\infty} D_{r,\geq m}(n)\frac{x^n}{n!}\right) =$$

$$x^{m-1}\sum_{n=0}^{\infty}\frac{x^n}{n!}\sum_{s=0}^{n}\binom{n}{s}D_{r,\geq m}(s)B(n-s,m-1).$$

The conclusion follows by comparing coefficients of $x^n/n!$. □

If $m = 2$, we obtain the relation for the r-derangement numbers and Lah numbers (see Theorem 9 of [21] and [16]):

$$(r+1)!L(n,r-1) = \sum_{s=0}^{n-1}\binom{n}{s}(n-s)D_r(s) = \sum_{s=0}^{n}\binom{n}{s}sD_r(n-s).$$

Acknowledgement. The first author is grateful to Mourad E. H. Ismail for introducing him to the references [1, 7, 8, 9], which led to the birth of the present paper. The authors would like to thank the anonymous referee for some useful comments.

References

[1] R. Askey, M. E. H. Ismail, Permutation problems and special functions, Canad. J. Math. **XXVIII**(4) (1976), 853–875.

[2] A. T. Benjamin, J. Ornstein, A bijective proof of a derangement recurrence, *Proceedings of the 17th International Conference on Fibonacci Numbers and Their Applications*, (2017), 28–29.

[3] F. Bowman, *Introduction to Bessel Functions*, Dover, New York, 1958.

[4] K. S. Briggs, J. B. Remmel, A p,q-analogue of the generalized derangement numbers, *Ann. Comb.* **13** (2009), 1–25.

[5] P. Flajolet, R. Sedgewick, *Analytic Combinatorics*, Cambridge University Press, Cambridge, 2009.

[6] W. Gautschi, *Orthogonal Polynomials, Computation and Approximation*, Oxford University Press, Oxford, 2004.

[7] J. Gillis, M. E. H. Ismail, T. Offer, An asymptotic problem in derangement theory, *SIAM J. Math. Anal.* **21**(1) (1990), 262–269.

[8] M. E. H. Ismail, A. Kasraoui, J. Zeng, Separation of variables and combinatorics of linearization coefficients of orthogonal polynomials, *J. Combin. Th. Ser. A* **120** (2013), 561–599.

[9] M. E. H. Ismail, P. Simeonov, Asymptotics of generalized derangements, *Adv. Comput. Math.* **39** (2013), 101–127.

[10] A. Ivić, Ž. Mijajlović, On Kurepa's problems in number theory, *Publ. Inst. Math. (Beograd) (N.S.)* **57**(71) (1995), 19–28.

[11] D. S. Kim, T. Kim, H. I. Kwon, T. Mansour, Barnes-type Narumi of the second kind and Poisson-Charlier mixed-type polynomials, *J. Comput. Anal. Appl.* **19**(5) (2015), 837–850.

[12] T. Kim, D. S. Kim, T. Mansour, S.-H. Rim, M. Schork, Umbral calculus and Sheffer sequences of polynomials, *J. Math. Phys.* **54** (2013), 083504.

[13] Đ. Kurepa, On the left factorial function $!n$, *Math. Balkan.* **1** (1971), 147–153.

[14] Đ. Kurepa, Left factorial function in complex domain, *Math. Balkan.* **3** (1973), 297–307.

[15] J. Labelle, Y.-N. Yeh, The combinatorics of Laguerre, Charlier and Hermite polynomials revisited, *Stud. Appl. Math.* **80** (1989), 25–36.

[16] J. Lindsay, T. Mansour, M. Shattuck, A new combinatorial interpretation of a q-analogue of the Lah numbers, *J. Comb.* **2**(2) (2011), 245–264.

[17] B. J. Malešević, Some considerations in connection with alternating Kurepa's function, *Integral Transforms Spec. Funct.* **19**(10) (2008), 747–756.

[18] M. Mihoubi, M. Rahmani, The partial r-Bell polynomials, *Afr. Mat.* **28** (2017), 1167–1183.

[19] V. Moll, *Numbers and Functions. From a Classical-Experimental Mathematician's Point of View.* Student Mathematical Library, Vol. 65, American Mathematical Society, 2012.

[20] V. Moll, J. L. Ramírez, D. Villamizar, Combinatorial and arithmetical properties of the restricted and associated Bell and factorial numbers, *J. Comb.* (accepted). See also on arXiv:1707.08138.

[21] Ch.-Y. Wang, P. Miska, I. Mező, The r-derangement numbers, *Discrete Math.* **340** (2017), 1681–1692.

[22] G. N. Watson, *A Treatise on the Theory of Bessel Functions* (second edition), Cambridge Univ. Press, 1944.

[23] H. S. Wilf, *Generatingfunctionology*, Academic Press, 1990.

7

ADDITIVE VOLUME OF SETS CONTAINED IN FEW ARITHMETIC PROGRESSIONS

Gregory A. Freiman
The Raymond and Beverly Sackler Faculty of Exact Sciences, School of Mathematical Sciences, Tel Aviv University, Tel Aviv, Israel
grisha@post.tau.ac.il

Oriol Serra[1]
Department of Mathematics, Universitat Politècnica de Catalunya and Barcelona Graduate School of Mathematics, Barcelona, Spain
oriol.serra@upc.edu

Christoph Spiegel[2]
Department of Mathematics, Universitat Politècnica de Catalunya and Barcelona Graduate School of Mathematics, Barcelona, Spain
christoph.spiegel@upc.edu

Abstract

A conjecture of Freiman gives an exact formula for the largest volume of a set of integers $A \subset \mathbb{Z}$ with given cardinality $k = |A|$ and doubling $T = |2A|$. The formula is known to hold when $T \leq 3k - 4$, for some small range over $3k - 4$ and for families of structured sets called chains. In this paper we extend the formula to sets of every dimension and prove it for sets composed of three segments, giving structural results for the extremal case. A weaker extension to sets composed of a bounded number of segments is also discussed.

1. Introduction

Let $A \subset \mathbb{Z}$ be a finite set of integers. The *Minkowski sum* of A is $A + A = \{a + a' : a, a' \in A\}$. The *doubling* of A is the cardinality of $2A = A + A$. The Freiman–Ruzsa theorem, giving the structure of sets of integers with small doubling, is one of the central results in Additive Number Theory. It states that a set A with doubling

[1]Supported by the Spanish Research Agency under project MTM2017-82166-P and MDM-2014-0445

[2]Supported by the Spanish Ministerio de Economía y Competitividad FPI grant under the project MTM2014-54745-P, the project MTM2017-82166-P and the María de Maetzu research grant MDM-2014-0445.

$|2A| \le c|A|$ is a dense subset of a multidimensional arithmetic progression P, where the density $|A|/|P|$ and dimension of P depend only on c, see Freiman [3], Bilu [1] and Ruzsa [13]. The estimation of the best lower bounds for the density of A in P was the object of a long series of papers and it was eventually brought to its essentially best values by Schoen [17].

At a conference in Toronto in 2008, Freiman proposed a precise formula for the largest possible volume of a set of integers $A \subset \mathbb{Z}$ with given doubling $T = |2A|$ in terms of a specific parametrisation of the value of T, see [5]. We start by recalling some definitions in order to state this conjecture.

Given abelian groups G and G', two sets $A \subset G$ and $B \subset G'$ are *Freiman isomorphic of order* 2 (F_2–isomorphic for short) if there is a bijection $\phi : A \to B$ such that, for every $x, y, z, t \in A$, we have

$$x + y = z + t \quad \Leftrightarrow \quad \phi(x) + \phi(y) = \phi(z) + \phi(t). \tag{1}$$

The *additive dimension* $\dim(A)$ of a set $A \subset \mathbb{Z}$ is the largest $d \in \mathbb{N}$ such that there exists a set $B \subset \mathbb{Z}^d$ not contained in a hyperplane of $\mathbb{Z}^d$ which is F_2–isomorphic to A. Note that any d–dimensional set A of cardinality k satisfies $d \le k - 1$ and that furthermore, by results of Freiman [2] as well as Konyagin and Lev [11], we have

$$(d+1)k - \binom{d+1}{2} \le |2A| \le \binom{k}{2} + d + 1. \tag{2}$$

The *volume* $\mathrm{vol}(A)$ of a d–dimensional set A is defined to be the minimum cardinality of the convex hull among all sets in $\mathbb{Z}^d$ that are F_2–isomorphic to A.

We say that a set of integers $A \subset \mathbb{Z}$ is in *normal form* if $\min(A) = 0$ and $\gcd(A) = 1$. We call

$$\tilde{A} = \big(A - \min(A)\big) / \gcd\big(A - \min(A)\big) \tag{3}$$

the *normalization* of A since it is a set in normal form and it is F_2–isomorphic to A. Note that for any 1–dimensional set A we have

$$\mathrm{vol}(A) = \max(\tilde{A}) + 1. \tag{4}$$

If $\min(A) = 0$, then the *reflection* of A is defined as $A^- = -A + \max(A)$. The reflection of A is certainly isomorphic to A.

We are interested in obtaining upper bounds for the volume of a set A of integers in terms of its cardinality $|A|$, the cardinality of its doubling $|2A|$ and its dimension $\dim(A)$. We denote the maximum volume of all sets A of integers with cardinality k, doubling T and dimension d by

$$\mathrm{vol}(k, T, d) = \max\{\mathrm{vol}(A) : A \subset \mathbb{N}\,, |A| = k, |2A| = T \text{ and } \dim(A) = d\}. \tag{5}$$

A set A is *extremal* if $\mathrm{vol}(A) = \mathrm{vol}(|A|, |2A|, \dim(A))$. The following is a more general and slightly reformulated version of the previously mentioned conjecture of

Freiman, which can be traced back to [2]. Its notable addition is that it takes the dimension of a set into consideration.

Conjecture 1. Given any $k, T, d \in \mathbb{N}$ such that

$$T = (d+c)k - \binom{d+c+1}{2} + d + b + 1, \tag{6}$$

where

$$1 \leq c \leq k-d-1 \text{ and } 0 \leq b \leq k-d-c-1, \tag{7}$$

we have

$$\mathrm{vol}(k,T,d) = 2^{c-1}(k-c+b) + 1. \tag{8}$$

Note that given any $k \in \mathbb{N}$, $1 \leq d \leq k-1$ and

$$T \in \left\{(d+1)k - \binom{d+1}{2}, \ldots, \binom{k}{2} + d + 1\right\},$$

there are uniquely defined

$$c = c(k,T,d) \quad \text{and} \quad b = b(k,T,d) \tag{9}$$

subject to the boundary conditions stated in the conjecture, such that T can be expressed as the right–hand side of (6). It follows that Conjecture 1 states a tight upper bound on the volume of any possible set of integers. If A has cardinality $|A| = k$, dimension $\dim(A) = d$ and doubling $|2A| = T$ then we call this uniquely determined $c = c(k,T,d)$ the *doubling constant* of A.

There are examples showing that the right–hand side in (8) is at least a lower bound for $\mathrm{vol}(k,T,d)$; see e.g., [2, 7]. Thus an extremal set A has volume

$$vol(A) \geq vol(k,T,d).$$

Furthermore, equality has been established in a few cases:

1. by Freiman [2] for one–dimensional sets satisfying $T \leq 3k-4$, that is either, $c = 1$ and any admissible b or $c = 2$ and $b = 0$, with an additional structural description of extremal sets given in [4],

2. by Freiman [2] as well as by Hamidoune and Plagne [8] for one–dimensional sets if $T = 3k-3$, that is $c = 2$ and $b = 1$, with a structural description of the extremal case due to Jin [9],

3. by Freiman [2] for two–dimensional sets satisfying $k \geq 10$ and $T \leq 10/3\,k - 6$, that is, $c = 1$ and $0 \leq b \leq k/3 - 2$, with a structural description of any such set,

4. by Jin [10] using tools from non-standard analysis in the case of large one–dimensional sets satisfying $T \leq (3+\epsilon)k$, that is, $c = 2$ and $0 \leq b \leq \epsilon k$, for some $\epsilon > 0$,

5. by Stanchescu [16] for any d–dimensional set satisfying $c = 1$ and $b = 0$,

6. by Freiman and Serra [7] for a class of one–dimensional sets called chains, which can be seen as extremal sets build by a greedy algorithm, and any admissible values of the doubling constant c.

In order to give further evidence towards the validity of this conjecture, we consider sets composed of a given number of segments. Throughout the paper we say that $A \subset \mathbb{Z}$ is the union of s segments if

$$A = P_1 \cup \cdots \cup P_s, \tag{10}$$

where each P_i is a segment of length k_i with $\max P_i + 1 < \min P_{i+1}$ for $1 \leq i < s$ and, moreover, $k_i > 1$ for some $1 \leq i \leq s$. Regarding the doubling of such a set, we have the upper bound

$$|2A| = |\cup_{1\leq i\leq j\leq s}(P_i + P_j)| \leq \sum_{1\leq i\leq j\leq s} (|P_i| + |P_j| - 1) = (s+1)|A| - \binom{s+1}{2}. \tag{11}$$

Equality holds if and only if the sums $P_i + P_j$ are pairwise distinct, in which case the set A has dimension s.

As previously mentioned, Conjecture 1 has been proved for sets A with doubling $|2A| \leq 3|A| - 3$, and the structure of extremal sets for this range of doubling is well understood. In spite of many efforts, not much is known about the exact maximum volume of sets with doubling at least $3|A| - 2$. This motivates us to consider sets composed of three segments and doubling larger than $3|A| - 3$. Our main result is to show that the statement of Conjecture 1 holds for sets A composed of three segments, also giving a structural description of the extremal cases.

Theorem 1. *Let A be an extremal set with cardinality $k > 7$ and doubling $|2A| > 3k - 4$ consisting of three segments. Then A is isomorphic to one of the following sets:*

(i) $([0, k+b-2] \setminus [1, b]) \cup \{2(k+b-2)\}$ *with* $\dim(A) = 1$, $\mathrm{vol}(A) = 2k + 2b - 3$ *and* $|2A| = 3k - 4 + b$,

(ii) $([0, k+b+i-3] \setminus [k-i-1, k+b-3]) \cup \{2(k+b-2)\}$ *for* $1 \leq i \leq k/3$ *and* $2 \leq b \leq k-2i-1$, *with* $\dim(A) = 1$, $\mathrm{vol}(A) = 2k+2b-3$ *and* $|2A| = 3k-4+b$,

(iii) $([0, k+b-2] \setminus [1, b]) \times \{0\} \cup \{(0,1)\} \subset \mathbb{Z}^2$ *with* $\dim(A) = 2$, $\mathrm{vol}(A) = k + b$ *and* $|2A| = 3k - 3 + b$, *or*

(iv) $([0,k_1-1]\times\{(0,0)\})\cup([0,k_2-1]\times\{(0,1)\})\cup([0,k_3-1]\times\{(1,0)\})\subset\mathbb{Z}^3$ *with* $k_1+k_2+k_3=k$, $k_1,k_2,k_3\geq 1$, $\dim(A)=3$, $\mathrm{vol}(A)=k$ *and* $|2A|=4k-6$.

The extremal sets composed of three segments described in Theorem 1 are illustrated for $k=11$ and $|2A|=3k-1$ in Figure 1.

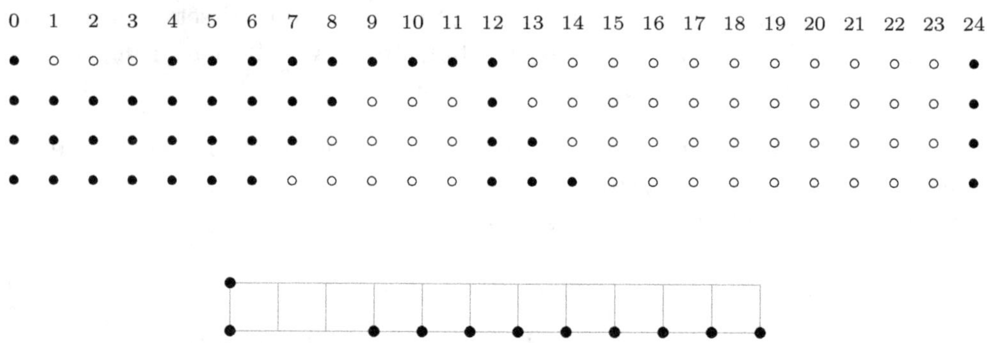

Figure 1: Extremal sets, up to isomorphism, for $k=11$ and $|2A|=3k-1$.

We also believe that the following statement, regarding an upper bound on the volume of 1–dimensional sets composed of few disjoint segments, should hold as well. It is independent of the doubling of such sets and in fact one may derive it from Conjecture 1 as the case with maximum doubling without too much effort.

Conjecture 2. Let A be a 1–dimensional set and let s be the minimum number of disjoint segments into which it can be decomposed. If $s\leq|A|-1$, then

$$\mathrm{vol}(A)\leq 2^{s-1}\big(|A|-s\big)+1. \tag{12}$$

The set $A_s=\{0,1,\ldots,k-s\}\cup\{2^i(k-s),i=1,\ldots,s-1\}$ shows that this conjectured upper bound would be tight. Note that, if we only know that a set A is 1–dimensional, Freiman [2] gave $\mathrm{vol}(A)\leq 2^{k-2}+1$ as an upper bound on the volume of A. In this paper, we show the validity of Conjecture 2 for some small values of s.

Proposition 1. *The statement of Conjecture 2 holds for* $s\leq 4$.

One may prove Conjecture 2 for further moderate values of s along the lines presented in the proof of Proposition 1, at the cost of a more involved case analysis. A general proof of this conjecture avoiding the increase of cases would be of interest.

The paper is organised as follows. In Section 2 we discuss the dimension of sets composed of few segments. We then prove Proposition 1 in Section 3 and Theorem 1 in Section 4.

2. The Dimension of Sets Contained in Few Segments

Konyagin and Lev [11] established a formula for the dimension of a given set $A \subset \mathbb{Z}^m$ of cardinality k. Let us write $A = \{a_1, \ldots, a_k\}$ and introduce some necessary notation. For $1 \leq i \leq k$, let $\mathbf{e}_i$ denote the vector in $\mathbb{R}^k$ that has a one at coordinate i and zero everywhere else. We denote by M_A the integer–valued matrix with k columns obtained by listing as its rows all vectors $\mathbf{e}_{i_1} + \mathbf{e}_{i_2} - \mathbf{e}_{i_3} - \mathbf{e}_{i_4}$ for which $a_{i_1} + a_{i_2} = a_{i_3} + a_{i_4}$ holds and for which we do not have $i_1 = i_2 = i_3 = i_4$. The dimension of A can be derived from the rank of M_A using the following result.

Theorem 2 (Konyagin and Lev [11]). *For any set $A \subseteq \mathbb{Z}^m$ we have*

$$\dim(A) = |A| - 1 - \mathrm{rank}(\mathrm{M_A}). \tag{13}$$

Now let $A \subset \mathbb{Z}$ be a set which is the union of s disjoint segments as in (10). Given such a set A, we denote by S_A the integer–valued matrix with s columns obtained by listing in its rows all vectors $\mathbf{e}_{j_1} + \mathbf{e}_{j_2} - \mathbf{e}_{j_3} - \mathbf{e}_{j_4}$ for which $(P_{j_1} + P_{j_2}) \cap (P_{j_3} + P_{j_4}) \neq \emptyset$ and for which we do not have $j_1 = j_2 = j_3 = j_4$. We derive the following Corollary from Theorem 2.

Corollary 1. *Any $A \subset \mathbb{Z}$ that is the union of $1 \leq s \leq |A| - 1$ disjoint segments satisfies*

$$\dim(A) = s - \mathrm{rank}(\mathrm{S_A}). \tag{14}$$

Proof. Every row in M_A is associated with up to four (not all equal) elements $a_{i_1}, a_{i_2}, a_{i_3}, a_{i_4}$ such that $a_{i_1} + a_{i_2} = a_{i_3} + a_{i_4}$. Let $a_{i_1} \in P_{j_1}$, $a_{i_2} \in P_{j_2}$, $a_{i_3} \in P_{j_3}$ and $a_{i_4} \in P_{j_4}$ where we may assume $j_1 \leq j_2$ and $j_3 \leq j_4$ as well as $\min\{j_1, j_2\} \leq \min\{j_3, j_4\}$. Furthermore, let $0 \leq y = \#\{j : |P_j| = 1\} < s$ denote the number of segments that are singletons. We distinguish the following cases.

`Case 1.` $\#\{j_1, j_2, j_3, j_4\} = 1$, that is, $j_1 = j_2 = j_3 = j_4 = j$ for some $1 \leq j \leq s$. Since P_j is one dimensional, by Theorem 2 there are a total of $\max\{|P_j| - 2, 0\}$ linearly independent equations of this type for each P_j. As these equations only involve elements in P_j and the segments are disjoint, it is clear that each equation is linearly independent from those of other segments, so we get a total of $|A| - 2s + y$ linearly independent equations of this type in M_A. On the other hand this case does not contribute to the rank of S_A

`Case 2.` $\#\{j_1, j_2, j_3, j_4\} = 2$. We distinguish two further cases.

`Case 2.1.` $j_1 = j_3 < j_4 = j_2$. Segments of length one can only give trivial equations of this type not contributing to the rank of M_A. If $P_{i_0} < P_{i_1} < \cdots < P_{i_{s-y}}$ are the $s - y$ segments which are not singletons, then equations of this type give us a total of $s - y - 1$ new linear independent ones on top of the ones given by Case 1, one for each pair P_{i_0}, P_{i_j} and

$j = 2, \ldots s - y$, the remaining ones being linearly dependent with these. Moreover, this case does not contribute to the rank of S_A.

`Case 2.2.` $j_1 < j_2 = j_3 = j_4$ or $j_1 = j_2 = j_3 < j_4$. Each pair $P_j, P_{j'}$ with $j \neq j'$ for which an equation of this type exists implies that $P_j \cap P_{j'}$ intersects either $2P_j$ or $2P_{j'}$, contributes one additional linear independent equation in M_A on top of the above ones and contributes to one additional linear equation in S_A as well.

`Case 3.` $\#\{j_1, j_2, j_3, j_4\} \geq 3$, that is, $j_1 < j_3 \leq j_4 < j_2$. This implies that $P_{j_1} + P_{j_2}$ intersects $P_{j_3} + P_{j_4}$. Each such intersection contributes with one additional linear independent equation in M_A and also on S_A on top of the above ones.

Taken together, it follows that $\text{rank}(M_A) = (|A| - 2s + y) + (s - y - 1) + \text{rank}(S_A)$ and therefore, by Theorem 2, $\dim(A) = s - \text{rank}(S_A)$. □

We note that for $s \geq 6$ there are sets for which $k_i = 1$ for all i that are not covered by Corollary 1 or Conjecture 2.

3. Proof of Proposition 1

Let $A \subset \mathbb{Z}$ again be a set which is the union of s segments. We denote the interval separating the two consecutive segments P_i and P_{i+1} by

$$L_i = [\max(P_i) + 1, \min(P_{i-1}) - 1], \tag{15}$$

and write $\ell_i = |L_i|$ for its cardinality. It follows that

$$\min P_i = \sum_{j<i}(k_j + l_j) \quad \text{and} \quad \max(P_i) = \min(P_i) + k_i - 1. \tag{16}$$

In order to prove Proposition 1, we will need the following lemma that gives us an inductive approach to Conjecture 2.

Lemma 1. *Let A be a 1–dimensional set of cardinality k that is composed of $s \leq k - 1$ disjoint segments $P_1, \ldots, P_s$ such that $\max P_i < \min P_{i+1}$ for $1 \leq i < s$. If Conjecture 2 holds for $s - 1$ and*

$$\text{vol}(A) > 2^{s-1}(|A| - s) + 1, \tag{17}$$

then we must have

$$P_i + P_j < P_i + P_{j+1}, \qquad 1 \leq i \leq s,\ 1 \leq j < s. \tag{18}$$

Proof. We observe that, for each $1 \le j < s$, the set $A \cup L_j$ is 1–dimensional, consists of $s-1$ disjoint segments and has the same volume as A. By the assumption that Conjecture 2 holds for $s-1$, we must have

$$\mathrm{vol}(A \cup L_j) \le 2^{s-2}(|A| + \ell_j - s + 1) + 1. \tag{19}$$

Hence, our assumption on vol(A) implies that

$$\ell_j \ge |A| - s \ge k_i - 1 \quad \text{for } 1 \le i \le s,\ 1 \le j < s. \tag{20}$$

In particular, we have

$$\begin{aligned}\min(P_i + P_{j+1}) - \max(P_i + P_j) =& (\min P_{j+1} - \max P_j) - (\max P_i - \min P_i)\\ =& \ell_j - k_i + 2 > 0,\end{aligned}$$

which implies the desired statement (18). □

It follows that, under the hypothesis of Lemma 1, the only possible intersections between sums of two segments are of the form

$$(P_i + P_j) \cap (P_{i'} + P_{j'}) \quad \text{where} \quad i < i' \le j' < j. \tag{21}$$

By using (21) we next prove Proposition 1.

Proof of Proposition 1. For $s = 1$ the conjecture trivially holds. For $s = 2$, suppose that A is composed of two segments and vol$(A) > 2k - 3$. By Lemma 1 we have $2P_1 < P_1 + P_2 < 2P_2$ and therefore rank$(S_A) = 0$, so that A must be 2–dimensional by Corollary 1.

Suppose now that $s = 3$ and that vol$(A) > 4k - 11$. By (21) the only possible intersections of sums of segments are $P_1 + P_3$ and $2P_2$. Again, A must be 2–dimensional by Corollary 1.

Finally, suppose that $s = 4$ and vol$(A) > 8k - 31$. By (21) the only possible intersections of sets of the type $P_i + P_j$ are

1. $P_1 + P_3$ can intersect with $2P_2$,
2. $P_1 + P_4$ can only intersect with at most one of $2P_2$, $P_2 + P_3$ or $2P_3$,
3. $P_2 + P_4$ can only intersect with $2P_3$.

By Corollary 1, the only case of interest is if three intersections occur, so let us distinguish the following two cases.

`Case 1.` $P_1 + P_4$ intersects $P_2 + P_3$. We note that the vector $(1, -1, -1, 1)$ can be written as the sum of $(1, -2, 1, 0)$ and $(0, 1, -2, 1)$ and therefore Corollary 1 again implies that A is 2–dimensional.

Case 2. $P_1 + P_4$ intersects $2P_2$ or $2P_3$. It is clear that these cases are identical by symmetry, so let us assume the former. We must have

$$\begin{aligned} P_1 + P_4 \cap 2P_2 \neq \emptyset &\quad\Leftrightarrow\quad k_1 + k_2 + \ell_1 \geq k_3 + \ell_2 + \ell_3 + 2, \\ P_1 + P_3 \cap 2P_2 \neq \emptyset &\quad\Leftrightarrow\quad k_2 + k_3 + \ell_2 \geq \ell_1 + 2, \\ P_2 + P_4 \cap 2P_3 \neq \emptyset &\quad\Leftrightarrow\quad k_3 + k_4 + \ell_3 \geq \ell_2 + 2. \end{aligned}$$

Combining the first two inequalities gives $\ell_3 \leq k_1 + 2k_2 - 4$ which, combined with the third inequality, gives $\ell_2 \leq |A| + k_2 - 6$ which inserted into the second inequality gives $\ell_1 \leq |A| + 2k_2 + k_3 - 8$. Taken together this would imply

$$\ell = \ell_1 + \ell_2 + \ell_3 \leq 2|A| + k_1 + 5k_2 + k_3 - 18 \leq 7|A| - 30$$

in contradiction to the assumption that $\ell = \ell_1 + \ell_2 + \ell_3 \geq 7|A| - 30$. □

4. Proof of Theorem 1

We quote explicitly the so–called $(3k-4)$–Theorem of Freiman [2] mentioned in the Introduction, which will be used throughout the proof.

Theorem 3 (Freiman). *Let $A \subset \mathbb{Z}$ in normal form with $a = \max(A)$ and $k = |A|$. We have*

$$|2A| \geq \min\{k + a, 3k - 3\}.$$

There are several versions of the above Theorem for the sum of distinct sets due to Freiman [2], Lev and Smeliansky [12] and Stanchescu [14]. We will use the following slightly weaker form of the one by Lev and Smeliansky [12].

Theorem 4 (Lev and Smelianski). *Any two finite sets $A, B \subset \mathbb{Z}$ in normal form with $\max(A) > \max(B)$ satisfy*

$$|A + B| \geq \min\{|A| + 2|B| - 2, \max(A) + |B|\}. \tag{22}$$

We make also use of the following result by Freiman [2].

Theorem 5. *Let A be a two dimensional set of cardinality $k > 6$ with $|2A| = 3|A| - 3 + b$. If A cannot be covered by a set consisting of two lines with volume at most $k + b$ then $b \geq |A|/3 - 2$.*

We will also use the following lemma which handles the case of two segments.

Lemma 2. *Let A be an extremal set with cardinality k composed by two segments. Then A is isomorphic to one of the following sets:*

(*i*) $[0, k+b-1] \setminus [1, b]$, *with* $\dim(A) = 1$, $\text{vol}(A) = k+b$ *and* $|2A| = 2k-1+b$ *for some* $1 \leq b \leq k-3$, *or*

(*ii*) $([0, k_1-1] \times \{0\}) \cup ([0, k_2-1] \times \{1\})$ *with* $k_1+k_2 = k$, $k_1, k_2 \geq 1$, $\dim(A) = 2$, $\text{vol}(A) = k$ *and* $|2A| = 3k-3$.

Proof. If $\dim(A) = 2$, then there must be no relation in the matrix S_A in Corollary 1 and we are led to Case (ii).

Suppose that $\dim(A) = 1$ and $A = P_1 \cup P_2$, say $P_1 = [0, k_1-1]$ and $P_2 = [k_1+\ell_1, k_1+\ell_1+k_2-1]$ for some $k_1, k_2 \geq 1$, where $k = k_1+k_2$ and $\ell_1 \geq 1$. Since $dim(A) = 1$, we may also assume that $(P_1+P_2) \cap 2P_2 \neq \emptyset$, so that $2A$ consists of the interval $[0, 2(k+\ell_1-1)]$ with a hole of some length $h \geq 0$. Since A is extremal and has volume $vol(A) = k+\ell_1$ we have $|2A| = 2k-1+\ell_1$, so that $h = \ell_1$. Therefore

$$\ell_1 = \min(P_1+P_2) - \max(2P_1) - 1 = \ell_1 + 1 - k_1,$$

which implies $k_1 = 1$ and gives Case (i). □

Proof of Theorem 1. Let A consist of three segments P_1, P_2, and P_3, separated by intervals of holes L_1 and L_2. We consider three cases according to the dimension of A.

`Case 1.` $\dim(A) = 3$. By Corollary 1 we must have $\text{rank}(S_A) = 0$, that is, all $P_i + P_j$ are disjoint for $1 \leq i, j \leq 3$, and we are led to Case (iv) of the Theorem.

`Case 2.` $\dim(A) = 2$. It follows from Corollary 1 that the matrix S_A has $\text{rank}(S_A) = 1$. Up to isomorphisms, we have two possibilities for the only independent relation in S_A.

`Case 2.1:` $(P_1+P_3) \cap 2P_2 \neq \emptyset$. In this case we may assume that

$$\begin{aligned} P_1 &= \{(0,0), \ldots, (0, k_1-1)\}, \\ P_2 &= \{(1,0), \ldots, (1, k_2-1)\} \text{ and} \\ P_3 &= \{(2,\ell), \ldots, (2, \ell+k_3-1)\} \end{aligned}$$

for some $\ell \in \mathbb{N}_0$. We know that $|2A| = 4k-6-|(P_1+P_3) \cap 2P_2|$ so that $b = k-3-|(P_1+P_3) \cap 2P_2|$. We also have $\text{vol}(A) = k+\max(\lfloor (k_1+\ell+k_3)/2 \rfloor - k_2, 0)$. Since $\dim(A) = 2$ we must have $|(P_1+P_3) \cap 2P_2| > 0$ and therefore $\ell \leq 2k_2-2$. Now if $0 \leq \ell \leq 2k_2-k_1-k_3$ then $b = k-3-(k_1+k_3-1) = k_2-2$ and $\text{vol}(A) = k$. If A is extremal, we therefore have $k_2 = 2$ so that $k_1+k_3 \leq 4$ and hence $k \leq 6$. If $\max(2k_2-k_1-k_3, 0) < \ell \leq 2k_2-2$ then $b = k-3-(2k_2-2-\ell+1) = k-2+2k_2-\ell > \max(\lfloor (k_1+\ell+k_3)/2 \rfloor - k_2, 0)$ so the set cannot be extremal.

`Case 2.2:` $2P_1 \cap (P_1 + P_2) \neq \emptyset$. The case $(P_1 + P_2) \cap 2P_2 \neq \emptyset$ works likewise. We may assume that

$$\begin{aligned} P_1 &= \{(0,0),(0,1),\cdots,(0,k_1-1)\}, \\ P_2 &= \{(0,k_1+\ell_1),(0,k_1+\ell_1+1),\ldots,(0,k_1+\ell_1+k_2-1)\} \text{ and} \\ P_3 &= \{(1,0),\ldots,(1,k_3-1)\} \end{aligned}$$

where $k_1 \geq \ell_1 + 2$. Let $A_0 = P_0 \cup P_1$ and $A_1 = P_2$, so that

$$2A = 2A_0 \cup (A_0 + A_1) \cup 2A_1,$$

the union being disjoint. We have $|2A_1| = 2k_3 - 1$ and, by Theorem 3, we also have $|2A_0| \geq 2(k_1 + k_2) - 1 + \ell_1$. Moreover, it can be readily checked that

$$|A_0+A_1| = \begin{cases} k_1 + \ell_1 + k_2 + (k_3 - 1) = k + \ell_1 - 1, & \text{if } k_3 > \ell_1 + 1, \\ (k_1 + k_3 - 1) + (k_2 + k_3 - 1) = k + k_3 - 2, & \text{otherwise.} \end{cases}$$

It follows that

$$|2A| \geq 3k - 3 + \ell_1 + \min\{k_3 - 1, \ell_1\}. \tag{23}$$

As $\mathrm{vol}(A) = k + \ell_1$, the set can only be extremal if $\min(k_3 - 1, \ell_1) = 0$, which implies $k_3 = 1$ (as $\ell_1 \geq 1$) and there is equality in (23), namely, if $|2A_0| = 2(k_1 + k_2) - 1 + \ell_1$. Applying Lemma 2 to A_0 leads to Case (iii) of the Theorem.

`Case 3.` $\dim(A) = 1$. We recall the notation

$$|P_1| = k_1, \quad |P_2| = k_2, \quad |P_3| = k_3, \quad |L_1| = \ell_1, \quad |L_2| = \ell_2, \tag{24}$$

and

$$a = \max(A) = k + \ell - 1. \tag{25}$$

where $\ell = \ell_1 + \ell_2$. The six segments in $2A$ are detailed below for further reference:

$$\begin{aligned} 2P_1 &= [0, 2k_1 - 2], \\ P_1 + P_2 &= (k_1 + \ell_1) + [0, k_1 + k_2 - 2], \\ 2P_2 &= 2(k_1 + \ell_1) + [0, 2k_2 - 2], \\ P_1 + P_3 &= k_1 + \ell_1 + k_2 + \ell_2 + [0, k_1 + k_3 - 2], \\ P_2 + P_3 &= 2(k_1 + \ell_1) + (k_2 + \ell_2) + [0, k_2 + k_3 - 2], \\ 2P_3 &= 2(k_1 + \ell_1 + k_2 + \ell_2) + [0, 2k_3 - 2]. \end{aligned}$$

Since A is extremal, we have

$$a \geq 2(k+b-2), \tag{26}$$

so that

$$\ell \geq k+2b-3. \tag{27}$$

We will use the following facts.

Claim 1. *If* $\max(k_1,k_2) < \ell_1+2$ *then* $2P_1$, P_1+P_2 *and* $2P_2$ *are pairwise disjoint. If* $\max(k_1,k_2) \geq \ell_1+2$ *then* $P_1 \cup P_2$ *is* 1*–dimensional and* $2(P_1 \cup P_2)$ *is a segment with a hole of length*

$$h = \max\{\ell_1 - \min(k_1,k_2)+1, 0\}.$$

Proof. We note that $2P_1$ does not intersect P_1+P_2 if and only if $\max(2P_1) < \min(P_1+P_2)$, which is equivalent to $k_1 < \ell_1+2$. Likewise, P_1+P_2 does not intersect $2P_2$ if and only if $k_2 < \ell_1+2$, establishing the first part of the claim.

Assume without loss of generality that $k_1 \leq k_2$ and $k_2 \geq \ell_1+2$. Then $(P_1+P_2) \cup 2P_2$ is a segment since the two parts intersect. In particular, $P_1 \cup P_2$ is 1–dimensional. Moreover, either $2P_1 \cup (P_1+P_2) \cup 2P_2$ is a segment or a segment with a hole of length $h = \min(P_1+P_2) - \max(2P_1) - 1 = \ell_1 - k_1 + 1$, establishing the second part of the claim. □

By the above Claim, if both $\max(k_1,k_2) < \ell_1+2$ and $\max(k_2,k_3) < \ell_2+2$, then the five segments in

$$2P_1 \cup (P_1+P_2) \cup 2P_2 \cup (P_2+P_3) \cup 2P_3$$

are pairwise disjoint. Using Corollary 1 it follows that $\operatorname{rank}(S_A) \leq 1$ and hence $\dim(A) \geq 2$, contradicting the assumption of this case.

We will therefore without loss of generality assume that $\max(k_1,k_2) \geq \ell_1+2$. In this case we have the following.

Claim 2. $\max\{k_2,k_3\} < \ell_2+2$.

Proof. Suppose on the contrary that $\max\{k_2,k_3\} \geq \ell_2+2$. Then, using (27),

$$k+2b-3 \leq \ell \leq \max\{k_1,k_2\} + \max\{k_2,k_3\} - 4$$

which implies $\max\{k_1,k_2\} = \max\{k_2,k_3\} = k_2$. It follows that $\max(2P_2) = 2(k_1 + l_1 + k_2 - 1) \geq 2(k_1+\ell_1) + k_2 + \ell_2 = \min(P_2+P_3)$. Hence, the sets $2(P_1 \cup P_2)$ and $2(P_2 \cup P_3)$ overlap and, by Claim 1, $2A$ consists of the interval $[0, 2a]$ with two holes of total length at most

$$\max\{\ell_1 - k_1 + 1, 0\} + \max\{\ell_2 - k_3 + 1, 0\} \leq \ell.$$

Therefore, by using $a = k+\ell-1$ and (27), we obtain $|2A| \geq 2a-\ell+1 \geq 3k+2b-4$ and therefore A is not extremal. □

It follows from Claim 1 and Claim 2 that the three segments $2P_2, P_2 + P_3, 2P_3$ are pairwise disjoint. Since A is one–dimensional, $2(P_1 \cup P_2)$ must intersect $P_1 + P_3$. In particular, $\max(2P_2) \geq \min(P_1 + P_3)$ which yields

$$k_1 + \ell_1 + k_2 \geq \ell_2 + 2. \tag{28}$$

Claim 3. $k_3 = 1$.

Proof. Suppose on the contrary that $k_3 > 1$. We then have $\ell_1 > 1$, since otherwise (28) and (27) give $k_1 + k_2 \geq \ell_2 + 1 \geq k + 2b - 3$ and we get $k_3 \leq 1$.

Let $B = 2(P_1 \cup P_2) \cup (P_1 + P_3) \cup (P_2 + P_3)$. We can write $2A$ as the disjoint union

$$2A = B \cup 2P_3.$$

Consider now the set A' obtained from A by replacing $\min(P_3)$ with $\max(P_1) + 1$ if $k_1 \geq k_2$ and with $\min(P_2) - 1$ otherwise. The resulting set is still composed of three disjoint segments, $A' = P_1' \cup P_2' \cup P_3'$ with $\ell_1' = \ell_1 - 1$, $\ell_2' = \ell_2 + 1$ and $\min\{k_1', k_2'\} = \min\{k_1, k_2\}$. We can write $2A'$ as the disjoint union

$$2A' = B' \cup 2P_3',$$

where $B' = 2(P_1' \cup P_2') \cup (P_1' + P_3') \cup (P_2' + P_3')$. We have $|2P_3'| = |2P_3| - 2$. Let us show that $|B'| \leq |B| + 1$.

By Claim 1, $|2(P_1' \cup P_2')| \leq |2(P_1 \cup P_2)| + 1$. If $k_1 \geq k_2$ then $P_2' = P_2$ and $|P_2' + P_3'| = |P_2 + P_3| - 1$, while $P_1' + P_3' = (P_1 + P_3) + 1$. If $P_1 + P_3$ and $P_2 + P_3$ are disjoint, then the two last modifications compensate each other, while if they intersect then there is no change in the cardinality of their union. Similarly, if $k_1 < k_2$ then $P_1' = P_1$ and we loose one unit in $P_1' + P_3'$ while $P_2' + P_3'$ is translated one unit to the right from $P_2 + P_3$, and again there is no change in the cardinality of the union of these two segments.

In either case, we get $|2A'| < |2A|$ so that, if A' is one–dimensional it would have the same volume as A, contradicting that A is extremal. It follows that A' must be 2–dimensional. This implies $\max(2P_2') < \min(P_1' + P_3')$. Since $\max(2P_2) \geq \min(P_1 + P_3)$, we have equality in the last inequality. Therefore,

$$|2A| = |2(P_1 \cup P_2)| + |P_3 + A| - 1. \tag{29}$$

By Theorem 4 we have

$$|P_3 + A| \geq |A| + 2|P_3| - 2 = k + 2k_3 - 2. \tag{30}$$

Therefore,

$$\begin{aligned} |2A| &= |2(P_1 \cup P_2)| + |P_3 + A| \\ &\geq (\max(2P_2) + 1 - \ell_1) + (k + 2k_3 - 2) \\ &= 2(k - k_3 + \ell_1 - 1) - \ell_1 + k + 2k_3 - 2 \\ &= 3k + \ell_1 - 4, \end{aligned}$$

so that $\ell_1 \leq b$. But then, by (28), we have

$$\ell = \ell_1 + \ell_2 \leq b + (k - k_3 + b - 2) = k - k_3 + 2b - 2, \tag{31}$$

contradicting (27). Therefore A could not have been extremal. □

We can therefore assume $P_3 = \{a\}$. It follows that

$$2A = 2(P_1 \cup P_2) \cup (a + A).$$

Moreover,

$$\min(P_2 + P_3) - \max(P_1 + P_3) = \ell_1 - k_3 + 2 = \ell_1 + 1 > 1.$$

We next consider two cases.

`Case 3.1:` $k_1 \leq k_2$.

The sumset $2A$ can be written as the disjoint union

$$2A = B \cup (P_2 + P_3) \cup 2P_3,$$

where $B = 2(P_1 \cup P_2) \cup (P_1 + P_3)$ is an interval with a hole of length $h = \max\{\ell_1 - k_1 + 1, 0\}$. Such a one–dimensional set with $k_1 > 1$ cannot be extremal since, by exchanging $\max(P_1)$ by $\min(P_2) - 1$ we get a one–dimensional set with the same volume and smaller doubling. It follows that $k_1 = 1$. By using (28), we get $\max(2P_2) - \max(P_1 + P_3) = \ell_1 + k_2 - \ell_2 - 1 \geq 0$. In this case $2(k + \ell_1 - 2) = \max(2P_2) \geq a = \max(P_1 + P_3)$ and, again by extremality, equality holds. We thus have $|2A| = (a - \ell_1 + 1) + (k - 2) + 1 = 3k + \ell_1 - 4$, leading to Case (i) of the Theorem.

`Case 3.2:` $k_1 > k_2$.

From

$$\begin{aligned} 3k - 4 + b = |2A| &= |2(P_1 \cup P_2)| + |a + A| - |2(P_1 \cup P_2) \cap (a + A)| \\ &= 2(k - 1 + \ell_1) - 1 - \max\{\ell_1 - k_2 + 1, 0\} + k \\ &\quad - |2(P_1 \cup P_2) \cap (a + A)|, \end{aligned}$$

we obtain

$$|2(P_1 \cup P_2) \cap (a + A)| = 2\ell_1 + 1 - \max\{\ell_1 - k_2 + 1, 0\} - b. \tag{32}$$

For this equality to hold, a necessary condition is

$$\max(2P_2) - a + 1 \geq 2\ell_1 + 1 - \max\{\ell_1 - k_2 + 1, 0\} - b. \tag{33}$$

By using $\max(2P_2) = 2(k-1) + 2\ell_1 - 2$ and $a = k + \ell - 1$ in (33), we obtain

$$\ell \leq k + b + \max\{\ell_1 - k_2 + 1, 0\} - 3. \tag{34}$$

By (27) we have $\ell_1 \geq k_2 + b - 1$. On the other hand, since $|2(P_1 \cup P_2) \cap (a + A)| \leq |2P_2| = 2k_2 - 1$, it follows from (32) that $\ell_1 \leq b + k_2 - 1$. Hence $\ell_1 = b + k_2 - 1$, there is equality in (34) and $2P_2$ must be included in $P_1 + P_3$, so that $2k_2 - 1 = |2P_2| \leq |P_1 + P_3| = k_1$. This gives Case (ii) of the Theorem.

□

References

[1] Y. Bilu, Structure of sets with small sumset, *Astérisque* **258.xi** (1999), 77–108.

[2] G.A. Freiman, Foundations of a structural theory of set addition, *Translations of Math. Monographs* **37**, AMS (1973).

[3] G.A. Freiman, What is the structure of K if $K + K$ is small?, *Number Theory*, Springer, Berlin, Heidelberg, 1987, 109–134.

[4] G.A. Freiman, Inverse additive number theory. XI. Long arithmetic progressions in sets with small sumsets, *Acta Arith.* **137** (2009), 325–331.

[5] G.A. Freiman, On the Additive Volume of Sets of Integers, *arxiv preprint* **1412.5082** (2014).

[6] G.A. Freiman, D. Grynkiewicz, O. Serra and Y.V. Stanchescu, Inverse additive problems for Minkowski sumsets I, *Collect. Math.* **63.3** (2012), 261–286.

[7] G.A. Freiman and O. Serra, On doubling and volume: Chains, *arXiv preprint* **1608.04916v2** (2017).

[8] Y.O. Hamidoune and A. Plagne, A generalization of Freiman's $3k - 3$ theorem, *Acta Arith.* **103.2** (2002), 147–156.

[9] R. Jin, Detailed structure for Freiman's 3k-3 theorem, *Integers* **15A** (2015), Article A9.

[10] R. Jin, Freiman's inverse problem with small doubling property, *Adv. Math.* **216.2** (2007), 711–752.

[11] S. V. Konyagin and V. F. Lev, Combinatorics and linear algebra of Freiman's isomorphism, *Mathematika* **47.1-2** (2000), 39–51.

[12] V. Lev and P. Smeliansky, On addition of two distinct sets of integers, *Acta Arith.* **70.1** (1995), 85–91.

[13] I. Ruzsa, Generalized arithmetic progressions and sumsets, *Acta Math. Hungar.* **65.4** (1994), 379–388.

[14] Y.V. Stanchescu, On addition of two distinct sets of integers, *Acta Arith.* **75.2** (1996), 191–194.

[15] Y.V. Stanchescu, On the simplest inverse problem for sums of sets in several dimensions, *Combinatorica* **18.1** (1998), 139–149.

[16] Y.V. Stanchescu, The structure of d–dimensional sets with small sumset, *J. Number Theory* **130.2** (2010), 289–303.

[17] T. Schoen, Near optimal bounds in Freiman's theorem, *Duke Math. J.* **158.1** (2011), 1–12.

8

MINIMUM COPRIME LABELINGS FOR OPERATIONS ON GRAPHS

John Asplund
Dept. of Technology and Mathematics, Dalton State College, Dalton, Georgia
jasplund@daltonstate.edu

N. Bradley Fox
Department of Mathematics and Statistics, Austin Peay State University, Clarksville, Tennessee
foxb@apsu.edu

Abstract
A prime labeling of a graph of order n is a labeling of the vertices with the integers 1 to n in which adjacent vertices have relatively prime labels. A coprime labeling maintains the same criterion on adjacent vertices using any set of distinct positive integers. In this paper, we consider several families of graphs or products of graphs that have been shown to not have prime labelings, and answer the natural question of how to label the vertices while minimizing the largest value in its set of labels.

1. Introduction

Consider G to be a simple graph with vertex set V in which $|V| = n$ and edge set E. Throughout this paper, we let p_i be the ith prime number. A *prime labeling* of G is a labeling of V using the distinct integers $\{1, \ldots, n\}$ such that the labels of any pair of adjacent vertices are relatively prime. If such a labeling exists, we call G a *prime graph*. More generally, a *coprime labeling* of G uses distinct labels from the set $\{1, \ldots, m\}$ for some integer $m \geq n$ such that adjacent labels are relatively prime. The minimum value m for which G has a coprime labeling is defined as the *minimum coprime number*, denoted as $\mathfrak{pr}(G)$, and a coprime labeling of G with largest label being $\mathfrak{pr}(G)$ is called a *minimum coprime labeling* of G. A prime graph therefore has $\mathfrak{pr}(G) = n$ as its minimum coprime number.

The concept of a prime labeling of a graph was first developed by Roger Entringer and introduced in [15] by Tout, Dabboucy, and Howalla. While most research has revolved around finding prime labelings for various classes of graphs, our focus is on the problem of determining the minimum coprime number for graphs that have been shown to not be prime, a question that was previously studied for complete bipartite graphs $K_{n,n}$ by Berliner et al. [1]. A dynamic survey of results on the 35 year history of prime labelings is given by Gallian in [4].

It was conjectured by Entringer that all trees have prime labelings, and many classes of trees such as paths, stars, complete binary trees, and spiders have been shown by Fu and Huang in [3] to be prime. Salmasian [10] showed that for every tree T with n vertices ($n \geq 50$), $\mathfrak{pr}(T) \leq 4n$. Pikhurko [7] improved this by showing that for any integer $c > 0$, there is an N such that for any tree T of order $n > N$, $\mathfrak{pr}(T) < (1+c)n$. Additionally, many graphs that are not trees have been proven to be prime, including cycles, helms, fans, flowers, and books for all sizes; see [2], [12], and [14]. There is a large collection of graphs whose primality depends on the size of its vertex set. The complete graph K_n, for example, is clearly prime only if $n \leq 3$. Additionally, the wheel graph W_n, which consists of a cycle of length n where each vertex on the cycle is adjacent to a central vertex, is prime if and only if n is even. Section 2 examines the minimum coprime number for complete graphs with at least 4 vertices and wheel graphs in which n is odd.

While paths and cycles on n vertices, denoted as P_n and C_n, respectively, are known to be prime, combinations of these through common graph operations often result in a graph that is not prime. Recall the *disjoint union* of graphs G and H is the graph $G \cup H$ with vertex set $V(G) \cup V(H)$ and edge set $E(G) \cup E(H)$. Deretsky et al. [2] proved that $C_{2k} \cup C_n$ is prime for all integers k and n. However, if both of the cycles are of odd length, their disjoint union is not prime. Determining the minimum coprime number in this case of the union of odd cycles will be our first focus in Section 3. We also consider the union of the complete graph with either a path or the star graph, denoted as S_n where n is the number of degree 1 vertices, also called *pendant* vertices. Youssef and El Sakhawi [16] studied these two union graphs, concluding that $K_m \cup P_n$ is prime if and only if $1 \leq m \leq 3$ or $m = 4$ with $n \geq 1$ being odd. They also determined $K_m \cup S_n$ is prime if and only if the number of primes less than or equal to $m + n + 1$ is at least m. Another graph operation that we will examine is the *corona* operation, which is defined as follows. The corona of a graph G with a graph H, in which $|V(G)| = n$, is denoted by $G \odot H$ and is obtained by combining one copy of G with n copies of H by attaching the ith vertex in G to every vertex within the ith copy of H. In particular, we examine the corona of a complete graph on n vertices with an empty graph on 1 or 2 vertices. We examine the non-prime cases for these unions and coronas to determine their minimum coprime number in Section 3.

Given a graph G, the kth *power* of G, denoted G^k, is defined as the graph with the same vertex set as G but with an edge between each $u, v \in V(G)$ for which $d(u,v) \leq k$ in G. Here the value $d(u,v)$ is the distance between u and v, or the length of the shortest path between the two vertices. The square of paths and cycles, denoted as P_n^2 and C_n^2, were shown not to be prime by Seoud and Youssef in [14]. This implies that higher powers of these are also not prime since G^k is a subgraph of G^ℓ for integers $k \leq \ell$. Section 4 explores the minimum coprime numbers for the square and cube of both the path and cycle graphs.

The *join* of two disjoint graphs G and H, denoted as $G+H$, consists of a vertex set $V(G) \cup V(H)$ with an edge added to connect each vertex in G to those in H, resulting in an edge set $E(G) \cup E(H) \cup \{uv : u \in V(G), v \in V(H)\}$. Seoud, Diab, and Elsahawi [12] studied the primality of the join of paths with the empty graph $\overline{K}_m$ on m isolated vertices. They proved that $P_n + \overline{K}_2$ is prime if and only $n = 2$ or n is odd, and that the join graph $P_n + \overline{K}_m$ is not prime for all $m \geq 3$. Since these two classes of graphs are subgraphs of $P_n + P_2$ and $P_n + P_m$, respectively, the join of paths follow similar criteria for not being prime. Analogous reasoning applies for the join of two cycles or of a path and a cycle. We will find the minimum coprime number for certain cases of these join graphs depending on the relationship between m and n within Section 5.

Our final section concludes with open problems for further research. While there are still many unanswered questions about the primality of graphs such as conjectures on trees and unicyclic graphs being prime, we focus on open questions regarding the minimum coprime number of particular classes of graphs.

2. Complete Graphs and Wheels

Consider the complete graph K_n on n vertices. It is easy to see that K_n is prime if and only if $n \leq 3$. The clearest way to determine when a graph is not prime is based on the *independence number* of the graph, defined as the size of the largest set of vertices S (called an independent set) in which no pair of these vertices are adjacent. In order to possibly have a prime labeling, a graph needs an independence number of at least $\left\lfloor \frac{|V|}{2} \right\rfloor$ in order for an independent set of vertices to exist for the even labels (a fact first noted in [3]). The graph K_n has an independence number of 1 since all vertices are adjacent to each other, hence we examine the minimum coprime number for the complete graph with 4 or more vertices.

Proposition 1. *Let $n \geq 4$. The minimum coprime number of K_n is* $\mathfrak{pr}(K_n) = p_{n-1}$.

Proof. Since each vertex is adjacent to every other vertex, only prime numbers and 1 can be used as vertex labels to keep each pair of vertices relatively prime. Thus, we can label the graph with a minimum coprime labeling by using the first $n-1$ primes along with 1. □

Next we consider the wheel graph W_n. We name the vertices $v_1, \ldots, v_n$ as shown in Figure 1 with v_1 representing the center vertex and the remaining v_i listed in clockwise order with v_2 being adjacent to v_{n+1}. Lee, Wui, and Yeh [5] demonstrated

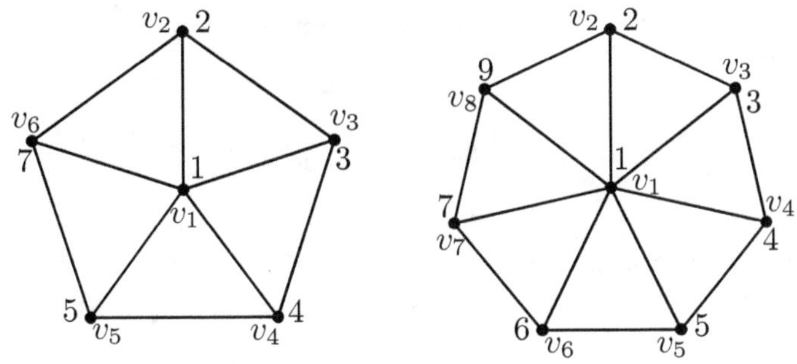

Figure 1: The graph W_n

that W_n is prime if and only if n is even. The following result determines the minimum coprime number for the odd case.

Proposition 2. *Let n be odd. Then the minimum coprime number of W_n is* $\mathfrak{pr}(W_n) = n + 2$.

Proof. We label the vertices v_i as i for each $i \in \{1, \ldots, n\}$ and v_{n+1} as $n + 2$. This labeling is coprime since each adjacent pair of labels falls into one of four cases: the center label 1 is in the pair, the labels are consecutive integers, the labels are n and $n + 2$ which are consecutive odd numbers, or the labels are 2 and the odd integer $n + 2$. The labeling is a minimum coprime labeling since it was proven in [5] to be not prime, hence a labeling with maximum label being the number of vertices, $n + 1$, is impossible to achieve. □

3. Disjoint Union and Corona Operations

The following observation is straightforward from the definitions of prime and coprime labelings and will be useful in some of our upcoming proofs. We first recall that if G and H are graphs with $V(G) = V(H)$ and $E(G) \subseteq E(H)$, then we say H is a *spanning supergraph* of G, and G is a *spanning subgraph* of H.

Observation 1. *If G is not prime, then a spanning supergraph of G is not prime. If G is prime, then any spanning subgraph of G is also prime.*

The disjoint union of two graphs has been shown to be prime for a variety of graphs under certain conditions. In [2], the disjoint union of cycles was shown to be prime when at least one of the cycles has an even number of vertices; that is, $C_{2k} \cup C_n$ is prime for all positive $k, n \in \mathbb{Z}$. Examining the case in which both cycles are odd, we see that, similarly to the wheel graph in Proposition 2, we only need the largest label to be $|V(G)| + 1$ to achieve a minimum coprime labeling.

Theorem 2. *For all $k, \ell \geq 1$, the minimum coprime number of the disjoint union of two odd-length cycles is*

$$\mathfrak{pr}(C_{2k+1} \cup C_{2\ell+1}) = 2(k + \ell) + 3.$$

Proof. We label the vertices of the graph, as shown in Figure 2, using the labels $1, 3, 4, \ldots, 2k+1, 2k+2$ in this order on the cycle C_{2k+1} and the labels $2, 2k+3, 2k+4, \ldots, 2(k+\ell)+1, 2(k+\ell)+3$ on the cycle $C_{2\ell+1}$. Most of the edges have endpoints with labels of the form $\{m, m+1\}$, which are relatively prime as consecutive integers. The remaining edges either include 1 as an endpoint, connect 2 to an odd label, or have consecutive odd labels $2(k+\ell)+1$ and $2(k+\ell)+3$ assigned to its vertices, which are all relatively prime pairs.

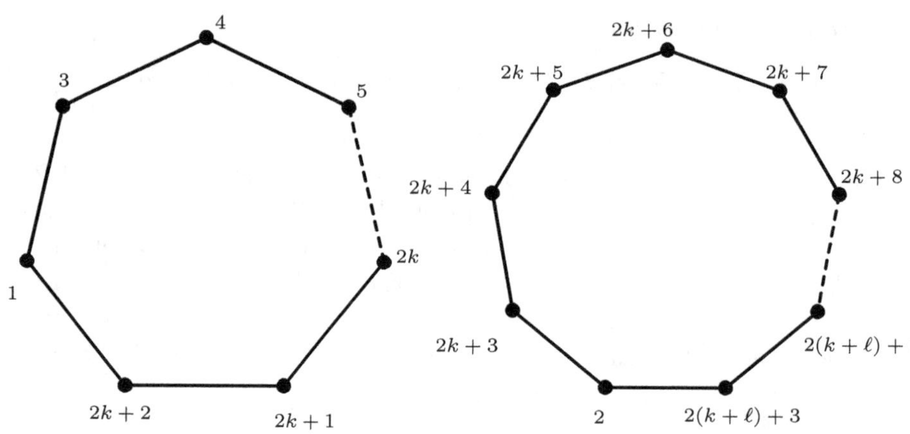

Figure 2: Minimum coprime labeling of $C_{2k+1} \cup C_{2\ell+1}$

It remains to show that this labeling is minimum, which is accomplished by demonstrating that the graph does not have a prime labeling since $2(k+\ell)+3$ is one larger than the size of the vertex set V of $C_{2k+1} \cup C_{2\ell+1}$. No prime labeling exists because the independence number for the graph is $k+\ell$, which is smaller than the required $\left\lfloor \frac{|V|}{2} \right\rfloor = k + \ell + 1$ independent vertices. Therefore, our labeling is a minimum coprime labeling. □

We now find a minimum coprime labeling for the union of the complete graph with a path or a star graph. These classes of graphs, $K_m \cup P_n$ and $K_m \cup S_n$, were investigated by Youssef and El Sakhawi in [16]. They proved that $K_m \cup P_n$ is prime if and only if $1 \leq m \leq 3$ or $m = 4$ and $n \geq 1$ is odd. See Figure 3 for an example of such a graph with a minimum coprime labeling.

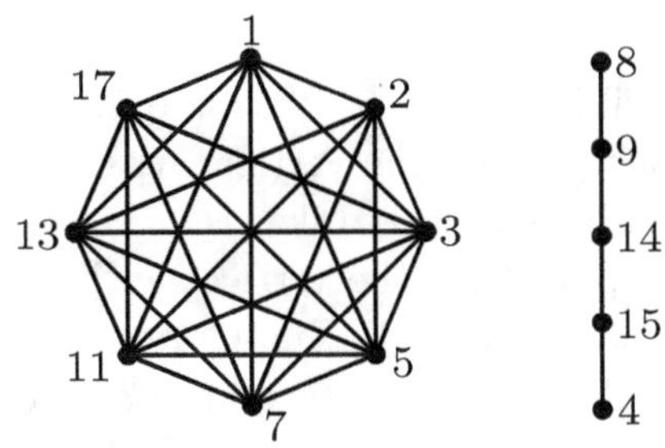

Figure 3: Minimum coprime labeling of $K_8 \cup P_5$ when $n \leq r$

Theorem 3. *Let m and n be positive integers, and if $m \geq 5$, we let $r = p_{m-1} - 2m + 4$. The minimum coprime number for $K_m \cup P_n$ is*

$$\mathfrak{pr}(K_m \cup P_n) = \begin{cases} m+n & \text{if } 1 \leq m \leq 3 \text{ or } m = 4 \text{ and } n \text{ is odd} \\ m+n+1 & \text{if } m = 4 \text{ and } n \text{ is even} \\ p_{m-1} & \text{if } m \geq 5 \text{ and } n \leq r \\ 2m+n-4 & \text{if } m \geq 5,\ n > r, \text{ and } n \text{ is odd} \\ 2m+n-3 & \text{if } m \geq 5,\ n > r, \text{ and } n \text{ is even.} \end{cases}$$

Proof. The case when $1 \leq m \leq 3$ or $m = 4$ and n is odd was already shown in [16] since $K_m \cup P_n$ is prime under those conditions. When $m = 4$ and n is even, we can label K_4 with $1, 2, 3, 5$ and use the integers $6, 7, \ldots, n+5$ to label P_n. Hence $\mathfrak{pr}(K_m \cup P_n) = m + n + 1$ since the graph is not prime.

Suppose that $m \geq 5$ and $n \leq r$. It is clear by our discussion on K_m that $\mathfrak{pr}(K_m \cup P_n) \geq p_{m-1}$ since K_m is a subgraph of $K_m \cup P_n$. As was done in Proposition 1, label the vertices of K_m with 1 and the first $m-1$ primes. Consider the vertices on the path as $V(P_n) = \{v_1, \ldots, v_n\}$, which we will label using the following sequence $X = \{x_i\}_{i=1}^{\infty}$. When n is even, we assign the value x_i to the vertex v_i for $i = 1, \ldots n$. When n is odd, we alter this slightly by assigning x_i to v_i for $i = 1, \ldots n-1$, and we label v_n by 4:

$$X = 8, 9, 14, 15, 16, 21, 22, 25, 26, 27, 28, 33, 34, 35, 38, 39, 44, 45, 46, 49, \ldots$$

The sequence is defined after the initial two terms by including each integer starting with 14 except we skip each prime p_i for all $i \geq 7$ along with the following three cases, determined partly by whether p_i and p_{i+1} are twin primes (i.e., $p_i + 2 = p_{i+1}$): we skip $p_i - 1$ if p_i is not the first of two twin primes and $3|(p_i - 1)$, as in the case of $p_{12} = 37$; skip $p_i - 1$ and $p_{i+1} - 1$ if p_i and p_{i+1} are twin primes and $5|(p_i - 1)$, such as the case of $p_{13} = 41$ and $p_{14} = 43$; and skip $p_i + 1$ for all other prime numbers.

It is clear that each pair of adjacent labels in K_m are relatively prime. The same is true for the labels on P_n since the distance between any such pair is at most 5, and we ensure, through the first two cases of values that were skipped, that no pair of adjacent labels are divisible by 3 or 5. This is because if $p_i - 1$ is divisible by 3 for p_i being a non-twin prime, then the vertices with labels $p_i - 2$ and $p_i + 1$ will be adjacent with $\gcd(p_i - 2, p_i + 1) = 1$. Likewise, if $p_i - 1$ is divisible by 5 when p_i is the first of a twin prime pair, then the vertices with labels $p_i - 2$ and $p_i + 3$ will be adjacent with $\gcd(p_i - 2, p_i + 3) = 1$. Therefore, this is a coprime labeling.

Notice that the largest label on P_n will be odd since the sequence X is increasing, and we use 4 as the final label if n is odd. There are $\frac{p_{m-1}+1}{2}$ odd numbers less than or equal to p_{m-1}. Of these, $m-1$ of them are odd prime numbers that we used as labels on K_m, leaving $\frac{p_{m-1}+1}{2} - m + 1$ odd integers that are smaller than p_{m-1}as labels for P_n. Then our path can include up to $2\left(\frac{p_{m-1}+1}{2} - m + 1\right) + 1 = p_{m-1} - 2m + 4 = r$ vertices while maintaining that all of its labels are smaller than p_{m-1}. By our assumption of $n \leq r$, $\mathfrak{pr}(K_m \cup P_n) = p_{m-1}$ in this case.

Next suppose that $m \geq 5$ and $n > r$. We label K_m as before, label the first $r - 1$ vertices of P_n using $x_1, \ldots, x_{r-1}$, and label v_r by 4. For the remaining vertices, if n is even, we label $v_{r+1}, \ldots, v_n$ by the sequence

$$p_{m-1} + 2, p_{m-1} + 3, \ldots, p_{m-1} + n - r + 1.$$

If n is odd, we label $v_{r+1}, \ldots, v_{n-1}$ by the above sequence up to $p_{m-1} + n - r$, but we label v_n by 8 and reassign the label for v_1 as 10 if $m > 5$. Note that if $m = 5$, this reassignment for n odd is not needed because $r = 1$ and hence the first label in the path is 4. The labeling is coprime based on the discussion in the last case about labeling the path using the sequence X, and the fact that any newly adjacent pairs of labels are consecutive, an odd integer adjacent to 4 or 8, or potentially the labels 10 and 9 at the beginning of the path. It is a minimum coprime labeling since our maximum label is an odd integer, either $p_{m-1}+n-r+1$ if n is even or $p_{m-1}+n-r$ if n is odd, and our labeling includes every odd value up to this maximum while using as many even labels as possible on the path. Our result for $\mathfrak{pr}(K_m \cup P_n)$ follows since $p_{m-1} + n - r + 1 = 2m + n - 3$ and $p_{m-1} + n - r = 2m + n - 4$. □

The last class of disjoint union graphs we will investigate is the union of a complete graph and a star, $K_m \cup S_n$, where S_n has n pendant vertices. See Figure 4 for an example of the union of K_9 and the star S_6 with a minimum coprime labeling. By [16], $K_m \cup S_n$ is prime when $\pi(m+n+1) \geq m$ where $\pi(m+n+1)$ is the number of prime numbers less than or equal to $m + n + 1$. To find the minimum coprime labeling of such a graph, we first define three numbers and provide examples of each.

Definition 4. For positive integers t and b, we let $\varphi(t, b)$ be the number of composite integers not equal to 1 that are less than or equal to b and relatively prime with t.

Example 5. We observe that $\varphi(25,31) = 14$, since there are eleven composite numbers less than 25 that are relatively prime with 25, and three such numbers that are between 25 and 31.

Next, we define a number k_b that will potentially be used in our minimum coprime labeling as the center of the star S_n.

Definition 6. Given positive integers m and n, and a composite integer b in the interval $[p_{m-1}, p_m]$, we let $k_b = \min(\{q^2 : q^2 < b, \varphi(q^2, b) \geq n,$ and q is prime$\})$. In the case that no such q exists in which q^2 fits the required inequalities, we define $k_b = \infty$.

Example 7. When $m = 10$ and $n = 10$, the smallest prime squares that fit the required inequalities for $b = 27$ is $k_{27} = 25$ yet when $b = 28$, we see that $k_{28} = 9$. On the other hand, when $m = 4$ and n is any positive integer, we find that when $b = 6$, $k_6 = \infty$ since the only $q^2 < b$ is 4 and $\varphi(4,6) = 0 < n$.

In the upcoming labeling, the integer b will represent the largest label assigned to a pendant vertex of the star S_n. The center of the star will be labeled by the integer k_b, resulting in the largest label among the vertices of S_n being the following value.

Definition 8. For $m, n \in \mathbb{N}$, we let $\alpha = \min(\{\max(k_b, b) : b \in (p_{m-1}, p_m)\})$.

Again, note it is possible for $k_b = \infty$ for each b value, particularly if n is sufficiently large compared to b, and in this case $\alpha = \infty$ as well. We now formalize our labeling in the following result.

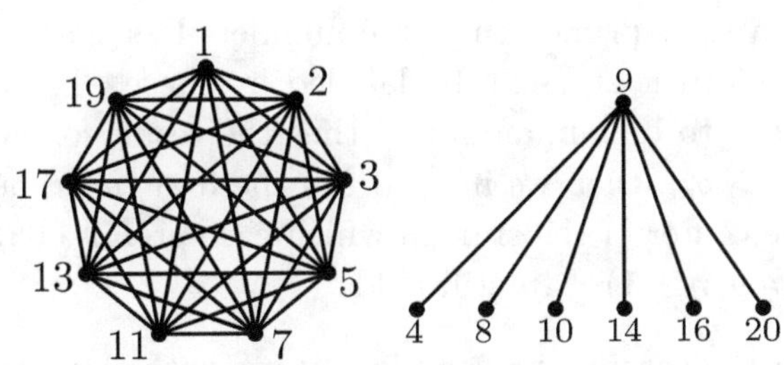

Figure 4: Minimum coprime labeling of $K_9 \cup S_6$ with $\pi(m+n+1) < m$, $n > r$, and $\alpha \leq p_m$

Theorem 9. *Let m, n be positive integers, p be the largest prime number such that $p^2 < p_{m-1}$, and $r = p_{m-1} - m - \left\lfloor \frac{p_{m-1}}{p} \right\rfloor + 1$. The minimum coprime number for*

$K_m \cup S_n$ *is*

$$\mathfrak{pr}(K_m \cup S_n) = \begin{cases} m+n+1 & \textit{if } \pi(m+n+1) \geq m, \\ p_{m-1} & \textit{if } \pi(m+n+1) < m \textit{ and } n \leq r \\ \alpha & \textit{if } \pi(m+n+1) < m,\ n > r, \textit{ and } \alpha \leq p_m \\ p_m & \textit{otherwise.} \end{cases}$$

Proof. By the work in [16], if $\pi(m+n+1) \geq m$ then $\mathfrak{pr}(K_m \cup S_n) = m+n+1$. Now suppose that $\pi(m+n+1) < m$. Since $\mathfrak{pr}(K_m) = p_{m-1}$ by Proposition 1, $\mathfrak{pr}(K_m \cup S_n) \geq p_{m-1}$. It is clear that $K_m \cup S_n$ will not be prime in this case. We first suppose that $n \leq r$. Again, label the vertices of K_m using 1 and the first $m-1$ primes. Let the center of the star be labelled as p^2. There are $p_{m-1} - m$ positive integers that are not used on the labels of K_m. Since the center of the star is labeled p^2, there are at most $\left\lfloor \frac{p_{m-1}}{p} \right\rfloor - 1$ positive integers less than p_{m-1} that we can use to label the pendant vertices of S_n. Since we assumed for this case that $n \leq r = p_{m-1} - m - \left\lfloor \frac{p_{m-1}}{p} \right\rfloor + 1$, we have $\mathfrak{pr}(K_m \cup S_n) = p_{m-1}$.

Now suppose that $n > r$. There are many options for the label of the center of the star, so we aim to use the smallest composite number less than p_m that is relatively prime with at least n composite numbers less than some fixed value $b \in (p_{m-1}, p_m)$. Since q^2 (q prime) is relatively prime with at least the same number of composite numbers as qt for some integer $t > 1$ with $t \neq q$, we would choose q^2 over qt as the label for the center of the star. Thus k_b is defined to be a candidate for the center of the star for each integer $b \in (p_{m-1}, p_m)$. If this were a minimum coprime labeling then its minimum coprime number is $\max(k_b, b)$. Hence we look for the minimum among all of these maximums and denote such a value as α. By construction, there are more than n relatively prime composite numbers less than α that are not 1, so the pendant vertices of the star can be labeled by these. Thus, this is a coprime labeling. We choose α to be minimum and thus the result follows if $\alpha \leq p_m$.

If we find that $\alpha > p_m$, then we instead use the first m primes on K_m, and use 1 as the label of the center of the star, in which case $\mathfrak{pr}(K_m \cup S_n) = p_m$ as long as the assumption $\pi(m+n+1) < m$ still holds. □

Next we consider the corona of a complete graph with the empty graph of one or two vertices. In [16], it was shown that $K_n \odot K_1$ and $K_n \odot \overline{K}_2$ are prime under certain conditions, particularly if $n \leq 7$ for K_1 and if $n \leq 16$ for $\overline{K}_2$. Later, El Sonbaty, Mahran, and Seoud [13] showed that $K_n \odot \overline{K}_m$ is not prime if $n > \pi(n(m+1)) + 1$ and conjectured that $K_n \odot \overline{K}_m$ is prime if $n < \pi(n(m+1)) + 1$. We give the minimum coprime number for $K_n \odot K_1$ and $K_n \odot \overline{K}_2$ below. An example of $K_8 \odot K_1$ is given in Figure 5.

Theorem 10. *If n is an integer with $n > 7$, then* $\mathfrak{pr}(K_n \odot K_1) = p_{n-1}$.

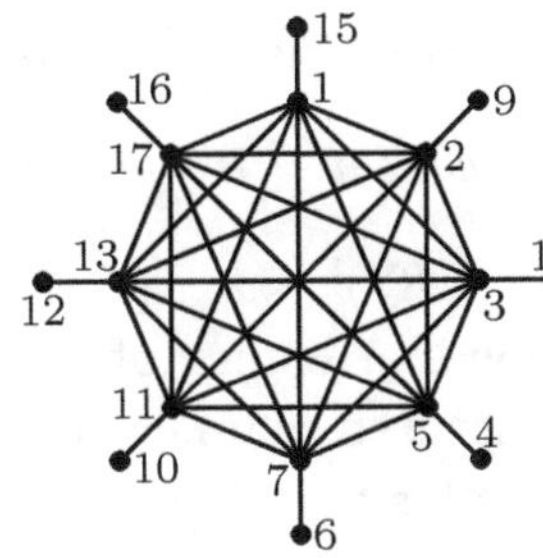

Figure 5: Minimum coprime labeling of $K_8 \odot K_1$

Proof. By Proposition 1, $\mathfrak{pr}(K_n \odot K_1) \geq p_{n-1} \geq 17$. Let $u_1, u_2, \ldots, u_n$ be the vertices in K_n, and for each $i \in \{1, \ldots, n\}$, let v_i be the vertices adjacent to u_i for the n copies of K_1. Label the vertices u_i with p_{i-1} for $i \in \{2, \ldots, n\}$ and label u_1 with 1. Then label v_i with $p_{i-1} - 1$ for $i \in \{4, \ldots, n\}$. We label v_1, v_2, v_3 with $15, 9, 14$, respectively. Since these three labels are specifically chosen to be relatively prime with their neighbors $1, 2$, and 3, and all other pendant edges connect labels that are consecutive, this labeling is coprime. This results in $\mathfrak{pr}(K_n \odot K_1) \leq p_{n-1}$, combining with the previous inequality to prove our claim of $\mathfrak{pr}(K_n \odot K_1) = p_{n-1}$. □

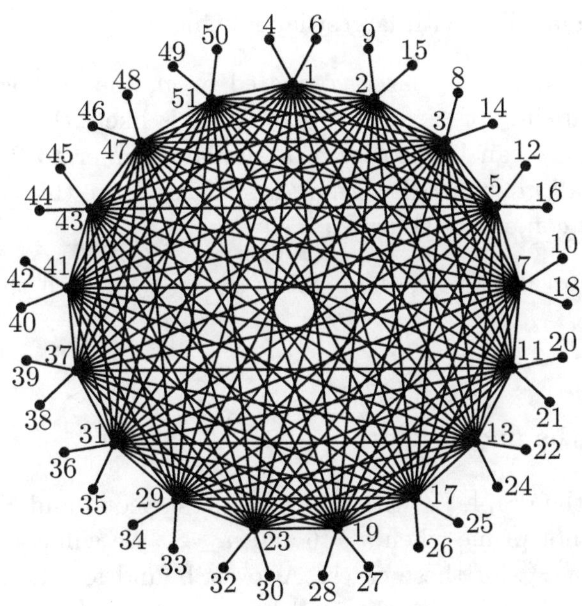

Figure 6: Minimum coprime labeling of $K_{17} \odot \overline{K}_2$

An example of $K_{17} \odot \overline{K}_2$ is given in Figure 6.

Theorem 11. *If n is an integer with $n > 16$, then $\mathfrak{pr}(K_n \odot \overline{K}_2) = p_{n-1}$.*

Proof. By Proposition 1, $\mathfrak{pr}(K_n \odot K_1) \geq p_{n-1} \geq 53$. We will label the vertices in K_n and the first 16 corresponding copies of $\overline{K}_2$, and then we will label the remaining vertices in a more structured manner. Let $u_1, u_2, \ldots, u_n$ be the vertices in K_n and $v_{i,1}, v_{i,2}$ be the vertices for the n copies of $\overline{K}_2$. Label the vertices u_i for $i \in \{2, 3, \ldots, n\}$ with p_{i-1} and label u_1 with 1. We label the sequence of vertices $v_{1,1}, v_{1,2}, v_{2,1}, v_{2,2}, v_{3,1}, v_{3,2}, \ldots, v_{16,2}$, respectively, with the following sequence of labels:

$$4, 6, 9, 15, 8, 14, 12, 16, 10, 18, 20, 21, 22, 24, 25, 26, 27, 28,$$
$$30, 32, 33, 34, 35, 36, 38, 39, 40, 42, 44, 45, 46, 48.$$

For $i > 16$, if p_{i-1} is not a twin prime, we label $v_{i,1}, v_{i,2}$ with $p_{i-1} - 2, p_{i-1} - 1$. If instead p_{i-1} and p_i are twin primes, we label $v_{i,1}, v_{i,2}$ with $p_{i-1} - 3, p_{i-1} - 2$ and label $v_{i+1,1}, v_{i+1,2}$ with $p_{i-1} - 1, p_{i-1} + 1$. One can see by inspection that the labels on the first 16 vertices in K_n are relatively prime with the corresponding pendant labels from the sequence, and the coprime condition is also upheld for pendant edges from the other vertices v_i with $i > 16$ since each label p_{i-1} is within distance 3 from each of its adjacent labels. Therefore, we have obtained our result since this is a minimum coprime labeling with largest label being p_{n-1}. $\square$

As long as n is sufficiently large compared to m, the authors believe that the minimum prime labeling of $K_n \odot \overline{K}_m$ is p_{n-1}. It is likely that several subsequent cases beyond $m = 2$ can be proven in a similar manner to Theorem 11, but a generalization for all m eludes discovery. As such, we leave this as an open problem in Section 6 and a conjecture below.

Conjecture 1. For all $m > 0$, there exists an $M > m$ such that for all $n > M$, $\mathfrak{pr}(K_n \odot \overline{K}_m) = p_{n-1}$.

4. Powers of Paths and Cycles

We next consider the graph P_n^2 with $n \geq 6$ vertices. Seoud and Youssef [14] proved that this graph is not prime when $n = 6$ and $n \geq 8$. We will construct a minimum coprime labeling of P_n^2 for these cases. A lower bound for the minimum coprime number for the graph would be obtained by using the maximum amount of even labels that can be used based on the independence number of P_n^2, shown in [14] to be $\left\lceil \frac{n}{3} \right\rceil$, along with the smallest possible odd labels. Figure 7 shows a minimum coprime labeling of the graph P_6^2 and P_{10}^2, and in the former case, the path is represented as $v_1, v_2, \ldots, v_6$ with the horizontal edges connecting vertices of distance 2. Since the independence number of P_6^2 is 2, we can only use two even labels, which prevents a prime labeling. Instead, a minimum coprime labeling can be achieved with $\mathfrak{pr}(P_6^2) = 7$.

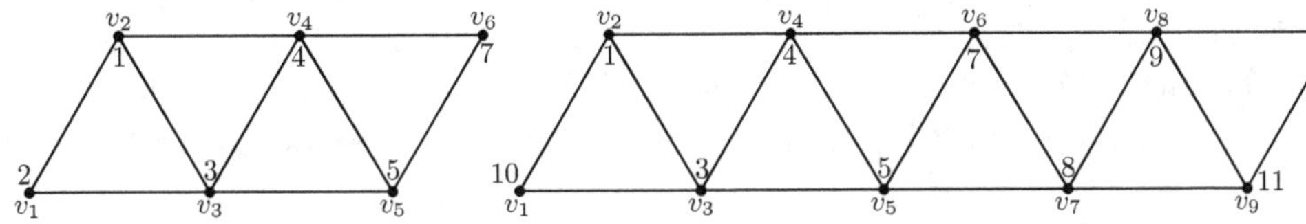

Figure 7: Minimum coprime labeling of P_6^2(left) and P_{10}^2 (right)

The following theorem regarding the minimum coprime number of the path squared was verified for the $n = 6$ case through the labeling in Figure 7, and the general case of $n \geq 8$ will be proven by a series of lemmas.

Theorem 12. *Let $n = 6$ or $n \geq 8$. The minimum coprime number of P_n^2 is given by*

$$\mathfrak{pr}(P_n^2) = \begin{cases} 4k-1 & \text{if } n = 3k \text{ or } 3k+1 \\ 4k+1 & \text{if } n = 3k+2. \end{cases}$$

Assume that $n \geq 8$ for the following lemmas that will prove the general case of Theorem 12. To construct a minimum coprime labeling of P_n^2, we define a sequence $X = \{x_i\}_{i=1}^{\infty}$ of integers for which the first n numbers will be used as labels for the vertices $\{v_1, \ldots, v_n\}$. The sequence X consists of a length 45 segment with the repeated pattern of even, odd, and odd integers, with the subsequent terms of the sequence defined by shifting the initial entries by multiples of 60. The first 30 odd numbers are included in the initial segment along with the first 15 even numbers that are not multiples of 3 or 5. The definition of the sequence is the following:

$$\begin{aligned}\{x_1, \ldots, x_{45}\} = \{&2, 1, 3, 4, 5, 7, 8, 9, 11, 14, 13, 15, 16, 17, 19, 22, 21, 23, 26, 25, 27, 28,\\ &29, 31, 32, 33, 35, 34, 37, 39, 38, 41, 43, 44, 45, 47, 46, 49, 51, 52, 53,\\ &55, 56, 57, 59\},\end{aligned}$$

and for $i > 45$, we recursively define $x_i = x_{i-45} + 60$.

In order to examine whether adjacent vertices will have relatively prime labels, the following fact about the distance between the labels of such vertices will be quite useful.

Lemma 1. *Given adjacent vertices v_i and v_j in P_n^2 for some $1 \leq i, j \leq n$, the labels in the sequence $\{x_1, \ldots, x_n\}$ satisfy $|x_i - x_j| \leq 5$.*

Proof. Based on the structure of P_n^2, the neighborhood of a vertex v_i with $i = 3, \dots, n-2$ is the set $\{v_{i-2}, v_{i-1}, v_{i+1}, v_{i+2}\}$. By inspection of the 45 labels in the initial segment of the sequence, we can verify that x_i is within distance 5 of the label of any adjacent vertex when $i \le 43$.

For adjacent vertices v_i and v_j where $v_i \in \{v_1, \dots, v_{45}\}$ and $v_j \in \{v_{46}, \dots, v_{90}\}$, we only need to consider the labels $x_{44} = 57$, $x_{45} = 59$, $x_{46} = 62$, and $x_{47} = 61$, where the latter two labels are the result of shifting x_1 and x_2 by 60. We see that our desired inequality $|x_i - x_j| \le 5$ is satisfied by each adjacent pair such as $|x_{44} - x_{46}| = 5$.

Adjacent vertices with indices that are both larger than 45 will also satisfy $|x_i - x_j| \le 5$ because $x_i = x_a + 60m$ and $x_j = x_b + 60m$ for some integers $1 \le a, b \le 47$ and positive integer m. Since x_a and x_b satisfy the inequality as shown above, the shifted values x_i and x_j also maintain a distance of 5 or less, which covers the remaining possible cases of indices i and j. □

Note that if we continued to define $x_{46}, \dots, x_{54}$ in the manner of the first 45 terms by including all even numbers that are not multiples of 3 or 5, then we would have $x_{52} = 64$ and $x_{54} = 71$. This would have contradicted Lemma 1 and eventually would result in the sequence containing adjacent labels that are both multiples of 7. Shifting the $x_1, \dots, x_{45}$ segment by 60 to have $x_{46} = 62$ skips the even number 58 and maintains that the distance between adjacent labels remain small enough to guarantee a coprime labeling, as shown in the following lemma.

Lemma 2. *The labels* $\{x_1, \dots, x_n\}$ *are a coprime labeling of* P_n^2.

Proof. By Lemma 1, the distance between the labels of adjacent vertices is at most 5; hence, no two adjacent vertices will have labels with a factor of 6 or higher in common. By design, even labels are not adjacent, eliminating the possibility of labels sharing a factor of 2. None of the even labels are divisible by 3 or 5, and any two odd labels both divisible by 3 or 5 are spaced out enough to avoid being adjacent, resulting in no adjacent labels having a 3 or 5 as a common factor. Thus our labeling is coprime. □

Lemma 3. *If* $\max(x_1, \dots, x_n)$ *is odd, then the labels* $\{x_1, \dots, x_n\}$ *are a minimum coprime labeling of* P_n^2.

Proof. Notice that even integer labels cannot be 1 or 2 indices apart, which corresponds to the independence number of our graph being $\left\lceil \frac{n}{3} \right\rceil$. Hence, we have used as few odd labels as possible, while also using all of the odd numbers from 1 to $\max(x_1, \dots, x_n)$. Thus we cannot make $\mathfrak{pr}(P_n^2)$ any smaller. □

For the case of the maximum label within $\{x_1, \dots, x_n\}$ being even, we alter the sequence of labels to achieve a minimum labeling. Through examination of the first 45 terms of the sequence and the fact that remaining terms are simply shifted from this initial segment, we see that this situation can only occur when $n = 3k+1$ or $3k+2$. Additionally, we observe that the largest even label in either case is in position $3k+1$. We create a new sequence $\{x_i^*\}_{i=1}^n$ by defining $x_1^* = 10$, $x_{3k+1}^* = 2$, and $x_i^* = x_i$ for $i \in \{2,\ldots,n\} \setminus \{3k+1\}$.

Lemma 4. *If* $\max(x_1,\ldots,x_n)$ *is even, then the labels* $\{x_1^*,\ldots,x_n^*\}$ *are a minimum coprime labeling of* P_n^2.

Proof. As described above, to be in the case of the maximum label being even, we know that $n = 3k+1$ or $3k+2$. By inspection of the initial segment of our sequence of labels, the first such $n \geq 8$ with an even label being the largest in the sequence $\{x_1,\ldots,x_n\}$ would be $n = 10$ in which $x_{10} = 14$. For this case, or for any larger such n value, replacing x_1 and x_{3k+1} (which we note is x_n or x_{n-1}) with $x_1^* = 10$ and $x_{3k+1}^* = 2$ would result in the maximum of the labels falling into the case of Lemma 3 since one can observe that the second largest label in $\{x_1,\ldots,x_n\}$ will always be odd if the largest one is even.

Following the reasoning of the previous lemma, the labeling $\{x_1^*,\ldots,x_n^*\}$ is a minimum coprime labeling as long as we show it is still a coprime labeling. This sequence matches the original sequence except for two values; hence, the only adjacent vertex pairs that need to be checked as still having relatively prime labels are the vertices v_1 and v_{3k+1} with their respective neighbors. In the case of $n = 3k+1$, $v_{3k+1} = v_n$ is only adjacent to v_{n-1} and v_{n-2}. Since both labels for these vertices are odd, the label $x_n^* = 2$ is relatively prime with its adjacent labels. Similarly if $n = 3k+2$, then $v_{3k+1} = v_{n-1}$ is only adjacent to v_{n-3}, v_{n-2}, and v_n. Again, all of these vertices will be labeled by odd numbers, which are relatively prime to 2. Likewise, the label $x_1^* = 10$ is only adjacent to the second and third vertices with labels $x_2^* = 1$ and $x_3^* = 3$, so the labels are once again relatively prime. Thus, the labeling $\{x_1^*,\ldots,x_n^*\}$ is a minimum coprime labeling of P_n^2. □

Proof of Theorem 12. The previous lemmas in this section have shown that the sequence $\{x_1,\ldots,x_n\}$ or $\{x_1^*,\ldots,x_n^*\}$ provides a minimum coprime labeling of P_n^2. It only remains to show that the maximum value in the labeling sequence, which was shown to be odd, is in fact $4k-1$ or $4k+1$ depending on the value of $n \pmod 3$. In the case of $n = 3k$, there are $\left\lceil \frac{n}{3} \right\rceil = k$ even labels used, so $\mathfrak{pr}(P_n^2)$ is the $(n-k)^{\text{th}}$ odd number, which is $2(n-k)-1 = 4k-1$. The other two cases follow similarly to give us the value of $\mathfrak{pr}(P_n^2)$. □

We next consider the square of the cycle, C_n^2, for $n \geq 4$. Seoud and Youssef [14] showed this graph is not prime when $n \geq 4$ and that its independence number is $\left\lfloor \frac{n}{3} \right\rfloor$, which will be the maximum allowable number of even labels. Note the only difference compared to the squared path graph is the additional edges v_1v_{n-1}, v_1v_n, and v_2v_n. These do require some alterations to our labeling of P_n^2 in some cases to maintain the coprime property, which results in the following for the minimum coprime number. See Figure 8 for an example of a minimum coprime labeling of C_8^2.

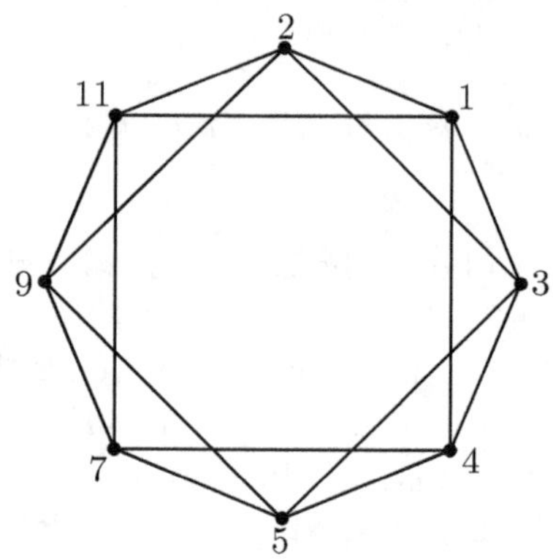

Figure 8: Minimum coprime labeling of C_8^2

Theorem 13. *Let $n \geq 4$. The minimum coprime number of C_n^2 is given by*

$$\mathfrak{pr}(C_n^2) = \begin{cases} 4k-1 & \text{if } n = 3k \\ 4k+1 & \text{if } n = 3k+1 \\ 4k+3 & \text{if } n = 3k+2. \end{cases}$$

Proof. For the case of $n = 3k$, the labeling of vertices $v_1, \ldots, v_n$ using the sequence $\{x_1, \ldots, x_n\}$ still provides a minimum coprime labeling as it did with the P_n^2. This is because $x_1 = 2$ and $x_2 = 1$ while the final two labels x_{n-1} and x_n are odd, so the additional three new edges maintain our coprime property.

The other two cases cannot simply use the same labeling as the squared path graph since the independence number is 1 smaller for C_n^2 compared to P_n^2. Also note that we will only need to adjust the sequence $\{x_1, \ldots x_n\}$ instead of $\{x_1^*, \ldots, x_n^*\}$ even in the cases of $n = 3k+1$ and $3k+2$ that needed the largest even label reassigned as 2 in the path squared labeling. This is because this largest even label will instead be reassigned as an odd number in the following cases.

When $n = 3k+1$, the edge v_1v_n would result in a common factor of 2 between the labels $x_1 = 2$ and x_n since x_n is even. Hence, we reassign $x_n = 4k+1$, which is the lowest available odd label. This label will now be relatively prime with its adjacent vertices whose labels are 1, 2, $4k-1$, and $4k-3$, making $\{x_1, \ldots, x_n\}$ a minimum coprime labeling.

Similarly, when $n = 3k+2$, the edge v_1v_{n-1} leads to adjacent vertices with even labels. We correct this while keeping the labeling minimum by reassigning

$x_{n-1} = 4k+1$ and $x_n = 4k+3$, which we note are the smallest available odd labels. The newly assigned x_{n-1} and x_n are coprime with any adjacent vertices' labels since their neighbors are only labeled by 1, 2, or an odd number of distance 2 or 4 from x_{n-1} or x_n. In each case, we obtain a minimum coprime labeling using the maximum amount of even labels and the smallest possible odd labels with the largest label being $4k-1$, $4k+1$, and $4k+3$ in the respective cases. □

Now we consider taking the third power of the path graph, P_n^3, which has additional edges from v_i to v_{i+3} for $i = 1, \ldots, n-3$. Since this graph contains P_n^2 as a spanning subgraph, it is also not prime when $n = 6$ and $n \geq 8$ by Observation 1. It is known that the independence number for the cubed path is $\left\lceil \frac{n}{4} \right\rceil$, which will determine how many even labels can be placed on its vertices. This fact results in P_n^3 not being prime for the additional cases of $n = 4$ and $n = 7$, but note that the path cubed on 4 vertices is simply the complete graph. It is, however, prime when $n = 5$, in which the path can be labeled using $2, 1, 3, 5, 4$.

Theorem 14. *Let $n \geq 6$. The minimum coprime number of P_n^3 is given by*

$$\mathfrak{pr}(P_n^3) = \begin{cases} 6k-1 & \text{if } n = 4k \text{ or } 4k+1 \\ 6k+1 & \text{if } n = 4k+2 \\ 6k+3 & \text{if } n = 4k+3. \end{cases}$$

Similar to our construction for the coprime labeling of P_n^2, we define a sequence of finite length, with the subsequent terms determined by shifting the initial sequence. For P_n^3, our initial sequence consists of 140 entries, and the shift is by 210 from y_i to y_{i+140}. Note that the sequence below is a repetition of even, odd, odd, and odd entries with the smallest possible odd numbers included and all even multiples of 3, 5, and 7 excluded. Additionally, some other even numbers were removed to maintain the inequality in Lemma 5 regarding the distance between adjacent labels. We define the labeling sequence as follows, where vertex v_n of P_n^3 is labeled by y_n:

$$\begin{aligned} \{y_1, \ldots, y_{140}\} = \{&2,1,3,5,4,7,9,11,8,13,15,17,16,19,21,23,22,25,27,29,26,31,\\ &33,35,32,37,39,41,38,43,45,47,44,49,51,53,52,55,57,59,58,\\ &61,63,65,62,67,69,71,68,73,75,77,74,79,81,83,82,85,87,89,\\ &86,91,93,95,92,97,99,101,104,103,105,107,106,109,111,113,\\ &116,115,117,119,118,121,123,125,122,127,129,131,128,133,\\ &135,137,134,139,141,143,142,145,147,149,146,151,153,155,\\ &152,157,159,161,158,163,165,167,164,169,171,173,172,175,\\ &177,179,176,181,183,185,184,187,189,191,188,193,195,197,\\ &194,199,201,203,202,205,207,209\} \end{aligned}$$

and $y_i = y_{i-140} + 210$ for $i > 140$.

We prove this theorem as was done for squared paths by using a sequence of lemmas.

Lemma 5. *Given adjacent vertices v_i and v_j in P_n^3 for some $1 \leq i, j \leq n$, the labels in the sequence $\{y_1, \ldots, y_n\}$ satisfy $|y_i - y_j| \leq 9$.*

Proof. The graph P_n^3 has a neighborhood for each vertex v_i consisting of $\{v_{i-3}, v_{i-2}, v_{i-1}, v_{i+1}, v_{i+2}, v_{i+3}\}$ with $i = 4, \ldots, n-3$. Through careful inspection of the 140 initial labels, the maximum distance between labels of a vertex v_i with $i \leq 137$ and any of its neighbors is 9, which is attained for example by vertices v_9 and v_{12} having labels 8 and 17, respectively.

We next examine adjacent vertices with labels with one from the initial sequence $\{y_1, \ldots, y_{140}\}$ and one from the shifted sequence $\{y_{141}, \ldots, y_{280}\}$. Since $y_{138} = 205$, $y_{139} = 207$, $y_{140} = 209$, $y_{141} = 212$, $y_{142} = 211$, and $y_{143} = 213$, the adjacent labels in this case with greatest distance apart are $|y_{138} - y_{141}| = 7$, which is within the desired distance.

As in Lemma 1, adjacent vertices with indices larger than 140 maintain the same distance for their labels as their corresponding vertices with indices between 1 and 143, making the inequality hold for all adjacent vertices. □

Lemma 6. *The labels $\{y_1, \ldots, y_n\}$ are a coprime labeling of P_n^3.*

Proof. By Lemma 5, the distance between adjacent labels is at most 9; thus, no adjacent labels have a common factor of 10 or higher. The construction of the sequence results in even labels being four indices apart and hence not adjacent, so no labels share a common factor of 2. The even labels were chosen to not contain a factor of 3, 5, or 7, and any pair of odd labels that both contain a multiple of 3, 5, or 7 have indices that are at least 4 apart, so they are not adjacent. Thus, each pair of adjacent labels are relatively prime. □

Lemma 7. *If $\max(y_1, \ldots, y_n)$ is odd, then the labels $\{y_1, \ldots, y_n\}$ are a minimum coprime labeling of P_n^3.*

Proof. Since the independence number of P_n^3 is $\left\lceil \frac{n}{4} \right\rceil$, our sequence of labels uses the maximum number of even labels. Therefore, the fact that we have used every odd number up to $\max(y_1, \ldots, y_n)$ implies that we have achieved a minimum coprime labeling. □

As with P_n^2, we define an altered labeling sequence for the case of $\max(y_1, \ldots, y_n)$ being even to switch out this largest even label to ensure the maximum is odd. We create our new sequence $\{y_i^*\}_{i=1}^n$ by defining $y_1^* = 14$, $y_m^* = 2$, and $y_i^* = y_i$ for all $i \in \{2, \ldots, n\} \setminus \{m\}$, where y_m was the maximum label of $\{y_1, \ldots, y_n\}$. It can be observed from the initial sequence $\{y_1, \ldots, y_{140}\}$ that $m = n$ or $n-1$.

Lemma 8. *If* $\max(y_1, \ldots, y_n)$ *is even, then the labels* $\{y_1^*, \ldots, y_n^*\}$ *are a minimum coprime labeling of* P_n^3.

Proof. Through examination of the sequence $\{y_1, \ldots, y_n\}$, the first n value such that an even integer is the largest value in the sequence is $n = 69$ and $n = 70$ with $y_{69} = 104$. Replacing the maximum even label with 2 and the first label with 14 will result in there now being an odd label as the maximum. This allows us to apply Lemma 7 if our labeling remains a coprime labeling. Reassigning the label of v_1 to be 14 maintains our coprime property since it is only adjacent to the labels $1, 3$, and 5. The vertex v_m, whose maximum label was reassigned as the label 2, is only adjacent to vertices with odd labels; hence, the adjacent labels remain relatively prime. □

Proof of Theorem 14. The preceding lemmas have proven that P_n^3 has a mimimum coprime labeling using either $\{y_1, \ldots, y_n\}$ or $\{y_1^*, \ldots, y_n^*\}$, leaving us to verify that the correct minimum coprime number was attained. The labels consist of $\left\lceil \frac{n}{4} \right\rceil$ even labels, so we consider the cases of $n \pmod 4$. For the case of $n = 4k$, $\left\lceil \frac{n}{4} \right\rceil = k$ even labels were used. Therefore, $\mathfrak{pr}(P_n^3)$ is the $(n-k)^{\text{th}}$ odd number, which is $2(n-k) - 1 = 6k - 1$. The other three cases follow similarly to find their minimum coprime numbers. □

We next demonstrate a minimum coprime labeling of C_n^3, which we note has an independence number of $\left\lfloor \frac{n}{4} \right\rfloor$. Also observe that $C_n^3 = K_n$ for $n \leq 7$.

Theorem 15. *Let* $n \geq 8$. *The minimum coprime number of* C_n^3 *is given by*

$$\mathfrak{pr}(C_n^3) = \begin{cases} 6k - 1 & \text{if } n = 4k \\ 6k + 1 & \text{if } n = 4k + 1 \\ 6k + 5 & \text{if } n = 4k + 2 \\ 6k + 7 & \text{if } n = 4k + 3. \end{cases}$$

Proof. We begin by considering the labeling sequence $\{y_1, \ldots, y_n\}$ that was used for P_n^3 on the vertices $v_1, \ldots, v_n$ of the cycle. Before considering each case, it is important to note that the vertex v_n is adjacent to v_3, which is labeled by $y_3 = 3$, in C_n^3 since their distance in the cycle graph is 3. Observe from our labeling sequence that y_n is a multiple of 3 if and only if $n = 4k + 3$.

For $n = 4k$, the labeling $\{y_1, \ldots, y_n\}$ is a minimum coprime labeling of C_n^3. The additional edges in C_n^3 that were not in P_n^3 have endpoints with relatively prime labels because $y_3 = 3$ with y_n not being a multiple of 3 as previously stated above, $y_2 = 1$, and $y_1 = 2$ with the three labels y_{n-2}, y_{n-1}, and y_n all being odd.

When $n = 4k+1$, the independence number being $\left\lfloor \frac{n}{4} \right\rfloor$ implies that the sequence $\{y_1, \ldots, y_n\}$ cannot be used for the labeling since it would include $k+1$ even labels.

Instead, we reassign $y_n = 6k + 1$, which is the smallest unused odd label. This label is not a multiple of 3, so it is relatively prime with $y_3 = 3$. This final label $y_n = 6k + 1$ is also clearly relatively prime with the labels 2 and 1 of the vertices v_1 and v_2, in addition to the odd labels of $y_{n-3} = 6k - 5$, $y_{n-2} = 6k - 3$, and $y_{n-1} = 6k - 1$ since it is not a multiple of 3, resulting in a minimum coprime labeling.

Assuming $n = 4k + 2$, as in the previous case, there are too many even labels which requires a reassignment of the last even label to be $y_{n-1} = 6k + 1$. This label is again relatively prime with y_1 and y_2. The last label, y_n, which originally was also $6k + 1$, cannot be reassigned to be the next smallest odd label of $6k + 3$, which is a multiple of 3. This is because v_n is adjacent to v_3, which is labeled as $v_3 = 3$. Furthermore, labels that are multiples of 3 cannot be shifted in any way to accommodate $6k + 3$ because of the independence number being $\left\lfloor \frac{n}{4} \right\rfloor = k$ and the fact that there are already k multiples of 3 in our labeling sequence. Thus, we set $y_n = 6k + 5$, the smallest possible label that is not even or a multiple of 3. Since it is coprime with the odd labels of $y_{n-3} = 6k - 3$, $y_{n-2} = 6k - 1$, and $y_{n-1} = 6k + 1$, as well as y_1 and y_2, we have a minimum coprime labeling.

We use the same reasoning for the $n = 4k+3$ to reassign the labels $y_{n-2} = 6k+1$, $y_{n-1} = 6k + 5$ (to avoid the multiple of 3), and $y_n = 6k + 7$. The labeling is again minimum because the independence number limits the number of even integers and multiples of 3 that can be used. □

The next logical step would be to generalize our constructions for P_n^k and C_n^k or at least continue with finding a minimum coprime labeling of P_n^4. However, to keep even labels spaced out enough, a sequence for P_n^4 would require repetition of the pattern of even, odd, odd, odd, and odd. Using the smallest possible odd numbers fails quickly though as it would begin $2, 1, 3, 5, 7, 4, 9$, resulting in the labels 3 and 9 being adjacent due to their vertices being distance 4 apart. Having to frequently skip odd numbers within our label sequence greatly increases the difficulty of finding minimum coprime labeling of P_n^k for $k \geq 4$, so we leave this as an open problem in Section 6.

5. The Join of Paths and Cycles

In this section we will establish the minimum coprime number for the join of two paths, two cycles, or a cycle and a path. As we did in Section 4, we will use the results on paths to find solutions for the join of cycles. In [12, 14], it was shown that $P_m + K_1 = P_m + P_1$ is prime, $P_m + \overline{K}_2$ is prime if and only if $m \geq 3$ is odd, and $P_m + \overline{K}_n$ is not prime for $n \geq 3$. We will consider similar join graphs in which edges are added to $\overline{K}_n$ to create a path P_n.

We will exploit the size of the gap between prime numbers in order to ensure we can form a minimum coprime labeling. We define the gap between two primes as $g(p_n) = p_{n+1} - p_n$. By the prime number theorem, for all $\varepsilon > 0$, there exists an integer N such that for all $n > N$, $g(p_n) < \varepsilon p_n$. More specifically, particular values for ε and N are mentioned in [8].

Theorem 16. [6, 9, 11] *Let $\varepsilon > 0$. For any positive integer $n > N$, $g(p_n) < \varepsilon p_n$ where $(N, \varepsilon) \in \{(9, 1/5), (118, 1/13), (2010760, 1/16597)\}$.*

We first investigate a minimum coprime labeling of $P_m + P_2$. We know from [14] that $P_m + \overline{K_2}$ is prime if and only if m is odd or $m = 2$. By Observation 1, adding an edge to create the second path results in $P_m + P_2$ not being prime when m is even and $m \geq 4$. Note that $P_2 + P_2$ would also not be prime, but this graph is simply K_4.

Theorem 17. *If $m \geq 4$ is even, then $\mathfrak{pr}(P_m + P_2) = m + 3$.*

Proof. For the case of $m = 4$, Figure 9 shows an example of a minimum coprime labeling of $P_4 + P_2$. Since $P_m + P_2$ is known to not be prime and the maximum label is $|V(P_4 + P_2)| + 1$, we see that $\mathfrak{pr}(P_4 + P_2) = 7$.

For $m = 6$, label the vertices of $V(P_6) = \{v_1, \ldots, v_6\}$ using the sequence $2, 9, 4, 5, 8, 3$, and label P_2 using 1 and 7. Then it is clear the result follows. Suppose that $m \geq 8$ is even. We label the vertices of $V(P_m) = \{v_1, \ldots, v_m\}$ using the sequence

$$2, m+3, 4, 3, 10, 9, 8, 5, 6, p_4, 12, p_5, 14, 15, 16, p_6, 18, \ldots.$$

Notice that p_i is the label of $v_{p_{i+1}-1}$ for $i \geq 4$. By Theorem 16, m is large enough so that $p_{i+1} < 1.2p_i$. Since the smallest number that is divisible by p_i that is not p_i is $2p_i$, we know p_i is relatively prime with the labels $p_{i+1} - 1$ and $p_{i+1} + 1$ that are adjacent to $v_{p_{i+1}-1}$ for all $i \geq 4$. Let p' be the largest prime that is less than or equal to $m + 1$, which we note is not a label on P_m based on our shift of the prime numbers within our labeling sequence. Then we label the vertices of P_2 using the labels 1 and p'.

It is clear that this is a coprime labeling based on our discussion of p_i being relatively prime to its adjacent labels in the sequence. It is a minimum coprime labeling with $\mathfrak{pr}(P_m + P_2) = m + 3$ since the graph is not prime and the maximum label is $|V(P_m + P_2)| + 1$. □

See Figure 9 for an example of the labeling in the following theorem when $m = 12$.

Theorem 18. *If $m \geq 3$, then the minimum coprime number of $P_m + P_3$ is*

$$\mathfrak{pr}(P_m + P_3) = \begin{cases} m + 4 & \textit{if } m \textit{ is odd} \\ m + 5 & \textit{if } m \textit{ is even.} \end{cases}$$

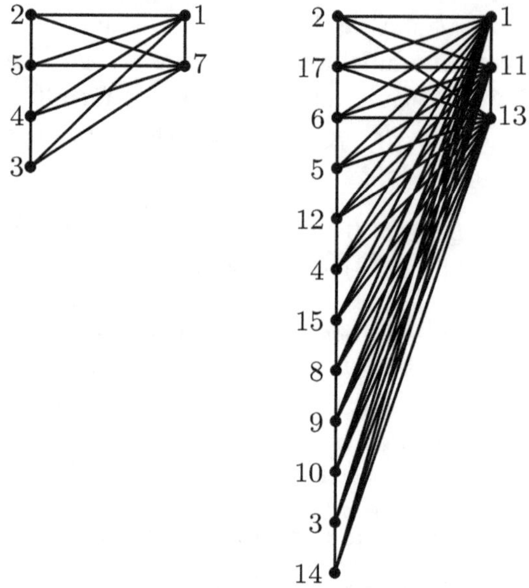

Figure 9: Minimum coprime labeling of $P_4 + P_2$ (left), $P_{12} + P_3$ (right)

Proof. Based on the independence number of $P_m + P_3$ being $\lceil \frac{m}{2} \rceil$, a minimum coprime labeling can at best use the smallest $\lfloor \frac{m}{2} \rfloor + 3$ odd integers. This would correspond to the largest label being $m+4$ if m is odd or $m+5$ if m is even. Thus, constructing labelings using these as the maximum label would prove our result.

Let $V(P_m) = \{v_1, \ldots, v_m\}$. For the initial cases of $m = 3, 4, 5$, and 6, by using the sequence $2, 3, 4, 9, 8, 11$ as the labels for $v_1, \ldots, v_6$, respectively, and labeling the vertices in P_3 with $1, 5, 7$, we see that $\mathfrak{pr}(P_3+P_3) = 7$, $\mathfrak{pr}(P_4+P_3) = \mathfrak{pr}(P_5+P_3) = 9$, and $\mathfrak{pr}(P_6 + P_3) = 11$. By using the sequence $6, 5, 4, 3, 2, 9, 8, 13, 10$ as the labels for $v_1, \ldots, v_9$, respectively, and labeling the vertices in P_3 with $1, 7, 11$, we have $\mathfrak{pr}(P_7 + P_3) = 11$ and $\mathfrak{pr}(P_8 + P_3) = \mathfrak{pr}(P_9 + P_3) = 13$. By using the sequence $6, 5, 4, 3, 2, 9, 10, 7, 8, 15, 14$ as the labels for $v_1, \ldots, v_{11}$, respectively, and labeling the vertices in P_3 with $1, 11, 13$, we see that $\mathfrak{pr}(P_{10} + P_3) = \mathfrak{pr}(P_{11} + P_3) = 15$.

Suppose that $m \geq 12$. We label the vertices of $V(P_m) = \{v_1, \ldots, v_m\}$ using the first m entries in the sequence

$$2, a, 6, 5, 12, 7, 4, b, 8, 9, 10, 3, 14, 15, 16, p_5, 18, p_6, 20, 21, 22, p_7, \ldots$$

where a and b are the values of m' and $m'+2$, with m' being the smallest odd number larger than $m + 1$. If m' is not a multiple of 3, we set $a = m'$ and $b = m' + 2$, but if $3 \mid m'$, we set $b = m'$ and $a = m' + 2$ to avoid a multiple of 3 being adjacent to the label 6. We know that m' will either be $m + 2$ or $m + 3$ depending on whether $m + 1$ is even or odd, implying the largest label is $m' + 2 = m + 4$ or $m + 5$, as described in our claim for $\mathfrak{pr}(P_m + P_3)$.

Notice that p_i is the label of $v_{p_{i+2}-1}$ for $i \geq 5$. By Theorem 16, m is large enough so that $p_{i+2} < 1.44p_i$, thus p_i is relatively prime with the labels $p_{i+2}-1$ and $p_{i+2}+1$ on either side of v_{p_i-1} for all $i \geq 5$. Let p' and p'' be the two largest primes in the sequence $2, 3, 4, \ldots, m+1$. Then we label the vertices in P_3 using the sequence $1, p', p''$. It is clear that this sequence is a coprime labeling, and since we achieved the goal of using a maximum label of $m+4$ and $m+5$ for the odd and even cases, respectively, it is also a minimum coprime labeling. □

Next, we consider the join of a path P_m with P_4, where we note the cases of $m < 4$ have already been covered by previous results.

Theorem 19. *If $m \geq 4$, then the minimum coprime number of $P_m + P_4$ is given by*

$$\mathfrak{pr}(P_m + P_4) = \begin{cases} m+6 & \text{if } m \text{ is odd} \\ m+7 & \text{if } m \text{ is even.} \end{cases}$$

Proof. Observe that the independence number of $P_m + P_4$ is once again $\lceil \frac{m}{2} \rceil$, as it was for $P_m + P_3$. We now have one additional odd label that must be used in our labeling, meaning $\mathfrak{pr}(P_m + P_4) \geq m+6$ for m odd and $m+7$ for m even. We now construct labelings that use this as the maximum label.

Let $V(P_m) = \{v_1, \ldots, v_m\}$. By using the sequence $2, 3, 4, 9, 8, 13, 6$ as the labels for $v_1, \ldots, v_7$, respectively, and labeling the vertices of P_4 with $1, 5, 7, 11$, we have $\mathfrak{pr}(P_m + P_4) = 11, 11, 13, 13$ for the cases of $m = 4, 5, 6, 7$, respectively. By using the sequence $6, 5, 4, 3, 10, 9, 8, 15, 2, 17, 16, 19, 18$ as the labels for $v_1, \ldots, v_{13}$, respectively, and labeling the vertices of P_4 with $1, 7, 11, 13$, we see that $\mathfrak{pr}(P_m + P_4) = m+6$ for each $m = 9, 11, 13$ and $m+7$ for $m = 8, 10, 12$.

Suppose that $m \geq 14$. We label the vertices of $V(P_m) = \{v_1, \ldots, v_m\}$ using the first m values of the sequence

$$2, a, 6, b, 10, 3, 4, c, 8, 9, 14, 5, 12, 15, 16, p_4, 18, p_5, 20, 21, \ldots$$

where $\{a, b, c\} = \{m', m'+2, m'+4\}$, with m' being the smallest odd number larger than $m+1$. Note that at most two of m', $m'+2$, and $m'+4$ can be divisible by 3 or 5, or at most one can be divisible by both 3 and 5. In the case of one of the three being a multiple of three and a different value being a multiple of 5, set a to equal the multiple of 5, c to be the multiple of 3, and b to be the other value. If one of m', $m'+2$, or $m'+4$ is a multiple of 3 and 5, set c to be that value, and set a and b to be the other two values. Again, whether $m+1$ is even or odd will determine if m' is $m+2$ or $m+3$, resulting in the largest label being $m'+4 = m+6$ or $m+7$.

Notice that p_i is the label for $v_{p_i+3} - 1$ for $i \geq 4$. By Theorem 16, m is large enough so that $p_{i+3} < 1.728p_i$ and so p_i is relatively prime with the numbers on either side of the position of p_{i+3} for all $i \geq 4$. Let q_1, q_2, q_3 be the three largest primes in the sequence $2, 3, 4, \ldots, m+1$. Then we label the vertices in P_4 using the sequence $1, q_1, q_2, q_3$. It is clear that this sequence is a coprime labeling. Since its maximum label matches the lower bound for the minimum coprime number, the result follows. $\square$

We now consider the more general join of paths P_m and P_n with restrictions set on m and n, particularly a larger lower bound for m. It is likely the case that $\mathfrak{pr}(P_m + P_n)$ satisfies the equality below for $m \leq 118$, but the task of completing these cases is better left to a computer.

Theorem 20. *Let $m > 118$ and $2 \leq n \leq 10$ be positive integers. Then the minimum coprime number of $P_m + P_n$ is given by*

$$\mathfrak{pr}(P_m + P_n) = \begin{cases} m + 2n - 2 & \text{if } m \text{ is odd} \\ m + 2n - 1 & \text{if } m \text{ is even.} \end{cases}$$

Proof. Since $m > n$, the independence number of the general $P_m + P_n$ is still $\lceil \frac{m}{2} \rceil$ as in the previous proofs. A coprime labeling could be minimized by using the smallest $\lfloor \frac{m}{2} \rfloor + n$ odd integers, resulting the smallest possible minimum coprime number being $m + 2n - 2$ if m is odd and $m + 2n - 1$ if m is even.

Theorems 17, 18, and 19 prove the smallest cases of n, so we let $5 \leq n \leq 10$. We label the vertices of $P_m = \{v_1, \ldots, v_m\}$ using the first n terms of the sequence

$$2, x_1, 4, x_2, 8, x_3, 16, x_4, 32, x_5, 64, x_6, 10, x_7, 20, x_8, 40, x_9, 50, 7, 6, 5, 12, 31, 18, 37, 24, 17, 30, 19, 36, 11, 42, 25, 48, 35, 54, 49, 60, 13, 66, 65, 72, 23, 78, 29, 70, 3, 14, 9, 28, 15, 56, 39, 22, 21, 26, 27, 34, 33, 38, 45, 44, 51, 46, 55, 52, 57, 58, 63, 62, 75, 68, 77, 74, 69, 76, p_{13}, 80, 81, 82, p_{12}, 84, 85, 86, 87, 88, p_{14}, 89, 90, 91, 92, 93, 94, 95, 96, p_{15}, 98, 99, 100, p_{16}, 102, p_{17}, 104, \ldots$$

where $x_1, x_2, \ldots, x_9$ are the 9 smallest odd integers larger than $m+1$. Observe that at most two of the integers in the sequence $x_1, \ldots, x_9$ are divisible by 5. Hence there are at least 4 integers not divisible by 5 that we will use to label the vertices $v_{12}, v_{14}, v_{16}, v_{18}$ to avoid adjacent pairs with a common factor of 5.

If $n < 10$ then the $10 - n$ largest of the labels in $\{x_1, x_2, \ldots, x_9\}$ will be reassigned to use $10 - n$ of the smallest primes in $\{p_{14}, p_{15}, \ldots, p_{22}\}$, and the placement of the primes will shift by $10 - n$ prime positions lower than they currently are in the sequence above. Whether $m + 1$ is even or odd will determine if $m' = \min(\{x_1, \ldots, x_9\})$ is $m + 2$ or $m + 3$. The largest label will be $m' + 2n - 4 = m + 2n - 2$ or $m + 2n - 1$ depending on the parity of m.

By Theorem 16, m is large enough so that $p_{i+n-10} < (14/13)^9 < p_i$ and hence p_i is relatively prime with the labels on either side of $v_{p_{i+n-10}-1}$ for all $i \geq 4$. Let $q_1, q_2, \ldots, q_{n-1}$ be the $n - 1$ largest primes in the sequence $2, 3, 4, \ldots, m + 1$. Then we label the vertices in $V(P_n)$ using the sequence $1, q_1, q_2, \ldots, q_{n-1}$. It is clear that this sequence is a coprime labeling using the desired minimum coprime label which is our claim for $\mathfrak{pr}(P_m + P_n)$. □

By applying similar methods as shown in the proofs of Theorem 19 and 20, the authors believe the following conjecture to be true.

Conjecture 2. For any positive integer N, there exists a positive integer M such that for all $m > M$ and $n \leq N$, the minimum coprime number of $P_m + P_n$ is

$$\mathfrak{pr}(P_m + P_n) = \begin{cases} m + 2n - 2 & \text{if } m \text{ is odd} \\ m + 2n - 1 & \text{if } m \text{ is even.} \end{cases}$$

Consequently, a stronger conjecture may be posed based solely on the size of $m + n$.

Conjecture 3. Let m and n be positive integers such that $m \geq n$. Then the minimum coprime numbers of $P_m + P_n$ is

$$\mathfrak{pr}(P_m + P_n) = \begin{cases} m + 2n - 2 & \text{if } m \text{ is odd} \\ m + 2n - 1 & \text{if } m \text{ is even.} \end{cases}$$

We now move our discussion to the join of two cycles. We will use our work on the join of two paths to find the following results with relative ease. Notice that by Observation 1, $C_m + C_n$, $C_m + P_n$ are not prime labelings when $m, n \geq 2$.

Theorem 21. *Let m and n be positive integers such that $m \geq n$, $n \leq 10$, and $m > 118$ when $n \geq 5$. Then the minimum coprime number of $C_m + C_n$ is given by*

$$\mathfrak{pr}(C_m + C_n) = \begin{cases} m + 2n & \text{if } m \text{ is odd} \\ m + 2n - 1 & \text{if } m \text{ is even.} \end{cases}$$

Proof. Consider the vertices of C_m to be $\{v_1, \ldots, v_m\}$ and those of C_n to be $\{u_1, \ldots, u_n\}$. Assume m is even. From the join of paths, two edges $v_1 v_m$ and $u_1 u_n$ are added to form $C_m + C_n$. For the labelings developed in Theorems 17, 18, 19, and 20, the label on v_1 (it is 2 in each case) is relatively prime with the odd integer label on v_m, and the label 1 on u_1 is relatively prime with any integer, so the result follows.

Suppose that m is odd. The labels on C_n still satisfy the coprime property. The labeling sequences from the previous $P_m + P_n$ results all have $m + 1$ as the label of v_m in the case of m being odd, but this label is even and thus is not relatively prime with the label of v_1. We can reassign the label on v_m to the next available odd number which will only increase the coprime labeling number by 2 from $\mathfrak{pr}(P_m + P_n)$, which is necessary since the independence number of C_m is $\frac{m-1}{2}$. □

The following corollary is directly from combining Observation 1 with Theorem 21 and from our reasoning in the previous result about the edge u_1u_n not violating the coprime condition.

Corollary 1. *Let m and n be positive integers such that $m \geq n$, $n \leq 10$, and $m > 118$ when $n \geq 5$. Then we have*

$$\mathfrak{pr}(C_m + P_n) = \begin{cases} m + 2n & \text{if } m \text{ is odd, } m \neq n \\ m + 2n - 1 & \text{if } m \text{ is even,} \end{cases}$$

and

$$\mathfrak{pr}(P_m + C_n) = \begin{cases} m + 2n - 2 & \text{if } m \text{ is odd} \\ m + 2n - 1 & \text{if } m \text{ is even.} \end{cases}$$

6. Concluding Remarks

We conclude by posing several open questions regarding minimum coprime numbers.

Question 1. Can the minimum coprime number be determined for P_n^k and C_n^k for $k \geq 4$?

Question 2. Trees and grid graphs are conjectured to be prime, meaning their minimum coprime number would match their order. Can the bound shown by Salmasian in [10] that $\mathfrak{pr}(T) \leq 4n$ for a tree T of order n be improved, and can a similar upper bound be found for the grid graph $P_m \times P_n$?

Question 3. Berliner et al. [1] investigated the minimum coprime number of $K_{n,n}$, but were not able to determine this number for all n. Does there exist a formula for $\mathfrak{pr}(K_{n,n})$, and can one determine the minimum coprime number more generally for all complete bipartite graphs $K_{m,n}$ in the cases in which there is no prime labeling?

Question 4. Many graphs that are not always prime are left to study in terms of minimum coprime labelings, such as Möbius ladders and $K_{1,n} + K_2$. Can their minimum coprime numbers be determined?

Question 5. Is the following true: If $mn \leq p_{n-1} - n - 1$ then $\mathfrak{pr}(K_n \odot \overline{K}_m) = p_{n-1}$?

Question 6. Is the following true: If $mn > p_{n-1} - n - 1$ then $\mathfrak{pr}(K_n \odot \overline{K}_m) = mn + n + 1$?

References

[1] A. Berliner, N. Dean, J. Hook, A. Marr, A. Mbirika, and C. McBee, Coprime and prime labelings of graphs, *Journal of Integer Sequences* **19** (2016), no. 2, 3–16.

[2] T. Deretsky, S.M. Lee, and J. Mitchem, On vertex prime labelings of graphs, *Graph Theory*, Combinatorics and Applications Vol. 1, 1991, pp. 359–369.

[3] H.L. Fu and K.C. Huang, On prime labelling, *Discrete Math.* **127** (1994), 181–186.

[4] J.A. Gallian, A dynamic survey of graph labeling, *Electron. J. Combin.* **DS6** (2014).

[5] S.M. Lee, I. Wui, and J. Yeh, On the amalgamation of prime graphs, *Bull. Malaysian Math. Soc.* (Second Series) **11** (1988), 59–67.

[6] J. Nagura, On the interval containing at least one prime number, *Proc. of the Japan Acad.* **28** (1952), no. 4, 177–181.

[7] O. Pikhurko, Trees are almost prime, *Discrete Math.* **307** (2007), no. 11, 1455–1462.

[8] P. Ribenboim, The new book of prime number records, *Springer-Verlag*, New York, 1996.

[9] H. Rohrbach and J. Weis, Zum finiten fall des bertrandschen postulats., *Journal für die reine und angewandte Mathematik* **214** (1964), 432–440.

[10] H. Salmasian, A result on prime labelings of trees, *Bull. Inst. Combin. Appl* **28** (2000), 36–38.

[11] L. Schoenfeld, Sharper bounds for the chebyshev functions $\theta(x)$ and $\psi(x)$. ii, *Math. of Comp.* **30** (1976), no. 134, 337–360.

[12] M.A. Seoud, A.T. Diab, and E.A. Elsahawi, On strongly-c harmonious, relatively prime, odd graceful and cordial graphs, *Proc. Math. Phys. Soc. Egypt*, vol. 73, 1998, pp. 33–55.

[13] M.A. Seoud, A. El Sonbaty, and A.E.A. Mahran, Primality of some graphs, *Ars Combin.* **112** (2013), 459–469.

[14] M.A. Seoud and M.Z. Youssef, On prime labelings of graphs, *Congr. Numer.* **141** (1999), 203–215.

[15] A. Tout, A.N. Dabboucy, and K. Howalla, Prime labeling of graphs, *Nat. Acad. Sci. Letters* **11** (1982), 365–368.

[16] M.Z. Youssef and E.A. El Sakhawi, Some properties of prime graphs, *Ars Combin.* **84** (2007), 129–140.

9

THE UNIT STRUCTURE IN A QUOTIENT RING OF A QUADRATIC NUMBER FIELD

Brian D. Sittinger
Department of Mathematics, CSU Channel Islands, Camarillo, California

Marina C. Morales
ACE Charter High School, Camarillo, California

Abstract

The unit group structure of $\mathbb{Z}_m$ is well-known in number theory, largely due to the significance of primitive roots modulo m whenever they exist. We investigate the analogous problem for a quadratic number ring $\mathcal{O}$, determining the unit group structure of $\mathcal{O}/\mathfrak{a}$ for some fixed nonzero ideal $\mathfrak{a}$ in $\mathcal{O}$ along with a set of generators.

1. Introduction

It is often useful to represent a finite abelian group as a direct product of cyclic groups. With this description, it is easy to deduce many important properties of the group, such as its order, subgroup lattice, and rank. In this paper, we investigate the structure of the group of units in any quotient ring of a quadratic number field.

As a motivating example, we start with $\mathbb{Z}_m$, the ring of integers modulo m. This is a finite commutative ring whose units (elements having multiplicative inverses) form a multiplicative group which is denoted by $(\mathbb{Z}_m)^*$. It is a standard fact that a nonzero element $x \in \mathbb{Z}_m$ is a unit if and only if $\gcd(x, m) = 1$. Moreover, the order of $(\mathbb{Z}_m)^*$ is given by the classic Euler phi function $\phi(m)$.

In order to find the group structure of $(\mathbb{Z}_m)^*$, recall that if $m = p_1^{n_1} p_2^{n_2} \ldots p_k^{n_k}$ is a prime factorization for m, then the Chinese Remainder Theorem induces the isomorphism

$$(\mathbb{Z}_m)^* \cong (\mathbb{Z}_{p_1^{n_1}})^* \times (\mathbb{Z}_{p_2^{n_2}})^* \times \cdots \times (\mathbb{Z}_{p_k^{n_k}})^*.$$

Thus, it suffices to understand $(\mathbb{Z}_{p^n})^*$ for any prime p. Its structure is given as follows [8]:

$$(\mathbb{Z}_{p^n})^* = \langle g \rangle \cong \mathbb{Z}_{p^n - p^{n-1}} \text{ for some } g \in \mathbb{Z}_{p^n} \text{ if } p \text{ is an odd prime,}$$

$$(\mathbb{Z}_{2^n})^* = \begin{cases} \{1\} & \text{if } n = 1, \\ \langle -1 \rangle \cong \mathbb{Z}_2 & \text{if } n = 2, \\ \langle -1 \rangle \times \langle 5 \rangle \cong \mathbb{Z}_2 \times \mathbb{Z}_{2^{n-2}} & \text{if } n \geq 3. \end{cases}$$

If $(\mathbb{Z}_m)^*$ is cyclic, then any of its generators is called a *primitive root modulo m*. For example, when p is odd, g is a primitive root modulo p^n for any $n \in \mathbb{N}$. We can say more; by using this fact in conjunction with the Chinese Remainder Theorem, it is straightforward to deduce that a primitive root modulo m exists if and only if $m = 2, 4, p^n, 2p^n$ for any odd prime p and positive integer n.

We consider the more general problem of finding the unit group structure and a set of generators for a quotient ring of any quadratic number field over $\mathbb{Q}$. The inspiration for this problem came from Cross, who investigated this problem in the ring of Gaussian integers [3]. Subsequently, Buçaj investigated this problem in the ring of Eisenstein integers [2]. The results on the unit group structure for a quotient ring of any quadratic number field exist in the literature; in [5], Kohler recently compiled them together (so he could use these results to explicitly compute characters on these groups), proving these results via intricate counting arguments *without giving the generators*. We follow Cross' approach to find the generators in any such quotient ring. After reviewing the pertinent facts about quadratic number fields, we spend the remainder of this paper deriving the unit group structure for a quotient ring of any quadratic number field. From here, we answer the corresponding question concerning the existence of primitive roots in a quotient ring of a quadratic number field.

2. Background and Terminology

2.1. Quadratic Number Rings

In this section, we recall some basic concepts from algebraic number theory, as found in [6] and [8]. Let K denote an algebraic field over $\mathbb{Q}$ (that is, $[K : \mathbb{Q}] < \infty$) with $\mathcal{O}$ being its corresponding ring of integers. In this paper, we are primarily interested in the quadratic number field $\mathbb{Q}(\sqrt{d})$ for any square-free integer d. Its ring of integers $\mathcal{O}$, called a *quadratic number ring*, is the set $\{a + b\omega : a, b \in \mathbb{Z}\}$, where

$$\omega = \begin{cases} \sqrt{d} & \text{if } d \not\equiv 1 \bmod 4, \\ \frac{1+\sqrt{d}}{2} & \text{if } d \equiv 1 \bmod 4. \end{cases}$$

Besides having the structure of a ring, $\mathbb{Q}(\sqrt{d})$ also has the operation of conjugation:

$$\overline{a + b\sqrt{d}} = a - b\sqrt{d} \text{ for any } a, b \in \mathbb{Q}.$$

With this operation, we define the norm of $\alpha \in \mathbb{Q}(\sqrt{d})$ by $N(\alpha) = \alpha\overline{\alpha}$. In particular, when $\alpha \in \mathcal{O}$, then $N(\alpha)$ is an integer.

Although we do not necessarily have unique factorization into irreducible elements in $\mathcal{O}$, we do have unique factorization into prime ideals in $\mathcal{O}$. We now give a description of the prime ideals in a quadratic number ring and how they arise from rational primes (primes in $\mathbb{Z}$).

Definition 1. Let p be a rational prime and $\mathcal{O}$ be a quadratic number ring.

(a) We say that p *splits* in $\mathcal{O}$ if $\langle p \rangle = \mathfrak{p}\overline{\mathfrak{p}}$ for distinct prime ideals $\mathfrak{p}, \overline{\mathfrak{p}}$ in $\mathcal{O}$.

(b) We say that p is *inert* in $\mathcal{O}$ if $\langle p \rangle$ is a prime ideal in $\mathcal{O}$.

(c) We say that p *ramifies* in $\mathcal{O}$ if $\langle p \rangle = \mathfrak{p}^2$ for some prime ideal $\mathfrak{p}$ in $\mathcal{O}$.

The resulting prime ideals are said to *lie above* p.

In the definition above, $\overline{\mathfrak{p}}$ is the ideal whose elements are conjugates to those in $\mathfrak{p}$. Next, we characterize the prime ideals in a quadratic number ring more precisely. A proof of this can be found in [6].

Theorem 1. *Let $\mathcal{O}$ be a quadratic number ring.*
For an odd rational prime p:

(a) p is inert in $\mathcal{O}$ if $(\frac{d}{p}) = -1$,

(b) p splits in $\mathcal{O}$ if $(\frac{d}{p}) = 1$,

(c) p ramifies in $\mathcal{O}$ if $p \mid d$.

For the rational prime 2:

(a) 2 is inert in $\mathcal{O}$ if $d \equiv 5 \bmod 8$,

(b) 2 splits in $\mathcal{O}$ if $d \equiv 1 \bmod 8$,

(c) 2 ramifies in $\mathcal{O}$ if d is even or $d \equiv 3 \bmod 4$.

It proves especially convenient to describe the generators of a prime ideal that lies above a rational prime p.

Lemma 1. *Suppose that $\mathfrak{p}$ lies above a ramifying prime p in a quadratic number ring $\mathcal{O}$.*

If p is odd, then $\mathfrak{p} = \langle p, \sqrt{d} \rangle$.

$$\text{If } p = 2, \text{ then } \mathfrak{p} = \begin{cases} \langle 2, \sqrt{d} \rangle & \text{if } d \equiv 2 \bmod 4, \\ \langle 2, \sqrt{d} - 1 \rangle & \text{if } d \equiv 3 \bmod 4. \end{cases}$$

Proof. First, suppose that p is odd. Note that

$$\mathfrak{p}^2 = \langle p, \sqrt{d} \rangle^2 = \langle p^2, p\sqrt{d}, d \rangle.$$

Since $p \mid d$ but $p^2 \nmid d$, we know that $d = kp$ for some integer k not divisible by p. Thus $\gcd(k,p) = 1$, and there exist integers x and y such that $kx + py = 1$. Then, since $p^2, d \in \mathfrak{p}^2$, we have $dx + p^2y = p(kx + py) = p \in \mathfrak{p}^2$. Therefore, we conclude that $\langle p, \sqrt{d}\rangle^2 = \langle p\rangle$.

Next, assume that $p = 2$. If $d \equiv 2 \bmod 4$, then $\mathfrak{p} = \langle 2, \sqrt{d}\rangle$ by using the same arguments as we did when p was odd. If $d \equiv 3 \bmod 4$, then

$$\mathfrak{p}^2 = \langle 2, \sqrt{d} - 1\rangle^2 = \langle 4, 2(\sqrt{d} - 1), d + 1 - 2\sqrt{d}\rangle.$$

Then, $2(\sqrt{d} - 1) + (d + 1 - 2\sqrt{d}) = d - 1 \in \mathfrak{p}^2$. Finally, since $d - 1 \equiv 2 \bmod 4$, we conclude that $2 \in \mathfrak{p}^2$, and the claim now immediately follows. □

In an effort to generalize the properties of $\mathbb{Z}$ to ideals in $\mathcal{O}$, we have a notion of *divides* in the context of ideals (which reduces to the usual definition of divisibility of an element in the case that all ideals are principal).

Definition 2. Given ideals $\mathfrak{a}, \mathfrak{b}$ in $\mathcal{O}$, we say that $\mathfrak{a}$ *divides* $\mathfrak{b}$, written $\mathfrak{a} \mid \mathfrak{b}$, if $\mathfrak{b} = \mathfrak{c}\mathfrak{a}$ for some ideal $\mathfrak{c}$ in $\mathcal{O}$.

In particular, any prime ideal in $\mathcal{O}$ divides the rational prime above which it lies. Moreover, it follows immediately from this definition that $\mathfrak{a} \mid \mathfrak{b}$ if and only if $\mathfrak{a} \supseteq \mathfrak{b}$.

2.2. Quotient Rings of a Quadratic Number Field

Fix a nonzero ideal $\mathfrak{a}$ in a number ring $\mathcal{O}$, and consider the quotient ring $\mathcal{O}/\mathfrak{a}$. We first state a version of the Chinese Remainder Theorem for $\mathcal{O}$ (as any two distinct prime ideals are comaximal in $\mathcal{O}$; see [4] for a proof).

Theorem 2. *Let $\mathfrak{a} = \mathfrak{p}_1^{n_1}\mathfrak{p}_2^{n_2}\ldots\mathfrak{p}_k^{n_k}$ be a prime factorization for the nonzero ideal $\mathfrak{a}$ in $\mathcal{O}$ where $\mathfrak{p}_1, \mathfrak{p}_2, \ldots, \mathfrak{p}_k$ are distinct prime ideals in $\mathcal{O}$. Then,*

$$\mathcal{O}/\mathfrak{a} \cong \mathcal{O}/\mathfrak{p}_1^{n_1} \times \mathcal{O}/\mathfrak{p}_2^{n_2} \times \cdots \times \mathcal{O}/\mathfrak{p}_k^{n_k}.$$

As we will be interested in the unit group $(\mathcal{O}/\mathfrak{a})^*$, the Chinese Remainder Theorem induces the isomorphism

$$(\mathcal{O}/\mathfrak{a})^* \cong (\mathcal{O}/\mathfrak{p}_1^{n_1})^* \times (\mathcal{O}/\mathfrak{p}_2^{n_2})^* \times \cdots \times (\mathcal{O}/\mathfrak{p}_k^{n_k})^*.$$

Hence, it suffices to study $(\mathcal{O}/\mathfrak{p}^n)^*$ for some fixed prime ideal $\mathfrak{p}$ in $\mathcal{O}$ and $n \in \mathbb{N}$.

As a first step in this endeavor, we first find a complete set of equivalence classes for $\mathcal{O}/\mathfrak{p}^n$. To assist us with this, we introduce the norm of an ideal.

Definition 3. The *norm* of a nonzero ideal $\mathfrak{a}$ in a number ring $\mathcal{O}$ is defined as $\mathfrak{N}(\mathfrak{a}) = |\mathcal{O}/\mathfrak{a}|$.

Note that the norm of an ideal is always finite and is related to the norm of an element $\alpha \in \mathcal{O}$ by $\mathfrak{N}(\langle\alpha\rangle) = N(\alpha)$. It can be shown that the norm of ideals is multiplicative: if $\mathfrak{a}, \mathfrak{b}$ are nonzero ideals in $\mathcal{O}$, then $\mathfrak{N}(\mathfrak{a}\mathfrak{b}) = \mathfrak{N}(\mathfrak{a})\mathfrak{N}(\mathfrak{b})$. Moreover, since a prime ideal $\mathfrak{p}$ is always maximal in $\mathcal{O}$, we know that $\mathcal{O}/\mathfrak{p}$ is a field. When $\mathcal{O}$ is a quadratic number ring, this implies that if $\mathfrak{p}$ is a prime ideal lying above the rational prime p, then $\mathfrak{N}(\mathfrak{p}) = p^2$ if p is inert, and $\mathfrak{N}(\mathfrak{p}) = p$ otherwise.

Theorem 3. *A complete set of congruence classes to a quadratic number ring $\mathcal{O}$ modulo a power of a prime ideal $\mathfrak{p}$ are given as follows:*

(a) If $\mathfrak{p}$ lies above a split prime p, then $\mathcal{O}/\mathfrak{p}^n = \{0, 1, \ldots, p^n - 1\}$.

(b) If p is inert, then $\mathcal{O}/\mathfrak{p}^n = \{a + b\omega \,:\, a, b = 0, 1, \ldots, p^n - 1\}$.

(c) If $\mathfrak{p}$ lies above a ramifying prime p, then for any $m \in \mathbb{N}$:
$\mathcal{O}/\mathfrak{p}^{2m} = \{a + b\omega \,:\, a, b = 0, 1, \ldots, p^m - 1\}$, *and*
$\mathcal{O}/\mathfrak{p}^{2m+1} = \{a + b\omega \,:\, a = 0, 1, \ldots, p^{m+1} - 1, \text{ and } b = 0, 1, \ldots, p^m - 1\}$.

Proof. By the remarks preceding this theorem, along with $\mathfrak{N}(\mathfrak{p}^n) = \mathfrak{N}(\mathfrak{p})^n$, we have the correct number of congruence classes for $\mathcal{O}/\mathfrak{p}^n$. Hence, it suffices to establish that the given congruence classes of $\mathcal{O}/\mathfrak{p}^n$ are distinct.

(a) Suppose that $\mathfrak{p}$ lies above a splitting rational prime p. If $a = b$ in $\mathcal{O}/\mathfrak{p}^n$ with $a, b \in \{0, 1, \ldots, p^n - 1\}$, then $\mathfrak{p}^n \mid \langle a - b\rangle$. By conjugation, this yields $\overline{\mathfrak{p}}^n \mid \langle a - b\rangle$. Since $\gcd(\mathfrak{p}, \overline{\mathfrak{p}}) = \langle 1\rangle$, this implies that $\mathfrak{p}^n\overline{\mathfrak{p}}^n = \langle p^n\rangle \mid \langle a - b\rangle$, which is equivalent to $p^n \mid (a - b)$. Since $a, b \in \{0, 1, \ldots, p^n - 1\}$, we conclude that $a = b$.

(b) Next, suppose that p is inert so that $\mathfrak{p} = \langle p\rangle$. If $a + b\omega = c + d\omega$ in $\mathcal{O}/\mathfrak{p}^n$ for some $a, b, c, d \in \{0, 1, \ldots, p^n - 1\}$, then $(a - c) + (b - d)\omega = 0$ in $\mathcal{O}/\mathfrak{p}^n$. This implies that $p^n|(a - c)$ and $p^n|(b - d)$. Since a, b, c, d are between 0 and $p^n - 1$ inclusive, we conclude that $a = c$ and $b = d$.

(c) Finally, suppose that $\mathfrak{p}$ lies above a ramifying prime p. If $n = 2m$, then the distinctness of the given congruence classes is proved as in the inert case. Now, suppose $n = 2m + 1$ and $a + b\omega = c + d\omega$ in $\mathcal{O}/\mathfrak{p}^n$, where $a, c \in \{0, 1, \ldots, p^{m+1} - 1\}$ and $b, d \in \{0, 1, \ldots, p^m - 1\}$. Since $\mathfrak{p}^n = \langle p^m\rangle\mathfrak{p}$, it follows that $p^m \mid (b - d)$ and thus $b = d$. Therefore, $\mathfrak{p}^n = \langle p^m\rangle\mathfrak{p} \mid \langle a - c\rangle$, or equivalently $a - c = p^m \cdot k$ for some integer k, since the only rational elements in $\mathcal{O}$ are integers. Then, $\mathfrak{p} \mid \langle k\rangle$. Taking norms, we find that $p \mid k^2$ and thus $p \mid k$. Hence, $p^{m+1} \mid (a - c)$, and we conclude that $a = c$. □

As in $\mathbb{Z}$, the following notational shorthand will prove useful.

Definition 4. Fix a nonzero ideal $\mathfrak{a}$ in $\mathcal{O}$. For any $\alpha, \beta \in \mathcal{O}$, we say that α *is congruent to β modulo* $\mathfrak{a}$, written $\alpha \equiv \beta \bmod \mathfrak{a}$, if $(\alpha - \beta) \in \mathfrak{a}$.

From this definition, we see that $\alpha \equiv \beta \bmod \mathfrak{a}$ if and only if α and β belong to the same coset in $\mathcal{O}/\mathfrak{a}$. It is straightforward to check that the fundamental properties of congruences over $\mathbb{Z}$ carry over to those in $\mathcal{O}$ unchanged.

Now we turn our attention to analyzing $(\mathcal{O}/\mathfrak{a})^*$. We first consider a generalization of Euler's phi function for number rings.

Definition 5. Let $\mathfrak{a}$ be a nonzero ideal of $\mathcal{O}$. Define the *phi function* as $\phi(\mathfrak{a}) = |(\mathcal{O}/\mathfrak{a})^*|$.

Its basic properties are reminiscent of its classical counterpart (see [1]), and we now quote its most essential facts for our purposes. First of all, if $\mathfrak{a}, \mathfrak{b}$ are relatively prime ideals in $\mathcal{O}$, then $\phi(\mathfrak{a}\mathfrak{b}) = \phi(\mathfrak{a})\phi(\mathfrak{b})$. Moreover, when $\mathfrak{p}$ is a prime ideal in $\mathcal{O}$ and $n \in \mathbb{N}$, we have

$$\phi(\mathfrak{p}^n) = \mathfrak{N}(\mathfrak{p})^n - \mathfrak{N}(\mathfrak{p})^{n-1}.$$

Although it is not crucial to the work that follows, we next give a complete set of congruence classes for $(\mathcal{O}/\mathfrak{p}^n)^*$ by suitably restricting the congruence classes from Theorem 3.

Theorem 4. *Let $\mathcal{O}$ be a quadratic number ring.*

(a) Suppose that $\mathfrak{p}$ lies above a split prime p. Then, $a \in (\mathcal{O}/\mathfrak{p}^n)^*$ if and only if $p \nmid a$.

(b) Suppose that p is inert. Then, $a + b\omega \in (\mathcal{O}/\mathfrak{p}^n)^*$ if and only if $p \nmid a$ or $p \nmid b$.

(c) Suppose that $\mathfrak{p}$ lies above a ramifying prime p.

(i) If either p is odd and $d \equiv 3 \bmod 4$, or $p = 2$ and $d \equiv 2 \bmod 4$, then $a + b\omega \in (\mathcal{O}/\mathfrak{p}^n)^*$ if and only if $p \nmid a$.

(ii) If p is odd and $d \equiv 1 \bmod 4$, then $a + b\omega \in (\mathcal{O}/\mathfrak{p}^n)^*$ if and only if $p \nmid (2a + b)$.

(iii) If $p = 2$ and $d \equiv 3 \bmod 4$, then $a+b\omega \in (\mathcal{O}/\mathfrak{p}^n)^*$ if and only if $2 \nmid (a+b)$.

Proof. Since $\mathcal{O}$ has unique factorization into prime ideals, $a \in \mathcal{O}/\mathfrak{p}^n$ is a unit if and only if $\mathfrak{p} \nmid \langle a \rangle$. With this observation, we now prove this theorem via contraposition.

(a) Suppose that p splits. By Theorem 3, any $a \in \mathcal{O}/\mathfrak{p}^n$ can be rewritten as an equivalence class from $\{0, 1, \ldots, p^n - 1\}$. Then, $a \in \mathcal{O}/\mathfrak{p}^n$ is a not a unit if and only if both $\mathfrak{p} \mid \langle a \rangle$ and $\overline{\mathfrak{p}} \mid \langle a \rangle$. Since $\gcd(\mathfrak{p}, \overline{\mathfrak{p}}) = \langle 1 \rangle$, and $\langle p \rangle = \mathfrak{p}\overline{\mathfrak{p}}$, this implies that $\langle p \rangle \mid \langle a \rangle$ and thus $p \mid a$.

(b) Now, suppose that p is inert. Then, $\mathfrak{p} = \langle p \rangle$, and $a + b\omega \in \mathcal{O}/\mathfrak{p}^n$ is not a unit if and only if $\langle p \rangle \mid \langle a + b\omega \rangle$. Hence, $p \mid (a + b\omega)$, and this is true if and only if both $p \mid a$ and $p \mid b$.

(c) Finally, suppose that p ramifies. Note that $a + b\omega \in \mathcal{O}/\mathfrak{p}^n$ is not a unit if and only if $\alpha \mid (a + b\omega)$ for some $\alpha \in \mathfrak{p}$. We have three cases to consider.

First, suppose that p is odd and $d \equiv 3 \bmod 4$, or $p = 2$ and $d \equiv 2 \bmod 4$. Then, taking norms and noting that $p \mid N(\alpha)$, we obtain $p \mid (a^2 - db^2)$. Finally, since $p \mid d$ due to p ramifying, we conclude that $p \mid a$.

Next, suppose that p is odd and $d \equiv 1 \bmod 4$ (as a reminder, this means that $\omega = \frac{1+\sqrt{d}}{2}$). Since $\mathfrak{p} = \langle p, \sqrt{d} \rangle$, we have that $\alpha = p(x + y\omega) + \sqrt{d}(z + t\omega)$ for some integers t, x, y, z. Without loss of generality, we may assume that $p \nmid (2z + t)$; otherwise we can also write $\sqrt{d}(z + t\omega)$ as a multiple of p. Since $\alpha \mid (a + b\omega)$, this implies that $N(\alpha) \mid (a + b\omega)\overline{\alpha}$. Then, because $p \mid N(\alpha)$ and $p \mid d$, it follows upon expanding $(a+b\omega)\overline{\alpha}$ in terms of $\sqrt{d}$ that $p \mid (-4az - 2at - 2bz - bt) = -(2z+t)(2a+b)$. Thus, $p \mid (2a + b)$ as required.

Finally, suppose that $p = 2$ and $d \equiv 3 \bmod 4$. Since $\mathfrak{p} = \langle 2, \sqrt{d} - 1 \rangle$, we have that $\alpha = p(x + y\sqrt{d}) + (\sqrt{d} - 1)(z + w\sqrt{d})$ for some integers w, x, y, z. Without loss of generality, we may assume that $2 \nmid (z - w)$; otherwise we can also write $(\sqrt{d} - 1)(z + w\sqrt{d})$ as a multiple of 2. Since $\alpha \mid (a + b\sqrt{d})$, this implies that $N(\alpha) \mid (a + b\sqrt{d})\overline{\alpha}$. Then, because $2 \mid N(\alpha)$ and d is odd, it follows upon expanding $(a + b\omega)\overline{\alpha}$ that $2 \mid (az + aw + bz + bw) = (w + z)(a + b)$. Thus $2 \mid (a + b)$ as required. $\square$

As a consequence of Theorems 3 and 4, it immediately follows for a quadratic number ring $\mathcal{O}$ that $(\mathcal{O}/\mathfrak{p}^n)^*$ has $p^{2n} - p^{2n-2}$ elements when $\mathfrak{p}$ lies above an inert prime, and $p^n - p^{n-1}$ elements when $\mathfrak{p}$ lies above a split or ramified prime, in accordance with the generalized phi function.

The following proposition is a variant of Hensel's lifting lemma in $\mathbb{Z}$ that will prove useful in the work that follows. This allows us to *lift* a solution to a polynomial congruence from one power of a prime ideal to the next power.

Proposition 1. *Suppose that $f(x) \in \mathcal{O}[x]$ and $\mathfrak{p}$ is a prime ideal in a number ring $\mathcal{O}$. If $x = \alpha \in \mathcal{O}$ is a solution to $f(x) \equiv 0 \bmod \mathfrak{p}^{n-1}$ for some $n \geq 2$ and $f(\alpha) \in \gcd(\langle f'(\alpha) \rangle, \mathfrak{p}^n)$, then $f(x) \equiv 0 \bmod \mathfrak{p}^n$ has a solution in $\mathcal{O}$.*

Proof. We first establish the following claim: If $\alpha, \beta \in \mathcal{O}$, then

$$f(\alpha + \beta) = f(\alpha) + \beta f'(\alpha) + \beta^2\gamma \text{ for some } \gamma \in \mathcal{O}.$$

To show this, note that for any $k \in \mathbb{Z}_{\geq 0}$, the binomial theorem yields

$$(\alpha + \beta)^k = \alpha^k + k\alpha^{k-1}\beta + \beta^2\delta_k \text{ for some } \delta_k \in \mathcal{O}.$$

Writing $f(x) = \sum_{k=0}^{n} \rho_k x^k$ for some $\rho_k \in \mathcal{O}$ and $n \in \mathbb{N}$, we have

$$\begin{aligned} f(\alpha+\beta) &= \sum_{k=0}^{n} \rho_k(\alpha+\beta)^k \\ &= \left[\sum_{k=2}^{n} \rho_k(\alpha^k + k\alpha^{k-1}\beta + \beta^2\delta_k)\right] + \rho_1(\alpha+\beta) + \rho_0 \\ &= \sum_{k=0}^{n} \rho_k\alpha^k + \beta \cdot \sum_{k=1}^{n} k\rho_k\alpha^{k-1} + \beta^2 \cdot \sum_{k=0}^{n} \rho_k\delta_k \\ &= f(\alpha) + \beta f'(\alpha) + \beta^2\gamma, \text{ where } \gamma = \sum_{k=0}^{n} \rho_k\delta_k. \end{aligned}$$

Now, we are ready to prove this proposition. Suppose that $f(\alpha) \equiv 0 \bmod \mathfrak{p}^{n-1}$. We want to solve $f(x) \equiv 0 \bmod \mathfrak{p}^n$ by using α. To do this, we write $x = \alpha + \beta$ for some $\beta \in \mathfrak{p}^{n-1}$. Substituting this into $f(\alpha) \equiv 0 \bmod \mathfrak{p}^n$ and using the claim yields

$$f(\alpha+\beta) = f(\alpha) + \beta f'(\alpha) + \beta^2\gamma \equiv 0 \bmod \mathfrak{p}^n \text{ for some } \gamma \in \mathcal{O}.$$

Since $\beta^2 \in \mathfrak{p}^{2n-2}$ and $n \geq 2$, the previous relation reduces to

$$f(\alpha) + \beta f'(\alpha) \equiv 0 \bmod \mathfrak{p}^n.$$

This is solvable for β if and only if $f(\alpha) \in \gcd(\langle f'(\alpha)\rangle, \mathfrak{p}^n)$. □

In the next two sections, we explicitly give the unit group structure of any quotient ring $\mathcal{O}/\mathfrak{p}^n$, where $\mathfrak{p}$ is a prime ideal of a quadratic number ring $\mathcal{O}$.

3. Unit Group Structure for an Unramified Prime

3.1. Split Case

Theorem 5. *Suppose that $\mathfrak{p}$ lies above a split rational prime p. Then,*

$$(\mathcal{O}/\mathfrak{p}^n)^* \cong (\mathbb{Z}_{p^n})^*.$$

Proof. By Theorem 4, the set of congruence classes comprising $(\mathcal{O}/\mathfrak{p}^n)^*$ is formally the same as $(\mathbb{Z}_{p^n})^*$. This leads us to consider $\psi : (\mathcal{O}/\mathfrak{p}^n)^* \to (\mathbb{Z}_{p^n})^*$ defined by $\psi(a) = a$. Note that ψ is well-defined, because $\mathfrak{p} \mid \langle p \rangle)$. Moreover since ψ is a bijection by construction, we deduce that ψ is an isomorphism. In other words, $(\mathcal{O}/\mathfrak{p}^n)^* \cong (\mathbb{Z}_{p^n})^*$ as required. □

3.2. Inert Case

In this section, we assume that p is inert in $\mathcal{O}$. Before stating the group structure theorem, we give a couple lemmas to expedite the proofs of the group structure theorems (Theorems 6 and 7) that follow.

Lemma 2. *Let p be a rational prime and $k \in \mathbb{N}$.*
(a) If p is odd, then $(1+p\omega)^{p^k} = 1 + p^{k+1}\omega + p^{k+2}\gamma$ for some $\gamma \in \mathcal{O}$.
(b) $(1+2\omega)^{2^k} = 1 + 2^{k+1} + 2^{k+2}\gamma$ for some $\gamma \in \mathcal{O}$.

Proof. Let $\beta \in \mathcal{O}$ and r be a rational prime. Then by using a straightforward proof by induction on k in conjunction with the binomial theorem, we find that

$$(1+\beta r)^{r^k} = 1 + \beta r^{k+1} + \frac{1}{2}\beta^2(r^k - 1)r^{k+2} + \delta r^{k+2} \text{ for some } \delta \in \mathcal{O}.$$

With this identity, we can readily prove the lemma. Part (a) of the lemma follows directly from this claim by letting $r = p$ where p is an odd prime, $\beta = \omega$, and collecting like terms.

For part (b), observe that 2 is inert precisely when $d \equiv 5 \bmod 8$. Thus, $\omega = \frac{1+\sqrt{d}}{2}$, and $\omega^2 + \omega = \frac{d-1}{4} \equiv 1 \bmod 2$. Hence, letting $r = 2$ and $\beta = \omega$ yields

$$\begin{aligned}(1+2\omega)^{2^k} &= 1 + 2^{k+1}(\omega + (2^k - 1)\omega^2) + 2^{k+2}\delta \\ &= 1 + 2^{k+1} + 2^{k+2}\gamma \text{ for some } \gamma \in \mathcal{O}.\end{aligned}$$

□

Lemma 3. *Suppose that p is inert in $\mathcal{O}$. If $n \in \mathbb{N}_{>1}$, then the order of $1 + p\omega$ in $(\mathcal{O}/\mathfrak{p}^n)^*$ equals p^{n-1}.*

Proof. First, consider the case when p is odd. Letting $k = n-1$ in Lemma 2 yields

$$(1+p\omega)^{p^{n-1}} = 1 + p^n\omega + p^{n+1}\gamma \text{ for some } \gamma \in \mathcal{O}.$$

Therefore, $(1+p\omega)^{p^{n-1}} \equiv 1 \bmod p^n$, and thus the order of $1+p\omega$ divides p^{n-1}. Next, letting $k = n-2$ in Lemma 2 yields $(1+p\omega)^{p^{n-1}} \not\equiv 1 \bmod p^n$. Therefore, the order of $1 + p\omega$ in $(\mathcal{O}/\mathfrak{p}^n)^*$ equals p^{n-1}.

Now, consider the case $p = 2$. Letting $k = n-1$ in Lemma 2 yields

$$(1+2\omega)^{2^{n-1}} = 1 + 2^n + 2^{n+1}\gamma \text{ for some } \gamma \in \mathcal{O}.$$

Hence, $(1+2\omega)^{2^{n-1}} \equiv 1 \bmod 2^n$, and the order of $1+2\omega$ divides 2^{n-1}. Next, letting $k = n-2$ in Lemma 2 and reducing modulo 2^n yields $(1+2\omega)^{2^{n-1}} \not\equiv 1 \bmod 2^n$. Therefore, the order of $1+2\omega$ in $(\mathcal{O}/\langle 2^n \rangle)^*$ equals 2^{n-1}. □

Now, we are able to give the group structure theorems for $(\mathcal{O}/\mathfrak{p}^n)^*$ in the case that p is inert. We start with the case when p is odd.

Theorem 6. *Let p be an inert odd prime and g be a primitive root modulo p^n. Then, there exists $\beta \in \mathcal{O}$ such that*

$$(\mathcal{O}/\mathfrak{p}^n)^* = \langle 1+p\omega\rangle \times \langle g^{p-1}\rangle \times \langle \beta^{p^{n-1}}\rangle \cong \mathbb{Z}_{p^{n-1}} \times \mathbb{Z}_{p^{n-1}} \times \mathbb{Z}_{p^2-1}.$$

Proof. We have already examined $\langle 1+p\omega\rangle$ in Lemma 3.

To construct g, the injective homomorphism $\psi : (\mathbb{Z}_{p^n})^* \to (\mathcal{O}/\mathfrak{p}^n)^*$ defined by $\psi(b) = b$ shows that we can view $(\mathbb{Z}_{p^n})^*$ as a subgroup of $(\mathcal{O}/\mathfrak{p}^n)^*$. Since p is odd, $(\mathbb{Z}_{p^n})^*$ is cyclic of order $\phi(p^n) = p^{n-1}(p-1)$ with generator g. Consequently, $\langle g^{p-1}\rangle$ is isomorphic to a cyclic group of order p^{n-1}.

Finally, we construct β. Since p is prime in $\mathcal{O}$, and prime ideals in $\mathcal{O}$ are maximal, $\mathcal{O}/\mathfrak{p}$ is a field with $\mathfrak{N}(\mathfrak{p}) = p^2$ elements. Thus, $(\mathcal{O}/\mathfrak{p})^*$ is a cyclic group of order p^2-1; let β be a generator. Since $\beta^{p^2-1} \equiv 1 \bmod p$, we have $\beta^{p^2-1} = 1+\gamma p$ for some $\gamma \in \mathcal{O}$. Using the techniques of Lemma 2, $(\beta^{p^2-1})^{p^n-1} = (1+\gamma p)^{p^{n-1}} = 1+\delta p^n$ for some $\delta \in \mathcal{O}$. Therefore, $(\beta^{p^2-1})^{p^n-1} \equiv (\beta^{p^{n-1}})^{p^2-1} \equiv 1 \bmod p^n$. Letting t be the order of $\beta^{p^{n-1}}$, we have $t \mid (p^2-1)$. Then, $\beta^{tp^{n-1}} \equiv 1 \bmod p$ implies that $(p^2-1) \mid tp^{n-1}$, and thus $(p^2-1) \mid t$. Therefore, $t = p^2-1$ and $\langle \beta^{p^{n-1}}\rangle \cong \mathbb{Z}_{p^2-1}$.

Next, we show that these cyclic groups have pairwise trivial intersections. Since all elements of $\langle 1+p\omega\rangle$ and $\langle g^{p-1}\rangle$ have orders that are powers of p, and $\langle \beta^{p^{n-1}}\rangle$ has order p^2-1, which is relatively prime to p, we can conclude that both $\langle 1+p\omega\rangle$ and $\langle g^{p-1}\rangle$ have trivial intersections with $\langle \beta^{p^{n-1}}\rangle$.

It remains to show that $\langle 1+p\omega\rangle$ and $\langle g^{p-1}\rangle$ have trivial intersection. Since both groups are cyclic with orders that are powers of the same prime p, $\langle 1+p\omega\rangle \cap \langle g^{p-1}\rangle$ is also cyclic of order p^k for some $k \in \{0,1,\dots,n-1\}$. If $k \geq 1$, then $\langle 1+p\omega\rangle \cap \langle g^{p-1}\rangle$ contains a cyclic subgroup of order p. Since $1+p\omega$ has order p^{n-1} in $(\mathcal{O}/\mathfrak{p}^n)^*$ by Lemma 2, it follows that $(1+p\omega)^{p^{n-2}} = 1+p^{n-1}\omega$ is an element in $\langle 1+p\omega\rangle$ that has order p. Hence, all other elements of order p in $\langle 1+p\omega\rangle$ have the form $(1+p^{n-1}\omega)^k = 1+p^{n-1}k\omega$, where $k \in \{1,2,\dots,p-1\}$. Since none of these are in $\langle g^{p-1}\rangle$, we conclude that $k=0$ and thus $\langle 1+p\omega\rangle \cap \langle g^{p-1}\rangle = \{1\}$.

Hence, we can construct the direct product $\langle 1+p\omega\rangle \times \langle g^{p-1}\rangle \times \langle \beta^{p^{n-1}}\rangle$, which is a subgroup of $(\mathcal{O}/\mathfrak{p}^n)^*$ having order $p^{2n-2}(p^2-1)$. However, since the order of $(\mathcal{O}/\mathfrak{p}^n)^*$ is $\phi(\mathfrak{p}^n) = p^{2n-2}(p^2-1)$, the two groups are equal. □

To complete our discussion of the inert prime case, we now address the case when $p = 2$.

Theorem 7. *Suppose that 2 is an inert prime in a quadratic number ring $\mathcal{O}$.*

(a) $(\mathcal{O}/\langle 2\rangle)^* = \langle \omega\rangle \cong \mathbb{Z}_3$.

(b) $(\mathcal{O}/\langle 2^2\rangle)^* = \langle 1+2\omega\rangle \times \langle -1\rangle \times \langle \alpha\rangle \cong \mathbb{Z}_2 \times \mathbb{Z}_2 \times \mathbb{Z}_3$ *for some* $\alpha \in \mathcal{O}$.

(c) For each $n \geq 3$, there exists $\alpha \in \mathcal{O}$ such that

$$(\mathcal{O}/\langle 2^n\rangle)^* = \langle 1+2\omega\rangle \times \langle 1+4\omega\rangle \times \langle -1\rangle \times \langle \alpha\rangle \cong \mathbb{Z}_{2^{n-1}} \times \mathbb{Z}_{2^{n-2}} \times \mathbb{Z}_2 \times \mathbb{Z}_3.$$

Proof. (a) Since $\mathcal{O}/\langle 2\rangle$ is a field with 2^2 elements, $(\mathcal{O}/\langle 2\rangle)^*$ is a cyclic group of order $2^2-1=3$. Since $\omega \neq 1$ in $\mathcal{O}/\langle 2\rangle$, we conclude that ω is a generator for $(\mathcal{O}/\langle 2\rangle)^*$.

(b) Clearly -1 is an element of order 2 in $(\mathcal{O}/\langle 2^2\rangle)^*$. Moreover since $d \equiv 5 \bmod 8$, we have $(1+2\omega)^2 = 1+4\omega+4\omega^2 \equiv 1 \bmod 4$, as $\omega^2+\omega+\frac{1-d}{4}=0$. Thus, $1+2\omega$ has order 2 in $(\mathcal{O}/\langle 2^2\rangle)^*$.

Next, we construct an element α having order 3 in $(\mathcal{O}/\langle 2^2\rangle)^*$. To this end, recall that $\omega^3 \equiv 1 \bmod 2$. By Proposition 1, we can use this to find a solution $\alpha \in \mathcal{O}$ to $x^3 \equiv 1 \bmod 4$ (by taking $f(x)=x^3-1$ and noting that $f'(\omega)=3\omega^2 \notin \mathfrak{p}$).

Plainly, these three subgroups have pairwise trivial intersections, and the product of their orders equals $\phi(\langle 2^2\rangle) = 2^4-2^2=12$ as required.

(c) Since $1+2\omega$ has order 2^{n-1} from Lemma 2, we have $\langle 1+2\omega\rangle \cong \mathbb{Z}_{2^{n-1}}$. Since -1 has order 2, we take $\langle -1\rangle \cong \mathbb{Z}_2$. By Proposition 1, we can inductively find an element $\alpha \in \mathcal{O}$ such that $\alpha^3 \equiv 1 \bmod 2^n$. Hence, $\langle \alpha\rangle \cong \mathbb{Z}_3$.

For the fourth cyclic subgroup, we claim that $\langle 1+4\omega\rangle \cong \mathbb{Z}_{2^{n-2}}$. This follows from the fact that $(1+4\omega)^{2^{n-3}} \equiv 1+2^{n-1}\omega \bmod 2^n$, which is easy to prove by induction in the style of the proof of Lemma 2.

We next observe that the four cyclic subgroups have pairwise trivial intersections. The only tricky case is showing that $\langle 1+4\omega\rangle \cap \langle 1+2\omega\rangle = \{1\}$. To this end, note that since $\langle 1+4\omega\rangle$ is cyclic of order 2^{n-2} and $\langle 1+2\omega\rangle$ is cyclic of order 2^{n-1}; if their intersection has a nontrivial element, then they both would share an element of order 2. This is impossible, since the elements of order 2 from the two cyclic subgroups are distinct, being $1+2^{n-1}\omega$ and $1+2^{n-1}$, respectively.

Finally, the product of the orders of the cyclic subgroups equals $\phi(\langle 2^n\rangle) = 2^{2n} - 2^{2n-2} = 2^{2n-2}\cdot 3$ as required. □

4. Unit Group Structure for a Ramified Prime

Now, suppose that p ramifies in $\mathcal{O}$. We first give a result that is repeatedly used in subsequent subsections.

Lemma 4. *Suppose that p is a ramifying rational prime so that $\langle p\rangle = \mathfrak{p}^2$ for some prime ideal $\mathfrak{p}$ in $\mathcal{O}$. Then for any fixed $m \in \mathbb{N}$ and $r \in \{0,1\}$, there exists a subgroup of $(\mathcal{O}/\mathfrak{p}^{2m+r})^*$ isomorphic to $(\mathbb{Z}_{p^{m+r}})^*$.*

Proof. Fix $m \in \mathbb{N}$ and $r \in \{0,1\}$, and consider the map

$$\psi : (\mathbb{Z}_{p^{m+r}})^* \to (\mathcal{O}/\mathfrak{p}^{2m+r})^* \text{ defined by } \psi(a) = a.$$

Since $\mathfrak{p}^{2m+r} \mid \langle p\rangle^{m+r}$, it follows that ψ is a well-defined injective group homomorphism. Thus, $(\mathbb{Z}_{p^{m+r}})^*$ is isomorphic to a subgroup of $(\mathcal{O}/\mathfrak{p}^{2m+r})^*$. □

4.1. Ramified Primes $p \geq 5$

As the cases for $p = 2$ and $p = 3$ ramifying are fairly involved, we first assume that $p \geq 5$ is a rational prime that ramifies in $\mathcal{O}$.

Theorem 8. *Suppose that $\mathfrak{p}$ lies above a ramifying prime $p \geq 5$.*

(a) $(\mathcal{O}/\mathfrak{p})^* = \langle g \rangle \cong \mathbb{Z}_{p-1}$, where g is a primitive root modulo p.

For any integer $m \geq 1$,

(b) $(\mathcal{O}/\mathfrak{p}^{2m})^* = \langle g \rangle \times \langle 1 + \sqrt{d} \rangle \cong (\mathbb{Z}_{p-1} \times \mathbb{Z}_{p^{m-1}}) \times \mathbb{Z}_{p^m}$,
where g is a primitive root modulo p^m.

(c) $(\mathcal{O}/\mathfrak{p}^{2m+1})^* = \langle g \rangle \times \langle 1 + \sqrt{d} \rangle \cong (\mathbb{Z}_{p-1} \times \mathbb{Z}_{p^m}) \times \mathbb{Z}_{p^m}$,
where g is a primitive root modulo p^{m+1}.

Proof. (a) Note that $\mathcal{O}/\mathfrak{p}$ is a field with $\mathfrak{N}(\mathfrak{p}) = p$ elements. Hence, $(\mathcal{O}/\mathfrak{p})^*$ is a cyclic group with $p - 1$ elements. Moreover, since $(\mathcal{O}/\mathfrak{p})^* = \{1, 2, \ldots, p - 1\}$, we can generate this group by using a primitive root g modulo p.

(b, c) By Lemma 4, there exists a subgroup of $(\mathcal{O}/\mathfrak{p}^{2m+r})^*$ that is an isomorphic copy of $(\mathbb{Z}_{p^{m+r}})^*$. Since p is odd, $(\mathbb{Z}_{p^{m+r}})^*$ is cyclic of order $p^{m+r-1}(p - 1)$ with generator g. (Note that $\langle g \rangle \cong \mathbb{Z}_{p^{m+r-1}(p-1)} \cong \mathbb{Z}_{p-1} \times \mathbb{Z}_{p^{m+r-1}}$.)

To show that $1 + \sqrt{d}$ has order p^m in $(\mathcal{O}/\mathfrak{p}^{2m+r})^*$ for both $r = 0$ and 1, we note that since $\sqrt{d} \in \mathfrak{p}$, the binomial theorem yields

$$(1 + \sqrt{d})^{p^k} = 1 + p^k\sqrt{d} + \gamma \text{ for some } \gamma \in \mathfrak{p}^{2k+2} \backslash \mathfrak{p}^{2k+3}.$$

By letting $k = m$, we see that $(1 + \sqrt{d})^{p^m} \equiv 1 \bmod \mathfrak{p}^{2m+1}$ (and thus mod $\mathfrak{p}^{2m}$ as well), because $p \in \mathfrak{p}^2$ and $\sqrt{d} \in \mathfrak{p}$. However, letting $k = m - 1$, we find that $(1 + \sqrt{d})^{p^{m-1}} \equiv 1 + p^{m-1}\sqrt{d} \not\equiv 1 \bmod \mathfrak{p}^{2m}$ (and thus mod $\mathfrak{p}^{2m+1}$ as well), because $p^{m-1}\sqrt{d} \in \mathfrak{p}^{2m-1} \backslash \mathfrak{p}^{2m}$. Therefore, $1 + \sqrt{d}$ has order p^m in $(\mathcal{O}/\mathfrak{p}^{2m+r})^*$.

Next, we show that $\langle 1 + \sqrt{d} \rangle$ and $\langle g \rangle$ have trivial intersection. Since both groups are cyclic whose orders are powers of the same prime p, $\langle 1 + \sqrt{d} \rangle \cap \langle g \rangle$ is also cyclic of order p^k for some $k \in \{0, 1, \ldots, m - 1\}$. We want to show that $k = 0$. If $k \geq 1$, then $\langle 1 + \sqrt{d} \rangle \cap \langle g \rangle$ contains a cyclic subgroup of order p. Since $1 + \sqrt{d}$ has order p^m in $(\mathcal{O}/\mathfrak{p}^{2m+r})^*$ for both $r = 0$ and 1, it follows that $(1 + p\sqrt{d})^{p^{m-1}} = 1 + p^{m-1}\sqrt{d} + \gamma$ where $\gamma \in \mathfrak{p}^{2m} \backslash \mathfrak{p}^{2m+1}$. Hence, $(1 + p\sqrt{d})^{p^{m-1}} \in \langle 1 + \sqrt{d} \rangle$ that has order p, and thus all other elements of order p in $\langle 1 + \sqrt{d} \rangle$ are powers of this element. Since $(1 + p^{m-1}\sqrt{d} + \gamma)^k \equiv 1 + kp^{m-1}\sqrt{d} + k\gamma \bmod \mathfrak{p}^{2m+1}$ for $k = 1, 2, \ldots, p - 1$, none of these elements are in $\langle g \rangle$, and we conclude that $\langle 1 + \sqrt{d} \rangle \cap \langle g \rangle = \{1\}$.

Finally for both $r = 0$ and 1, since $\langle g \rangle \times \langle 1 + \sqrt{d} \rangle$ is a subgroup of $(\mathcal{O}/\mathfrak{p}^{2m+r})^*$ that has the same order as $(\mathcal{O}/\mathfrak{p}^{2m+r})^*$, we are done. $\square$

4.2. Ramified Prime 3

Now, suppose that 3 ramifies in $\mathcal{O}$ and $\mathfrak{p}$ lies above 3. First, we address the lower powers of $\mathfrak{p}$.

Theorem 9. *Suppose that $\mathfrak{p}$ lies above the ramifying prime 3.*

(a) $(\mathcal{O}/\mathfrak{p})^* = \langle -1 \rangle \cong \mathbb{Z}_2$.

(b) $(\mathcal{O}/\mathfrak{p}^2)^* = \langle -1 \rangle \times \langle 1+\sqrt{d} \rangle \cong \mathbb{Z}_2 \times \mathbb{Z}_3$.

(c) $(\mathcal{O}/\mathfrak{p}^3)^* = \langle 2 \rangle \times \langle 1+\sqrt{d} \rangle \cong (\mathbb{Z}_2 \times \mathbb{Z}_3) \times \mathbb{Z}_3$.

Proof. Since part (a) is trivial, we concentrate on the remaining parts.

(b) Clearly, -1 is an element of order 2. Next, we show that $1+\sqrt{d}$ has order 3 in $(\mathcal{O}/\mathfrak{p}^2)^*$. To do this, observe that $(1+\sqrt{d})^k \equiv 1+k\sqrt{d} \bmod \mathfrak{p}^2$ by induction on k. From this claim, it follows that $(1+\sqrt{d})^3 \equiv 1 \bmod \mathfrak{p}^2$. However, since $\sqrt{d} \neq 0$ in $(\mathcal{O}/\mathfrak{p}^2)^*$, we conclude that the order of $1+\sqrt{d}$ is equal to 3 in $(\mathcal{O}/\mathfrak{p}^2)^*$.

Finally, the two cyclic groups from these generators have trivial intersection, because their orders are relatively prime. Since the order of $\langle g \rangle \times \langle 1+\sqrt{d} \rangle$ equals $\phi(\mathfrak{p}^2) = 3^2 - 3 = 6$, we are done.

(c) By Lemma 4, we can view $(\mathbb{Z}_{3^2})^*$ as a subgroup of $(\mathcal{O}/\mathfrak{p}^3)^*$. Note $(\mathbb{Z}_{3^2})^*$ is cyclic of order $3 \cdot 2 = 6$ with generator 2. As in part (b), we see that $1+\sqrt{d}$ has order 3 in $(\mathcal{O}/\mathfrak{p}^3)^*$. Finally, the cyclic subgroups have trivial intersection; otherwise, they would share a cyclic subgroup of order 3. However, since $(1+\sqrt{d})^k \equiv 1+k\sqrt{d} \bmod \mathfrak{p}^2$ for both $k = 1, 2$ have nonzero $\sqrt{d}$ component modulo $\mathfrak{p}^2$, the claim immediately follows.

Since the order of $\langle g \rangle \times \langle 1+\sqrt{d} \rangle$ equals $\phi(\mathfrak{p}^3) = 3^2(3-1) = 18$, we are done. □

For higher powers of $\mathfrak{p}$, we must be more careful. First of all, note that since $3|d$ but $3^2 \nmid d$, $\frac{d}{3}$ is an integer congruent to 1 or 2 mod 3. It turns out that the group structure of $(\mathcal{O}/\mathfrak{p}^n)^*$ depends on d in the aforementioned manner.

Theorem 10. *Suppose that $\mathfrak{p}$ lies above the ramifying prime 3. For any $m \geq 2$:*

(a) If $\frac{d}{3} \equiv 1 \bmod 3$, *then*

(i) $(\mathcal{O}/\mathfrak{p}^{2m})^* = \langle g \rangle \times \langle 1+\sqrt{d} \rangle \cong (\mathbb{Z}_2 \times \mathbb{Z}_{3^{m-1}}) \times \mathbb{Z}_{3^m}$, *where g is a primitive root modulo 3^m.*

(ii) $(\mathcal{O}/\mathfrak{p}^{2m+1})^* = \langle g \rangle \times \langle 1+\sqrt{d} \rangle \cong (\mathbb{Z}_2 \times \mathbb{Z}_{3^m}) \times \mathbb{Z}_{3^m}$, *where g is a primitive root modulo 3^{m+1}.*

(b) If $\frac{d}{3} \equiv 2 \bmod 3$, *then for some $\alpha \in \mathcal{O}$:*

(i) $(\mathcal{O}/\mathfrak{p}^{2m})^* = \langle g\rangle \times \langle 1+3\sqrt{d}\rangle \times \langle \alpha\rangle \cong (\mathbb{Z}_2 \times \mathbb{Z}_{3^{m-1}}) \times \mathbb{Z}_{3^{m-1}} \times \mathbb{Z}_3$, *where g is a primitive root modulo 3^m.*

(ii) $(\mathcal{O}/\mathfrak{p}^{2m+1})^* = \langle g\rangle \times \langle 1+3\sqrt{d}\rangle \times \langle \alpha\rangle \cong (\mathbb{Z}_2 \times \mathbb{Z}_{3^m}) \times \mathbb{Z}_{3^{m-1}} \times \mathbb{Z}_3$, *where g is a primitive root modulo 3^{m+1}.*

Proof. By Lemma 4, we can view $(\mathbb{Z}_{3^{m+r}})^*$ as a subgroup of $(\mathcal{O}/\mathfrak{p}^{2m+r})^*$. Since 3 is odd, $(\mathbb{Z}_{3^{m+r}})^*$ is cyclic of order $2\cdot 3^{m+r-1}$ with generator g. In particular, note that $\langle g\rangle \cong \mathbb{Z}_{2\cdot 3^{m+r-1}} \cong \mathbb{Z}_2 \times \mathbb{Z}_{3^{m+r-1}}$.

Now, it remains to address the remaining cyclic subgroups.

(a) First of all, suppose that $\frac{d}{3} \equiv 1 \bmod 3$. We claim that $1+\sqrt{d}$ has order 3^m in both $(\mathcal{O}/\mathfrak{p}^{2m})^*$ and $(\mathcal{O}/\mathfrak{p}^{2m+1})^*$.

For any positive integer k, the binomial theorem yields

$$(1+\sqrt{d})^{3^k} = 1 + \sum_{j=1}^{3^k} \binom{3^k}{j} (\sqrt{d})^j.$$

Letting $k=m$, we find that $(1+\sqrt{d})^{3^m} \equiv 1 \bmod \mathfrak{p}^{2m+1}$ since $\sqrt{d} \in \mathfrak{p}$ and $3^m \in \mathfrak{p}^{2m}$. Since this implies that $(1+\sqrt{d})^{3^m} \equiv 1 \bmod \mathfrak{p}^{2m}$, the order of $1+\sqrt{d}$ divides 3^m in $(\mathcal{O}/\mathfrak{p}^{2m+r})^*$ for both $r=0$ and 1.

Next, we let $k=m-1$ and rearrange terms to find that

$$(1+\sqrt{d})^{3^{m-1}} = \left(1+\binom{3^{m-1}}{2}d\right) + \left(3^{m-1} + \binom{3^{m-1}}{3}d\right)\sqrt{d} + \sum_{j=4}^{3^{m-1}} \binom{3^{m-1}}{j}(\sqrt{d})^j.$$

Now, we carefully analyze each term. First of all, since $\frac{d}{3} \equiv 1 \bmod 3$, we have $d = 3(1+3j)$ for some $j \in \mathbb{N}$. In particular, $d \in \mathfrak{p}^2\backslash\mathfrak{p}^3$. Moreover, since $\sqrt{d} \in \mathfrak{p}$ and $3 \in \mathfrak{p}^2$, we have the following consequences: (i) $1+\binom{3^{m-1}}{2}d \in \mathfrak{p}^{2m}\backslash\mathfrak{p}^{2m+1}$; (ii) since $1+\frac{1}{2}(3^{m-1}-1)(3^{k-1}-2)(1+3j) \equiv 2 \bmod 3$, $\left(3^{m-1}+\binom{3^{m-1}}{3}d\right)\sqrt{d} \in \mathfrak{p}^{2m-1}\backslash\mathfrak{p}^{2m}$; and (iii) $\sum_{j=4}^{3^{m-1}} \binom{3^{m-1}}{j}(\sqrt{d})^j \in \mathfrak{p}^{2m+2}$. Hence,

$$(1+\sqrt{d})^{3^{m-1}} \equiv 1+\alpha+\beta \bmod \mathfrak{p}^{2m+2}$$

for some nonzero $\alpha \in \mathfrak{p}^{2m-1}\backslash\mathfrak{p}^{2m}$ and $\beta \in \mathfrak{p}^{2m}\backslash\mathfrak{p}^{2m+1}$.

Thus, $(1+\sqrt{d})^{3^{m-1}} \not\equiv 1 \bmod \mathfrak{p}^{2m}$ and therefore $(1+\sqrt{d})^{3^{m-1}} \not\equiv 1 \bmod \mathfrak{p}^{2m+1}$ as well. Hence, we can conclude that the order of $1+\sqrt{d}$ in both $(\mathcal{O}/\mathfrak{p}^{2m})^*$ and $(\mathcal{O}/\mathfrak{p}^{2m+1})^*$ equals 3^m.

(b) Now, suppose that $\frac{d}{3} \equiv 2 \bmod 3$.

First, we show that the order of $1+3\sqrt{d}$ equals 3^{m-1}. By the binomial theorem, $(1+3\sqrt{d})^{3^{m-1}} \equiv 1+3^m\sqrt{d} \bmod \mathfrak{p}^{2m+1}$. Moreover, since $\sqrt{d} \in \mathfrak{p}$ and $3 \in \mathfrak{p}^2$, it follows that $(1+3\sqrt{d})^{3^{m-1}} \equiv 1 \bmod \mathfrak{p}^{2m+1}$, and the order of $1+3\sqrt{d}$ in $(\mathcal{O}/\mathfrak{p}^{2m+1})^*$ (and thus in $(\mathcal{O}/\mathfrak{p}^{2m})^*$) divides 3^{m-1} as well.

A similar calculation shows $(1+3\sqrt{d})^{3^{m-2}} \equiv 1 + 3^{m-1}\sqrt{d} \bmod \mathfrak{p}^{2m}$. However, $3^m\sqrt{d} \in \mathfrak{p}^{2m-1}$. Therefore, $(1+3\sqrt{d})^{3^{m-2}} \not\equiv 1 \bmod \mathfrak{p}^{2m}$ (and also mod $\mathfrak{p}^{2m+1}$). So, the order of $1+3\sqrt{d}$ in both $(\mathcal{O}/\mathfrak{p}^{2m})^*$ and $(\mathcal{O}/\mathfrak{p}^{2m+1})^*$ equals 3^{m-1}.

Note that $\langle 1+3\sqrt{d}\rangle \cap \langle g\rangle = \{1\}$, since any positive power of $1+3\sqrt{d}$ less than 3^{m-1} has a nonzero $\sqrt{d}$ coefficient.

For the other generator, given any integer $k \geq 4$, we need to find $\alpha \in \mathcal{O}\backslash\{1\}$ such that $\alpha^3 \equiv 1 \bmod \mathfrak{p}^k$. By direct calculation, we can take $\alpha = 1+\sqrt{d}$ for $k = 4$. For $k > 4$, we inductively invoke Proposition 1. Suppose we have found $\alpha \in \mathcal{O}\backslash\{1\}$ such that $\alpha^3 \equiv 1 \bmod \mathfrak{p}^k$. Letting $f(x) = x^3 - 1$, note that $f'(x) = 3x^2$. Although $f'(\alpha) \in \mathfrak{p}$, we have $f(\alpha) \in \mathfrak{p}^k$ and $f'(\alpha) \in \mathfrak{p}^2\backslash\mathfrak{p}^3$ (since α is a unit in $\mathcal{O}/\mathfrak{p}^k$). Thus, indeed $f(\alpha) \in \gcd(f'(\alpha), \mathfrak{p}^k) = \mathfrak{p}^{k-2}$, and we can lift α to a solution to the congruence $x^3 \equiv 1 \bmod \mathfrak{p}^{k+1}$.

It remains to show $\langle\alpha\rangle \times (\langle 1+3\sqrt{d}\rangle \cap \langle g\rangle) = \{1\}$. Again, we prove this inductively on $k \geq 4$. This is true for $k = 4$, since $1+\sqrt{d}$ and $(1+\sqrt{d})^2 \equiv 1+2\sqrt{d} \bmod \mathfrak{p}^4$ are not in $\langle 1+3\sqrt{d}\rangle \times \langle g\rangle$ (as their coefficients of $\sqrt{d}$ are not divisible by 3). Next, assume that there exist $x, y \in \mathbb{Z}$ such that $(x+y\sqrt{d})^3 \equiv 1 \bmod \mathfrak{p}^k$ with $3 \nmid y$. Then, $x+y\sqrt{d}$ lifts to a solution modulo $\mathfrak{p}^{k+1}$ of the form $(x+y\sqrt{d}) + (r+s\sqrt{d})$ where $(r+s\sqrt{d}) \in \mathfrak{p}^k$. Then $3 \mid s$ and thus $3 \nmid (y+s)$, establishing the inductive step. □

4.3. Ramified Prime 2

Finally, we address the case where $\mathfrak{p}$ lies above the ramifying prime 2. Not so surprisingly, this is the most involved case. Recall that $\langle 2\rangle$ ramifies if and only if $d \equiv 2, 3 \bmod 4$. This gives some indication how the group structure of $(\mathcal{O}/\mathfrak{p}^n)^*$ behaves. We first address the group structure of $(\mathcal{O}/\mathfrak{p}^n)^*$ for small powers of $\mathfrak{p}$.

Theorem 11. *Suppose that $\mathfrak{p}$ lies above the ramifying prime 2.*

(a) $(\mathcal{O}/\mathfrak{p})^* = \{1\}$.

(b) $(\mathcal{O}/\mathfrak{p}^2)^* = \langle 1+\sqrt{d}\rangle \cong \mathbb{Z}_2$.

(c) $$(\mathcal{O}/\mathfrak{p}^3)^* = \begin{cases} \langle 1+\sqrt{d}\rangle \cong \mathbb{Z}_4 & \text{if } d \equiv 2 \bmod 4, \\ \langle \sqrt{d}\rangle \cong \mathbb{Z}_4 & \text{if } d \equiv 3 \bmod 4. \end{cases}$$

(d) $$(\mathcal{O}/\mathfrak{p}^4)^* = \begin{cases} \langle 1+2\sqrt{d}\rangle \times \langle 1+\sqrt{d}\rangle \cong \mathbb{Z}_2 \times \mathbb{Z}_4 \text{ if } d \equiv 2 \bmod 4, \\ \langle 1+2\sqrt{d}\rangle \times \langle \sqrt{d}\rangle \cong \mathbb{Z}_2 \times \mathbb{Z}_4 \text{ if } d \equiv 3 \bmod 4. \end{cases}$$

(e) $$(\mathcal{O}/\mathfrak{p}^5)^* = \begin{cases} \langle -1\rangle \times \langle 1+2\sqrt{d}\rangle \times \langle 1+\sqrt{d}\rangle \cong \mathbb{Z}_2 \times \mathbb{Z}_2 \times \mathbb{Z}_4 \text{ if } d \equiv 2 \bmod 4, \\ \langle -1\rangle \times \langle 1+2\sqrt{d}\rangle \times \langle \sqrt{d}\rangle \cong \mathbb{Z}_2 \times \mathbb{Z}_2 \times \mathbb{Z}_4 \text{ if } d \equiv 3 \bmod 4. \end{cases}$$

Proof. (a) This follows from $\mathcal{O}/\mathfrak{p}$ being a field of order 2.

(b) By Theorem 4, $(\mathcal{O}/\mathfrak{p}^2)^*$ is a cyclic group with 2 elements. One such generator is $1+\sqrt{d}$, because $1+\sqrt{d}\neq 1$ and $(1+\sqrt{d})^2\equiv 1+2\sqrt{d}+d^2\equiv 1 \bmod \mathfrak{p}^2$. The last congruence follows from $2\mid d$ and $\mathfrak{p}^2=\langle 2\rangle$.

(c) By Theorem 4, we know that $(\mathcal{O}/\mathfrak{p}^2)^*$ has order 4. In fact, it is cyclic.

If $d\equiv 3 \bmod 4$, observe that $(\sqrt{d})^4\equiv 1 \bmod 4$ and thus $(\sqrt{d})^4\equiv 1 \bmod \mathfrak{p}^3$, because $\mathfrak{p}^3\mid\langle 4\rangle$. Since $(\sqrt{d})^2=d\not\equiv 1 \bmod \mathfrak{p}^3$, we conclude that $\sqrt{d}$ is a generator.

If $d\equiv 2 \bmod 4$, then although $(1+\sqrt{d})^2=(1+d)+2\sqrt{d}\not\equiv 1 \bmod \mathfrak{p}^3$, we have $(1+\sqrt{d})^4=1+4\sqrt{d}+6d+4d\sqrt{d}+d^2\equiv 1 \bmod 4$. Thus $(1+\sqrt{d})^4\equiv 1 \bmod \mathfrak{p}^3$, and we conclude that $1+\sqrt{d}$ is a generator.

(d) Since $\mathfrak{p}^4=\langle 4\rangle$, it follows that $(1+2\sqrt{d})^2\equiv 1 \bmod \mathfrak{p}^4$. Thus, $1+2\sqrt{d}$ has order 2 in $(\mathcal{O}/\mathfrak{p}^4)^*$. Next if $d\equiv 2 \bmod 4$, then $(1+\sqrt{d})^2=(1+d)+2\sqrt{d}\not\equiv 1 \bmod \mathfrak{p}^4$, but $(1+\sqrt{d})^4\equiv 1 \bmod \mathfrak{p}^4$; thus $1+\sqrt{d}$ has order 4 in $(\mathcal{O}/\mathfrak{p}^4)^*$. If $d\equiv 3 \bmod 4$, we have $(\sqrt{d})^2=d\not\equiv 1 \bmod \mathfrak{p}^4$, but $(\sqrt{d})^4\equiv 1 \bmod \mathfrak{p}^4$; thus $\sqrt{d}$ has order 4 in $(\mathcal{O}/\mathfrak{p}^4)^*$. For both cases of d, both $(\sqrt{d})^2$ and $(1+\sqrt{d})^2$ are not equal to $1+2\sqrt{d}$ in $(\mathcal{O}/\mathfrak{p}^4)^*$. This, combined with $\phi(\mathfrak{p}^4)=8$, gives the desired group structure for $(\mathcal{O}/\mathfrak{p}^4)^*$.

(e) Plainly -1 is an element of order 2 for both $d\equiv 2,3 \bmod 4$. Moreover, we claim that $1+2\sqrt{d}$ has order 2. When $d\equiv 2 \bmod 4$, $(1+2\sqrt{d})^2=1+4\sqrt{d}+4d\equiv 1 \bmod \mathfrak{p}^5$ (since $\sqrt{d}\in\mathfrak{p}$ and $2\in\mathfrak{p}^2$), and when $d\equiv 3 \bmod 4$, we note that $1-\sqrt{d}\in\mathfrak{p}$ and $2\in\mathfrak{p}^2$ to deduce that $(1+2\sqrt{d})^2=1+4\sqrt{d}(1-\sqrt{d})\equiv 1 \bmod \mathfrak{p}^5$.

Next, we find a generator having order 4. When $d\equiv 2 \bmod 4$, we claim that $1+\sqrt{d}$ has order 4. To this end, $(1+\sqrt{d})^2=(1+d)+2\sqrt{d}\not\equiv 1 \bmod \mathfrak{p}^5$, but since $8\mid(6d+d^2)$ $(1+\sqrt{d})^4\equiv 1+(6d+d^2)+4\sqrt{d}(1+d)\equiv 1 \bmod \mathfrak{p}^5$. When $d\equiv 3 \bmod 4$, we use $\sqrt{d}$, because $(\sqrt{d})^2=d\not\equiv 1 \bmod \mathfrak{p}^5$, but $(\sqrt{d})^4\equiv 1 \bmod \mathfrak{p}^5$. In both cases, the cyclic subgroups generated by this element of order 4 have trivial intersections with those generated by -1 and $1+2\sqrt{d}$.

Finally, since the order of the direct product of these three cyclic groups equals $\phi(\mathfrak{p}^5)=16$ for both cases of d, we have the desired group structure for $(\mathcal{O}/\mathfrak{p}^5)^*$. ☐

Now, we consider higher powers of $\mathfrak{p}$. However, when $d\equiv 3 \bmod 4$, it turns out that we have to investigate $d\equiv 3$ and $7 \bmod 8$ separately.

Theorem 12. *Suppose that $\mathfrak{p}$ lies above the ramifying prime 2, and let $m\geq 3$.*

(a) If $d\equiv 2 \bmod 4$, then

$$(\mathcal{O}/\mathfrak{p}^{2m})^*=\langle -1\rangle\times\langle 5\rangle\times\langle 1+\sqrt{d}\rangle\cong\mathbb{Z}_2\times\mathbb{Z}_{2^{m-2}}\times\mathbb{Z}_{2^m},$$

$$(\mathcal{O}/\mathfrak{p}^{2m+1})^*=\langle -1\rangle\times\langle 5\rangle\times\langle 1+\sqrt{d}\rangle\cong\mathbb{Z}_2\times\mathbb{Z}_{2^{m-1}}\times\mathbb{Z}_{2^m}.$$

(b) If $d\equiv 7 \bmod 8$, then for some $\alpha\in\mathcal{O}$

$$(\mathcal{O}/\mathfrak{p}^{2m})^* = \langle\alpha\rangle \times \langle 5\rangle \times \langle 1+2\sqrt{d}\rangle \cong \mathbb{Z}_4 \times \mathbb{Z}_{2^{m-2}} \times \mathbb{Z}_{2^{m-1}},$$
$$(\mathcal{O}/\mathfrak{p}^{2m+1})^* = \langle\alpha\rangle \times \langle 5\rangle \times \langle 1+2\sqrt{d}\rangle \cong \mathbb{Z}_4 \times \mathbb{Z}_{2^{m-1}} \times \mathbb{Z}_{2^{m-1}}.$$

(c) If $d \equiv 3 \bmod 8$, *then for some* $\alpha \in \mathcal{O}$

$$(\mathcal{O}/\mathfrak{p}^{2m})^* = \langle -1\rangle \times \langle 1+2\sqrt{d}\rangle \times \langle\alpha\rangle \cong \mathbb{Z}_2 \times \mathbb{Z}_{2^{m-1}} \times \mathbb{Z}_{2^{m-1}},$$
$$(\mathcal{O}/\mathfrak{p}^{2m+1})^* = \langle -1\rangle \times \langle 1+2\sqrt{d}\rangle \times \langle\alpha\rangle \cong \mathbb{Z}_2 \times \mathbb{Z}_{2^{m-1}} \times \mathbb{Z}_{2^{m}}.$$

Proof. (a) Suppose that $d \equiv 2 \bmod 4$. By Lemma 4, we can view $(\mathbb{Z}_{2^{m+r}})^*$ as a subgroup of $(\mathcal{O}/\mathfrak{p}^{2m+r})^*$. Since $(\mathbb{Z}_{2^{m+r}})^* = \langle -1\rangle \times \langle 5\rangle \cong \mathbb{Z}_2 \times \mathbb{Z}_{2^{m+r-2}}$, we have our generators for $\mathbb{Z}_2 \times \mathbb{Z}_{2^{m+r-2}}$.

It remains to find an element of order 2^m for both $(\mathcal{O}/\mathfrak{p}^{2m})^*$ and $(\mathcal{O}/\mathfrak{p}^{2m+1})^*$. We claim that one such element is $1+\sqrt{d}$. We prove this by induction on $m \geq 3$ by showing that

$$\text{(i) } (1+\sqrt{d})^{2^m} \equiv 1 \bmod \mathfrak{p}^{2m+1}, \text{ and (ii) } (1+\sqrt{d})^{2^{m-1}} \equiv 1+\gamma \not\equiv 1 \bmod \mathfrak{p}^{2m}$$

for some $\gamma \in \mathfrak{p}^{2m-1} \backslash \mathfrak{p}^{2m}$. Note these readily imply that $(1+\sqrt{d})^{2^m} \equiv 1 \bmod \mathfrak{p}^{2m}$, and $(1+\sqrt{d})^{2^{m-1}} \not\equiv 1 \bmod \mathfrak{p}^{2m+1}$ as well.

For $m=3$, since $\sqrt{d} \in \mathfrak{p}$ and $2 \in \mathfrak{p}^2$,

$$\begin{aligned}(1+\sqrt{d})^{2^3} &\equiv 1+8\sqrt{d}+28(\sqrt{d})^2+56(\sqrt{d})^3+70(\sqrt{d})^4+0\\ &\equiv 1+28(\sqrt{d})^2+70(\sqrt{d})^4\\ &\equiv 1+7\cdot 2(\sqrt{d})^2(2+5d)\\ &\equiv 1+7\cdot 2^3(\sqrt{d})^2(3+5k) \text{ since } d=2+4k \text{ for some integer } k\\ &\equiv 1 \bmod \mathfrak{p}^7.\end{aligned}$$

This establishes (i). For (ii), working modulo $\mathfrak{p}^6$ yields

$$\begin{aligned}(1+\sqrt{d})^{2^2} &\equiv 1+4\sqrt{d}+6(\sqrt{d})^2+4(\sqrt{d})^3+(\sqrt{d})^4\\ &\equiv 1+4\sqrt{d}+6(\sqrt{d})^2+(\sqrt{d})^4\\ &\equiv 1+4\sqrt{d} \text{ since } d \equiv 2 \bmod 4\\ &\not\equiv 1 \bmod \mathfrak{p}^6.\end{aligned}$$

Moreover, $4\sqrt{d} \in \mathfrak{p}^5 \backslash \mathfrak{p}^6$, thereby finishing the inductive step.

Now we assume the claim is true for m and show it is true for $m+1$. To show (i), note that the inductive hypothesis yields $(1+\sqrt{d})^{2^m} = 1+\alpha$ for some $\alpha \in \mathfrak{p}^{2m+1}$. Then, $(1+\sqrt{d})^{2^{m+1}} = (1+\alpha)^2 = 1+2\alpha+\alpha^2$. Reducing modulo $\mathfrak{p}^{2m+3}$ immediately yields $(1+\sqrt{d})^{2^{m+1}} \equiv 1 \bmod \mathfrak{p}^{2m+3}$.

To establish (ii), by the inductive hypothesis, $(1+\sqrt{d})^{2^{m-1}} \equiv 1+\gamma \bmod \mathfrak{p}^{2m}$ for some $\gamma \in \mathfrak{p}^{2m-1} \backslash \mathfrak{p}^{2m}$. So, we have $(1+\sqrt{d})^{2^{m-1}} = 1+\gamma+\delta$ for some $\delta \in \mathfrak{p}^{2m}$.

Then $(1+\sqrt{d})^{2^m} = (1+\gamma+\delta)^2 \equiv 1+2\gamma \not\equiv 1 \bmod \mathfrak{p}^{2m+2}$ as required, because $2\gamma \in \mathfrak{p}^{2m+1}\backslash\mathfrak{p}^{2m+2}$. This concludes the induction.

Since no nontrivial power of $1+\sqrt{d}$ is an integer, $\langle 1+\sqrt{d}\rangle \cap (\langle -1\rangle \times \langle 5\rangle) = \{1\}$. Finally, since the order of $\langle -1\rangle \times \langle 5\rangle \times \langle 1+\sqrt{d}\rangle$ equals $\mathfrak{N}(\mathfrak{p}^n)$ when $n \geq 6$, we have the desired unit group structure result.

(b) Now, suppose that $d \equiv 7 \bmod 8$.

We know that 5 has order 2^{m-2}, being one of the generators of $(\mathbb{Z}_{2^m})^*$.

It remains to find an element of order 2^{m-1} in $(\mathcal{O}/\mathfrak{p}^{2m+r})^*$ for both $r = 0, 1$. We claim that one such element is $1+2\sqrt{d}$. We prove this by induction on $m \geq 3$ by showing that

$$\text{(i) } (1+2\sqrt{d})^{2^{m-1}} \equiv 1 \bmod \mathfrak{p}^{2m+1}, \text{ and (ii) } (1+\sqrt{d})^{2^{m-2}} \equiv 1+\gamma \not\equiv 1 \bmod \mathfrak{p}^{2m}$$

for some $\gamma \in \mathfrak{p}^{2m-1}\backslash\mathfrak{p}^{2m}$.

For $m = 3$, since $(\sqrt{d}-1) \in \mathfrak{p}$ and $2 \in \mathfrak{p}^2$, we have

$$\begin{aligned}(1+\sqrt{d})^{2^{3-1}} &\equiv 1+4(2\sqrt{d})+6(2\sqrt{d})^2+4(2\sqrt{d})^3+(2\sqrt{d})^4\\ &\equiv 1+16d+8(\sqrt{d}-1)\\ &\equiv 1 \bmod \mathfrak{p}^7.\end{aligned}$$

This establishes (i). For (ii), working modulo $\mathfrak{p}^6 = \langle 8\rangle$ yields

$$\begin{aligned}(1+\sqrt{d})^{2^{3-2}} &\equiv 1+2(2\sqrt{d})+(2\sqrt{d})^2\\ &\equiv 1+4\sqrt{d}(\sqrt{d}-1)\\ &\not\equiv 1 \bmod \mathfrak{p}^6.\end{aligned}$$

The last line follows from $4\sqrt{d}(\sqrt{d}-1) \in \mathfrak{p}^5\backslash\mathfrak{p}^6$, thereby finishing the inductive step.

Now we assume the claim is true for m and show it is true for $m+1$.
To show (i), by the inductive hypothesis, $(1+2\sqrt{d})^{2^{m-1}} \equiv 1 \bmod \mathfrak{p}^{2m+1}$. Thus, we can write $(1+2\sqrt{d})^{2^{m-1}} = 1+\alpha$ for some $\alpha \in \mathfrak{p}^{2m+1}$. Then, $(1+2\sqrt{d})^{2^m} = (1+\alpha)^2 = 1+2\alpha+\alpha^2$. Reducing modulo $\mathfrak{p}^{2m+3}$ yields $(1+2\sqrt{d})^{2^m} \equiv 1 \bmod \mathfrak{p}^{2m+3}$.

To establish (ii), we have $(1+2\sqrt{d})^{2^{m-2}} \equiv 1+\gamma \bmod \mathfrak{p}^{2m}$ for some $\gamma \in \mathfrak{p}^{2m-1}\backslash\mathfrak{p}^{2m}$ by the inductive hypothesis. Then, $(1+2\sqrt{d})^{2^{m-1}} = 1+\gamma+\delta$ for some $\delta \in \mathfrak{p}^{2m}$. Therefore, $(1+2\sqrt{d})^{2^m} = (1+\gamma+\delta)^2 \equiv 1+2\gamma \not\equiv 1 \bmod \mathfrak{p}^{2m+2}$ as required, because $2\gamma \in \mathfrak{p}^{2m+1}\backslash\mathfrak{p}^{2m+2}$. This concludes the induction.

For the third generator, we find an element of order 4. To make sure the cyclic subgroup generated by this element has trivial intersection with those generated by 5 and $1+2\sqrt{d}$, we make sure that its square equals -1. Hence, it suffices to solve $\alpha^2 \equiv -1 \bmod \mathfrak{p}^k$ for $k \geq 6$. We begin with $k = 4$, because we can let $\alpha = \sqrt{d}$. For $k > 4$, we inductively invoke Hensel lifting. Suppose we have found $\alpha \in \mathcal{O}$ such that $\alpha^2 \equiv -1 \bmod \mathfrak{p}^k$. Letting $f(x) = x^2 - 1$, then $f(\alpha) \in \mathfrak{p}^k$ and $f'(\alpha) \in \mathfrak{p}^2 \backslash \mathfrak{p}^3$ (since α is a unit in $\mathcal{O}/\mathfrak{p}^k$). Thus, indeed $f(\alpha) \in \gcd(\langle f'(\alpha)\rangle, \mathfrak{p}^k) = \mathfrak{p}^{k-2}$ and we can lift α to a solution to $x^2 \equiv -1 \bmod \mathfrak{p}^{k+1}$.

Now, it is routine to show that the order of $\langle\alpha\rangle \times \langle 5\rangle \times \langle 1 + 2\sqrt{d}\rangle$ equals $\phi(\mathfrak{p}^n)$ for $n \geq 6$.

(c) Suppose that $d \equiv 3 \bmod 8$.

Clearly, -1 has order 2 and it can be checked using the binomial theorem that $1 + 2\sqrt{d}$ has order 2^{m-1}.

It remains to find a generator for $\mathbb{Z}_{2^{m-1}}$ and $\mathbb{Z}_{2^m}$, depending on m being even or odd, respectively. Since -5 has order 2^{m-2} and 2^{m-1} (as m is even or odd) respectively, we can construct a generator α that satisfies $\alpha^2 \equiv -5 \bmod \mathfrak{p}^n$ when $n \geq 6$. (Note that this cyclic subgroup generated by α necessarily has trivial intersection with those generated by -1 and $1 + 2\sqrt{d}$.)

Hence, it suffices to solve $\alpha^2 \equiv -1 \bmod \mathfrak{p}^k$ for $k \geq 6$. We begin with $k = 6$, because letting $\alpha = 4 + \sqrt{d}$ yields $(4 + \sqrt{d})^2 = 16 + 8\sqrt{d} + d \equiv -5 \bmod 8$. For $k > 6$, we inductively invoke Hensel lifting. Suppose we have found $\alpha \in \mathcal{O}$ such that $\alpha^2 \equiv -5 \bmod \mathfrak{p}^k$. Letting $f(x) = x^2 - 1$, we see that $f(\alpha) \in \mathfrak{p}^k$ and $f'(\alpha) \in \mathfrak{p}^2 \backslash \mathfrak{p}^3$ (since α is a unit in $\mathcal{O}/\mathfrak{p}^k$). Thus, $f(\alpha) \in \gcd(\langle f'(\alpha)\rangle, \mathfrak{p}^k) = \mathfrak{p}^{k-2}$ and we can lift α to a solution to $x^2 \equiv -5 \bmod \mathfrak{p}^{k+1}$.

As before, the order of $\langle -1\rangle \times \langle 1 + 2\sqrt{d}\rangle \times \langle\alpha\rangle$ equals $\phi(\mathfrak{p}^n)$ for any $n \geq 6$. □

5. Primitive Roots in Quadratic Number Rings

As an application of our work, we give a quadratic number ring generalization of primitive roots modulo m in this section.

Definition 6. Fix an algebraic number ring $\mathcal{O}$ and an ideal $\mathfrak{a}$ in $\mathcal{O}$. Then we say that $\alpha \in \mathcal{O}$ is a *primitive root* modulo $\mathfrak{a}$ if and only if $\gcd(\langle\alpha\rangle, \mathfrak{a}) = \langle 1\rangle$ and α has order $\phi(\mathfrak{a})$ in $(\mathcal{O}/\mathfrak{a})^*$.

Plainly, a primitive root modulo $\mathfrak{a}$ exists if and only if $(\mathcal{O}/\mathfrak{a})^*$ is a cyclic group. The following theorem catalogs when primitive roots exist.

Theorem 13. *Suppose that $\mathcal{O}$ is a quadratic number ring.*

(a) There exists a primitive root modulo $\mathfrak{p}^n$ for any prime ideal $\mathfrak{p}$ lying above a split odd rational prime with $n \in \mathbb{N}$, or lying above the split rational prime 2 with $n \in \{1, 2\}$.

(b) There exists a primitive root modulo $\langle p\rangle$ for any inert rational prime p.

(c) *There exists a primitive root modulo* $\mathfrak{p}^n$ *for any prime ideal* $\mathfrak{p}$ *lying above a ramifying odd rational prime with* $n \in \{1, 2\}$, *or lying above the ramifying rational prime* 2 *with* $n \in \{1, 2, 3\}$.

(d) *If* 2 *splits in* $\mathcal{O}$ *with* $\mathfrak{p}$ *lying above* 2, *then there exists a primitive root modulo:*

- $\mathfrak{p}\langle q\rangle$ *when* q *is an inert odd rational prime,*
- $\mathfrak{p}\mathfrak{q}^n$ *when* $\mathfrak{q}$ *lies over a ramifying odd rational prime and* $n \in \{1, 2\}$,
- $\mathfrak{p}\mathfrak{q}^n$ *when* $\mathfrak{q}$ *lies over a split odd rational prime and* $n \in \mathbb{N}$.

(e) *If* 2 *is inert, then there exists a primitive root modulo:*

- $\langle 2\rangle\mathfrak{p}^n$ *when* $\mathfrak{p}$ *lies over a split odd rational prime* $p \neq 3$ *and* $n \in \mathbb{N}$,
- $\langle 2\rangle\mathfrak{p}^n$ *when* $\mathfrak{p}$ *lies over a ramifying odd rational prime* $p \neq 3$ *and* $n \in \{1, 2\}$, *and*
- $\langle 6\rangle$ *when* 3 *is also inert.*

(f) *If* 2 *ramifies in* $\mathcal{O}$ *with* $\mathfrak{p}$ *lying above* 2, *then there exists a primitive root modulo:*

- $\mathfrak{p}\mathfrak{q}^n$ *when* $\mathfrak{q}$ *lies over a split odd rational prime and* $n \in \mathbb{N}$,
- $\mathfrak{p}\langle q\rangle$ *when* q *is an inert odd rational prime,*
- $\mathfrak{p}\mathfrak{q}^n$ *when* $\mathfrak{q}$ *lies over a ramifying odd rational prime and* $n \in \{1, 2\}$.

Proof. The assertions (a)-(c) follow directly from our unit group structure theorems from Sections 3 and 4, while the assertions (d)-(f) follow from these same theorems, in conjunction with the fact that $\mathbb{Z}_m \times \mathbb{Z}_n$ is cyclic if and only if $\gcd(m, n) = 1$. □

We give two corollaries of this theorem. The first of these gives the existence of primitive roots modulo an ideal in the Gaussian integers $\mathbb{Z}[i]$ (also given in [3]). Note that since $\mathbb{Z}[i]$ is a PID, any such ideal has the form $\langle\gamma\rangle$ for some $\gamma \in \mathbb{Z}[i]$.

Corollary 1. *A primitive root in* $\mathbb{Z}[i]$ *modulo* $\langle\gamma\rangle$ *exists if and only if*

$$\gamma = \pi^n, (1+i)\pi^n, q, (1+i)q, \text{ or } (1+i)^k,$$

where π *is a factor of an rational prime* $p \equiv 1 \bmod 4$, $q \equiv 3 \bmod 4$ *is a rational prime,* $n \in \mathbb{N}$, *and* $k \in \{1, 2\}$.

Proof. This follows immediately from the previous theorem, along with the characterization of primes in $\mathbb{Z}[i]$: A rational prime p is inert in $\mathbb{Z}[i]$ if $p \equiv 3 \bmod 4$, split in $\mathbb{Z}[i]$ if $p \equiv 1 \bmod 4$, and ramifies if $p = 2$. □

The second corollary gives a companion result to the case of the Eisenstein integers $\mathbb{Z}[\omega]$. Again since $\mathbb{Z}[\omega]$ is a PID, any such ideal has the form $\langle\gamma\rangle$ for some $\gamma \in \mathbb{Z}[\omega]$.

Corollary 2. *A primitive root in $\mathbb{Z}[\omega]$ modulo $\langle\gamma\rangle$ exists if and only if*

$$\gamma = \pi^n, 2\pi^n, q, \text{ or } (1-\omega)^k,$$

where π is a factor of a rational prime $p \equiv 1 \bmod 3$, $q \equiv 2 \bmod 3$ is a rational prime, $n \in \mathbb{N}$, and $k \in \{1,2\}$.

Proof. This follows immediately from the previous theorem, along with the characterization of primes in $\mathbb{Z}[\omega]$: a rational prime p is inert in $\mathbb{Z}[\omega]$ if $p \equiv 2 \bmod 3$, split in $\mathbb{Z}[\omega]$ if $p \equiv 1 \bmod 3$, and ramifies if $p = 3$. □

References

[1] Z. Borevich and I. Shafarevich, *Number Theory*, Academic Press, New York, 1966.

[2] V. Buçaj, Finding Factors of Factor Rings over Eisenstein Integers, *Int. Math. Forum*, **9(31)** (2014) 1521-1537.

[3] J. Cross, The Euler ϕ Function in the Gaussian Integers, *Amer. Math. Monthly*, **90** (1983) 518-528.

[4] D. Dummit and R. Foote, *Abstract Algebra*, 3rd ed., John Wiley and Sons, New York, 2003.

[5] G. Kohler, *Eta Products and Theta Series Identities*, Springer-Verlag, Berlin, 2011.

[6] D.A. Marcus, *Number Fields*, 2nd ed., Springer-Verlag, Berlin, 1995.

[7] I. Niven, *An Introduction to the Theory of Numbers*, John Wiley and Sons, New York, 1991.

[8] I. Stewart and D. Tall, *Algebraic Number Theory and Fermat's Last Theorem*, 3rd ed., AK Peters/CRC Press, Boca Raton, 2001.

10

ON ALGORITHMS TO CALCULATE INTEGER COMPLEXITY

Katherine Cordwell
Department of Mathematics, University of Maryland, College Park, Maryland
Computer Science Dept., Carnegie Mellon University, Pittsburgh, Pennsylvania
kcordwel@cs.cmu.edu

Alyssa Epstein
Dept. of Math. and Statistics, Williams College, Williamstown, Massachusetts
Stanford Law School, Stanford, California
alye@stanford.edu

Anand Hemmady
Dept. of Math. and Statistics, Williams College, Williamstown, Massachusetts
ash6@williams.edu

Steven J. Miller
Dept. of Math. and Statistics, Williams College, Williamstown, Massachusetts
Dept. of Math. Sciences, Carnegie Mellon University, Pittsburgh, Pennsylvania
sjm1@williams.edu, stevenm1@andrew.cmu.edu

Eyvindur Palsson
Dept. of Math., Virginia Polytechnic Institute and State University, Virginia
palsson@vt.edu

Aaditya Sharma
Dept. of Math. and Statistics, Williams College, Williamstown, Massachusetts
DPMMS, University of Cambridge
as2718@cam.ac.uk

Stefan Steinerberger
Department of Mathematics, Yale University, Connecticut
stefan.steinerberger@yale.edu

Yen Nhi Truong Vu
Dept. of Mathematics and Statistics, Amherst College, Amherst, Massachusetts
ytruongvu17@amherst.edu

Abstract

We consider a problem first proposed by Mahler and Popken in 1953 and later developed by Coppersmith, Erdős, Guy, Isbell, Selfridge, and others. Let $f(n)$ be the complexity of $n \in \mathbb{Z}^+$, where $f(n)$ is defined as the least number of 1's needed to represent n in conjunction with an arbitrary number of $+$'s, $*$'s, and parentheses. Several algorithms have been developed to calculate the complexity of all integers up to n. Currently, the fastest known algorithm runs in time $\mathcal{O}(n^{1.230175})$ and was given by J. Arias de Reyna and J. van de Lune in 2014. This algorithm makes use of a recursive definition given by Guy and iterates through products, $f(d)+f\left(\frac{n}{d}\right)$, for $d \mid n$, and sums, $f(a)+f(n-a)$, for a up to some function of n. The rate-limiting factor is iterating through the sums. We discuss potential improvements to this algorithm via a method that provides a strong uniform bound on the number of summands that must be calculated for almost all n. We also develop code to run J. Arias de Reyna and J. van de Lune's analysis in higher bases and thus reduce their runtime of $\mathcal{O}(n^{1.230175})$ to $\mathcal{O}(n^{1.222911236})$. All of our code can be found online at `https://github.com/kcordwel/Integer-Complexity`.

1. Introduction

1.1. Background

In this paper, log denotes ln, and $\log_b$ denotes the logarithm in base b. Given $n \in \mathbb{N}$, the complexity of n, which we denote $f(n)$, is defined as the least number of 1's needed to represent n using an arbitrary number of additions, multiplications, and parentheses. For example, because 6 may be represented as $(1+1)(1+1+1)$, $f(6) \leq 5$. Calculating $f(n)$ for arbitrary n is a problem that was posed in 1953 by Mahler and Popken [10]. Guy [8] drew attention to this problem in 1986 when he discussed it and several other simply stated problems in an Amer. Math. Monthly article. The following recursive expression for integer complexity highlights the interplay of additive and multiplicative structures:

$$f(n) = \min_{\substack{d \mid n \\ 2\leq d\leq\sqrt{n} \\ 1\leq a\leq n/2}} \left\{ f(d) + f\left(\frac{n}{d}\right), f(a) + f(n-a) \right\}. \tag{1}$$

Some unconditional bounds on $f(n)$ are known. In particular, [8] attributes a lower bound of $f(n) \geq 3\log_3(n)$ to Selfridge. Also, an upper bound of $f(n) \leq 3\log_2(n)$ is attributed to Coppersmith. Extensive numerical investigation (see [9]) suggests that $f(n) \sim 3.3\log_3(n)$ for n large but it is not even known whether $f(n) \geq (3+\varepsilon_0)\log_3 n$ for some $\varepsilon_0 > 0$. As a step towards understanding these problems Altman and Zelinsky [4] introduced the discrepancy $\delta(n) = f(n) - 3\log_3(n)$ and provided a way to classify those numbers with a small discrepancy. This classification was taken further by Altman [1, 2] where he obtained a finite set of polynomials that represent precisely the numbers with small defects. As a consequence Altman [3] was able to calculate the integer complexity of certain classes of numbers. Any progress on these difficult questions likely requires a substantial new idea; the main difficulty, the interplay between additive and multiplicative structures, is at the core of a variety of different open problems, which we believe adds to its allure.

1.2. Algorithms

Much of the progress on this problem has been algorithmic. Using the above recursive definition, it is possible to write algorithms to calculate $f(n)$ for large values of n where the rate-limiting step of the algorithm is iterating through the summands, $f(a) + f(n-a)$, for many values of a. In particular, the brute-force algorithm that iterates over all $a's$ such that $1 \leq a \leq n/2$ runs in time $\mathcal{O}(n^2)$, but there are ways to bound the number of summands that must be checked so as to significantly decrease the computational complexity. Srinivar and Shankar [13] used the unconditional upper and lower bounds on $f(n)$ to bound the number of summands, obtaining an algorithm that runs in time $\mathcal{O}(n^{\log_2(3)}) < \mathcal{O}(n^{1.59})$.

The fastest known algorithm runs in time $\mathcal{O}(n^{1.230175})$ and is due to J. Arias de Reyna and J. van de Lune [5]. Also, the experimental data in [9] is based on an algorithm that calculates $f(n)$ for n up to $n = 10^{12}$. They derive many interesting results from their data, but they do not analyze the runtime of their algorithm. We obtain both an overall improvement on the runtime of the J. Arias de Reyna and J. van de Lune algorithm and a potential internal improvement to the workings of the algorithm. The overall improvement is derived from running the analysis of [5] in much higher bases, while the internal improvement gives a strong uniform bound on the number of summands $f(a) + f(n-a)$ that must be calculated for almost all n. We detail the overall improvement in Section 2. We introduce the potential internal improvement in Section 3 and test it in Section 4. We end the paper by proposing a new approach for improving the current unconditional upper bound on $f(n)$.

2. Algorithmic Aspects

2.1. The de Reyna & van de Lune Algorithm

J. Arias de Reyna and J. van de Lune [5] developed code in Python to perform the analysis of their algorithm, which they have generously shared with us. Additionally, Fuller has published open-source code [7] written in C to calculate integer complexities. Using these, we have developed code[1] in C that is comparable to J. Arias de Reyna and J. van de Lune's Python code. The heart of the code is the calc_count method, which calculates $D(b, r)$ for varying values of b and r, where $D(b, r)$ is an upper bound on how much multiplying by b and adding r increases the complexity of any given number. More precisely, we define $D(b, r)$ to be the

[1]See the "calculate_complexities.c" file at https://github.com/kcordwel/Integer-Complexity

smallest integer such that

$$f(r+bn) \leq f(n) + D(b,r) \tag{2}$$

for all n. As an example, notice that $D(b,0) \leq f(b)$, because we can always represent b with $f(b)$ 1's and n with $f(n)$ 1's and then multiply these two representations to achieve a representation of bn—and thus $f(bn) \leq f(n) + f(b)$. Similarly, $D(1,r) \leq f(r)$ because we can represent r with $f(r)$ 1's and n with $f(n)$ 1's, and then add these two representations to achieve a representation of $n+r$ that uses $f(n)+f(r)$ 1's.

These integers $D(b,r)$ are useful for bounding $f(n)$ in the following way: [5] defined C_{avg} as the infimum of all C such that $f(n) \leq C\log(n)$ for a set of natural numbers of density 1 and showed that

$$C_{avg} \leq \frac{1}{b\log(b)} \sum_{r=0}^{b-1} D(b,r). \tag{3}$$

In this calculation, we refer to b as the base in which we are working. Our code closely follows the logic of J. Arias de Reyna and J. van de Lune's program, making the following slight optimization.

Theorem 1. *Take $b = 2^i 3^j$ where $b < 10^{12}$ and $i + j > 0$. If $r \mid b$ for $2 \leq r < b$, then $D(b,r) = f(b) + 1$.*

Proof. From the equality $r + bn = r(1 + n \cdot \frac{b}{r})$, notice that:

$$f(bn+r) \leq 1 + f(r) + f\left(n \cdot \frac{b}{r}\right) \leq 1 + f(r) + f(n) + f(\frac{b}{r}) \tag{4}$$

From [9], we know that $f(2^v 3^w) = 2v + 3w$ for $2^v 3^w < 10^{12}$ and $v + w > 0$. We know that $b = 2^i 3^j$, and since r divides b, r is of the form $2^x 3^y$ for $x \leq i$, $y \leq j$. This means that b/r is of the form $2^{i-x}3^{j-y}$. Since $r \geq 2$, we have $x + y > 0$ and since $r < b$, we have $(i-x) + (j-y) > 0$.

Then applying the result of [9] to both r and $\frac{b}{r}$, we obtain

$$\begin{aligned} f(bn+r) &\leq 1 + f(r) + f(n) + f\left(\frac{b}{r}\right) \\ &= 1 + 2x + 3y + f(n) + 2(i-x) + 3(i-y) \\ &= 1 + 2i + 3j + f(n) \\ &= 1 + f(b) + f(n). \end{aligned} \tag{5}$$

This shows that $D(b,r) \leq f(b) + 1$.

Now we wish to argue that $D(b,r) \geq f(b)+1$. While this makes sense intuitively, in order to be rigorous we do a computer aided proof[2]. Our computer calculations

[2]See the code in the "Thm2.1" folder at https://github.com/kcordwel/Integer-Complexity

work as follows: First, we calculate the complexities of $b+r$ for all b, r as in the theorem. For the vast majority of the b, r, $f(b+r) = f(b)+1$, meaning that $D(b,r) \geq f(b)+1$ from the definition of $D(b,r)$. However, there are 372 pairs of b, r such that $f(b+r) \neq f(b)+1$. For these pairs, we do a second pass and calculate the complexities of $2b+r$. For all of the pairs, $f(2b+r) = f(b)+3 = f(b)+1+f(2)$, meaning that $D(b,r) \geq f(b)+1$. □

J. Arias de Reyna and J. van de Lune [5] suggest that their algorithms will be more powerful when implemented in C and Pascal. [5] proved that their algorithm has running time $\mathcal{O}(N^\alpha)$ where

$$\alpha = -1 + \frac{1}{\log b} \log \left(\sum_{d=0}^{b-1} 3^{\frac{1}{3}D(b,d)} \right). \tag{6}$$

They calculated the runtime of their algorithm for bases $2^n 3^m$ up to 3188646, and found the best value of α as

$$\begin{aligned} \alpha &= \frac{\log(48399164638047 + 3^{1/3} \cdot 33606823799088 + 3^{2/3} \cdot 23231513379231)}{\log(2^{10} \cdot 3^7)} - 1 \\ &\approx 1.230175 \end{aligned}$$

in base $2^{10}3^7 = 2239488$. Using C is advantageous because it runs much faster than Python, and so we are able to calculate values for higher bases. We calculated values for bases $2^n 3^m \leq 57395628$. In base $2^{13}3^8 = 53747712$, we find that the runtime is $\mathcal{O}(n^\alpha)$ where the value of α is

$$\frac{\log(50903564566217859 + 35271975106952037 \cdot 3^{1/3} + 24493392174530898 \cdot 3^{2/3})}{\log(2^{13}3^8)} - 1,$$

so that the runtime is $\mathcal{O}(n^{1.222911236})$.

2.2. Improved Asymptotic Results

Probably Guy [8] was the first who remarked that while pointwise bounds seem difficult, it is possible to establish bounds that are true for almost all (in the sense of asymptotic density 1) numbers. His method showed that $f(n) \leq 3.816 \log_3 n$ for a subset of integers with density 1.

Using their definition of C_{avg} as the infimum of all C where $f(n) \leq C \log(n)$ for a set of natural numbers of density 1, [5] showed that for any base $b \geq 2$

$$C_{avg} \leq \frac{1}{b \log b} \sum_{r=0}^{b-1} D(b,r). \tag{7}$$

In base $b = 2^9 3^8$, they obtain

$$C_{avg} \leq \frac{166991500}{2^9 3^8 \log(2^9 3^8)}, \tag{8}$$

so that $f(n) \leq 3.30808 \log(n)$, or $f(n) \leq 3.63430 \log_3(n)$, for a set of natural numbers of density 1.

We find that in base $2^{11}3^9$,

$$C_{avg} \leq \frac{2326006662}{2^{11}3^9 \log(2^{11}3^9)}, \tag{9}$$

so that $f(n) \leq 3.29497 \log(n)$, or $f(n) \leq 3.61989 \log_3(n)$, for a set of natural numbers of density 1.

3. Possible Improvements via Balancing Digits

3.1. Balancing Digits

Our goal is to improve the algorithm for calculating complexity given in [5]. The rate-limiting factor in this algorithm is checking, for all $n \leq N$, $f(a) + f(n-a)$ for all $1 \leq a \leq$ kMax for some kMax that is specially calculated for each n. We will show that we can give a strong uniform bound on the number of summands that must be checked for almost all n.

We say that $n \in \mathbb{Z}$ is *digit-balanced* in base b if each of the digits $1, \ldots, b-1$ occurs roughly $1/b$ times in the base b representation of n, or *digit-unbalanced* if some digits occur significantly more often than others. We will show that almost all numbers are digit-balanced, although the exact threshold of variation that we allow will depend on the base b. Finally, assuming that we have a set S of digit-balanced numbers in base b, we will use Guy's method to find that for any $n \in S$, $f(n) \leq c \log_3(n)$ for some c. Then, using this bound on $f(n)$ and assuming that $f(n) = f(a) + f(n-a)$, we are able to bound a, which, in turn, narrows the search space that a reasonable algorithm has to cover.

3.2. Bounds on Digit-Balanced Numbers

Our main result is as follows.

Proposition 1. *There exists a constant $c_b > 0$, depending only on the base b, such that*

$$\#\left\{1 \leq n \leq N : \max_{1 \leq i \leq b} \left| \frac{\text{number of digits of } n \text{ in base } b \text{ that are } i}{\text{number of digits of } n \text{ in base } b} - \frac{1}{b} \right| \geq \varepsilon \right\} \leq N^{1-c_b\varepsilon^2}.$$

Proof. The main idea behind the argument is to replace a combinatorial counting argument by the probabilistic large deviation theory. Let $N = b^k$, consider all k-digit numbers in base b, and let X_i be a random variable such that $X_i = 1$ with probability $1/b$ and 0 otherwise for $1 \leq i \leq k$. For any given digit $0 \leq d < b$, each X_i gives the probability that this digit will appear in a fixed position i in the base b representation of a number. Since we are considering k-digit numbers, we need to understand the average value of $X_1 + \cdots + X_k$ and to analyze how close this average is to $\frac{1}{b}$. Let $\overline{X} = \frac{1}{k}(X_1 + \cdots + X_k)$. Next, we can use Hoeffding's inequality, which gives

$$P\left(\overline{X} - \frac{1}{b} \geq \epsilon\right) \leq e^{-2k\epsilon^2}. \tag{10}$$

We know that $k \approx \log_b(N) = \frac{\log(N)}{\log(b)}$, so:

$$e^{-2k\epsilon^2} = e^{-2\epsilon^2 \frac{\log(N)}{\log(b)}} = \left(e^{\log(N)}\right)^{-2\epsilon^2 \frac{1}{\log(b)}} = N^{\frac{-2\epsilon^2}{\log(b)}}. \tag{11}$$

So, the probability that a number with k digits in its base b representation has some digits that appear more often than the average is less than or equal to $N^{\frac{-2\epsilon^2}{\log(b)}}$, meaning that $|S| \leq N \cdot N^{\frac{-2\epsilon^2}{\log(b)}} = N^{1-\frac{2\epsilon^2}{\log(b)}}$. □

3.3. Bound on Number of Summands

Assume now that $f(n) = f(n-a)+f(a)$ and that this is the optimal representation using the least number of 1's. We assume that $f(n) = c\log_3(n)$ for some $c > 0$. Our goal is to derive a bound on a. The main idea is to show that the logarithmic growth implies that a cannot be very large (otherwise the growth of $f(n)$ would be closer to linear). Using the lower bound due to Selfridge [8], we attain:

$$c\log_3(n) \geq 3(\log_3(n-a) + \log_3(a)). \tag{12}$$

This is equivalent to:

$$\log_3(n^{c/3}) \geq \log_3(n-a) + \log_3(a). \tag{13}$$

Say that $a = qn$, where necessarily $q \leq \frac{1}{2}$. Then we have:

$$\log_3(n^{c/3}) \geq \log_3((1-q)n \cdot a). \tag{14}$$

Exponentiating both sides and simplifying gives

$$\frac{n^{c/3-1}}{1-q} \geq a. \tag{15}$$

Since $q \leq \frac{1}{2}$, then $1-q \geq \frac{1}{2}$, and so

$$\frac{n^{c/3-1}}{1/2} \geq \frac{n^{c/3-1}}{1-q} \geq a, \tag{16}$$

or:

$$2n^{c/3-1} \geq \frac{n^{c/3-1}}{1-q} \geq a. \tag{17}$$

Thus, we need only check for values of a at most $2n^{c/3-1}$.

3.4. Binary Analysis

To see how our result works, we analyze it in the simplest possible base, which is binary. Consider k-digit numbers less than N (so that $k \approx \log_2(N)$). The average case in Guy's method, illustrated in [8] and based on Horner's scheme of representing binary numbers, gives $f(n) \leq 5\log_2(n)/2$, or $f(n) < 3.962407\log_3(n)$. "Bad" numbers in base 2 are those that have many 1's, as that is when the representation is rather inefficient. If we move away from the average case to numbers which have, say, 75% 1's and 25% 0's, then the constant in Guy's method is

$$\frac{1}{\log(2)}(3 \cdot .75 + 2 \cdot .25)\log(3) \; < \; 4.358647. \tag{18}$$

This is already much worse than the original average case constant of 3.962407, and so we need to stay much closer to the average case. In particular, the following percentages of 1's and 0's give the following values for the constant in Guy's method:

Percent 0's	Percent 1's	Constant
46	54	4.02581
47	53	4.00997
48	52	3.99411
49	51	3.97826
49.9	50.1	3.96399
49.99	50.01	3.962565

Consider numbers with at most 46% 0's and 54% 1's. The previous section affords a bound of $a \leq 2n^{4.02581/3-1} \leq 2n^{0.342}$ for such numbers. We want to understand how often this case occurs. Recall that we are considering k-digit numbers. We need to bound the number of times that 0 occurs at most $\frac{46k}{100}$ times, or the number of times that 1 occurs at least $\frac{54k}{100}$ times. Say that X_i is the Bernoulli variable corresponding to digit i, $1 \leq i \leq k$. Then $P(X_i = 1) = \frac{1}{2}$. Let $S_k = X_1 + \cdots + X_k$, so that S_k represents the total number of 1's in our number. Since $\frac{1}{2} < \frac{54}{100} < 1$, we may apply Theorem 1 from [6] to achieve the following bound:

$$P\left(S_k \geq \frac{54k}{100}\right) \leq e^{-kD\left(\frac{54}{100}||\frac{1}{2}\right)} \tag{19}$$

where

$$D\left(\frac{54}{100} \mid\mid \frac{1}{2}\right) = \frac{54}{100}\log\left(2\left(\frac{54}{100}\right)\right) + \left(1-\frac{54}{100}\right)\log\left(2\left(1-\frac{54}{100}\right)\right). \tag{20}$$

Because $k \approx \log_2(N)$, we get that

$$P\left(S_k \geq \frac{54k}{100}\right) \leq N^{-D\left(\frac{54}{100} \,||\, \frac{1}{2}\right)\frac{1}{\log 2}}. \tag{21}$$

In particular, then, there are at most $N^{1-\frac{1}{\log(2)}\cdot D\left(\frac{54}{100}||\frac{1}{2}\right)} < N^{1-.004622}$ "bad" numbers, i.e. we have the desired bound $a \leq 2n^{0.342}$ for the other $> N^{.004622}$ numbers, which is significant as N grows large. Call this set of numbers for which we have this bound $\mathcal{U}$.

Following the analysis in [5], Arias de Reyna and van de Lune's algorithm has a runtime of n^{α} in base 2 where

$$\alpha = -1 + \frac{\log(3^{D(2,0)/3} + 3^{D(2,1)/3})}{\log(2)} = -1 + \frac{\log(3^{2/3}+3)}{\log(2)} \approx 1.3448. \tag{22}$$

Recall that in their complexity proof, Arias de Reyna and van de Lune denote the number of summands that must be checked for each n by kMax. Our bound on the numbers in $\mathcal{U}$ compares well to [5]'s bound in that if kMax were uniform for all numbers in [5], our bound would be lower on all $u \in \mathcal{U}$. More explicitly, in binary, if kMax were uniform, then [5] would require checking summands up to $\approx n^{0.3448}$ whereas we require checking summands up to $\approx 2n^{0.342}$ for numbers in $\mathcal{U}$.

Unfortunately, kMax is not uniform in this way, and so we cannot claim a definitive improvement with our uniform bound on α. It is possible that some of the $u \in \mathcal{U}$ have a low value of kMax to begin with, and for such numbers our bound may not afford an improvement. Conversely, it is possible that our bound will improve some numbers that are not in $\mathcal{U}$. Overall, since kMax is not uniform, it is not easy to theoretically compare our bound to [5]. Given this, and given that the ideal bases are much larger than binary (which significantly complicates theoretical analysis), we performed a number of empirical tests to understand how our algorithm compares to [5] in the general case.

4. Empirical Calculations

To see whether our method improves J. Arias de Reyna and J. van de Lune's algorithm in practice, we modified J. Arias de Reyna and J. van de Lune's code by adding various precomputations and calculating how many numbers would be improved with these precomputations[3].

The first precomputation uses a greedy algorithm due to Steinerberger [12], which gives that $f(n) \leq 3.66 \log_3(n)$ for most n. The recursive algorithm works as follows: if $n \equiv 0 \mod 6$ or $n \equiv 3 \mod 6$, take $n = 3(n/3)$ and run the algorithm on $n/3$. If $n \equiv 2 \mod 6$ or $n \equiv 4 \mod 6$, take $n = 2(n/2)$ and run the algorithm on $n/2$. If $n \equiv 1 \mod 6$, take $n = 1 + 3(n-1)/3$ and run the algorithm on $(n-1)/3$. If $n \equiv 5 \mod 6$, take $n = 1 + 2(n-1)/2$ and run the algorithm on $(n-1)/2$.

The method is as follows: First, run the greedy algorithm on all of the numbers up to some limit and store the results in a dictionary. Then, use these values to compute a bound on the number of summands for each number (using the formula derived in Section 3.3). Store a counter that is initialized to 0. Next, run J. Arias de Reyna and J. van de Lune's algorithm. For each number, test whether the precomputed summand bound is better than the summand bound in the original algorithm. If an improvement is found, increment the counter. When we use this algorithm to precompute summands, we improve 7153 numbers out of the first 200000, or less than 3.6% of numbers. If we compute complexities further, up to 2000000, we improve 60864 numbers, or less than 3.05% of numbers.

We can also combine Steinerberger's algorithm with a stronger algorithm, due to Shriver [11]. Shriver developed a greedy algorithm in base 2310. If we use the best upper bound on complexities from Shriver and Steinerberger's greedy algorithm, we improve 11188 numbers out of 200000, or about 5.6% of numbers. If we compute complexities up to 2000000, we improve 107077 numbers, or less than 5.36% of numbers.

Shriver conjectures that his best algorithm, which uses simulated annealing, produces a bound of $f(n) \leq 3.529 \log_3(n)$ for generic integers. In fact, only 824 numbers up to 2000000 would be improved by assuming a uniform bound of $f(n) \leq 3.529 \log_3(n)$. Of course, this is a purely theoretical result—if we were to actually introduce a uniform bound, then we would not be able to accurately calculate complexities. If we become even more optimistic and use a uniform bound of $f(n) \leq 3.5 \log_3(n)$, we would only potentially improve 4978 numbers out of the first 2000000. Similarly, using $f(n) \leq 3.4 \log_3(n)$ would improve 124707 numbers of 2000000, which is about 6.23%. If we venture significantly below Shriver's conjecture of $3.529 \log_3(n)$ and use $f(n) \leq 3.3 \log_3(n)$ uniformly, then we start to see a significant difference—we would improve 726756 numbers of 2000000, or about 36%.

Overall, it seems that Arias de Reyna and van de Lune's algorithm already has a strong bound on the number of summands that are computed. It is possible that we are encountering difficulties because kMax is not uniform, or it is possible that the complexity of J. Arias de Reyna and J. van de Lune's algorithm is significantly lower

[3] See the "ExperimentalResults" folder at https://github.com/kcordwel/Integer-Complexity

than $\mathcal{O}(n^{1.223})$. Thus, while summand precomputing improves the complexity computation for some numbers, given the overhead for performing precomputations and the current speed of J. Arias de Reyna and J. van de Lune's algorithm, introducing a precomputation does not seem to yield an overall improvement to the algorithm.

5. Progress Towards an Unconditional Upper Bound

The current unconditional upper bound on complexity, $f(n) \leq 3\log_2(x)$, is derived from applying Guy's method in base 2 to n. In particular, the most complex numbers have binary expansions of the form $11\cdots1_2$ so that at each step, Guy's method requires three 1's. The resulting representation is of the form $1+(1+1)[1+(1+1)[\cdots]]$.

Say that $n \bmod 3 \equiv k$. Instead of applying Guy's method to n, what if write $n = k + (1+1+1)(n-k)/3$ and then apply Guy's method to $(n-k)/3$? Then in the case where $n = 11\cdots1_2$, $(n-k)/3$ is either of the form $1010\cdots1$ or $1010\cdots0$, and applying Guy's method to $(n-k)/3$ gives $f((n-k)/3) \leq 1+2.5\log_2(n)$. Using this, we find that $f(n) \leq 6 + 2.5\log_2(n)$, which is a significant improvement over $f(n) \leq 3\log_2(n)$.

This suggests the following method: If the binary representation of n contains more than a certain percentage of 1's, then write n as $k + (1+1+1)\cdot(n-k)/3$ and apply Guy's method instead to $(n-k)/3$. Empirically, in most cases, when the binary expansion of n contains a high percentage of 1's, $(n-k)/3$ has a significantly lower percentage of 1's. However, there are some examples where this fails. For example, if $n = 2^{102} - 2^{100} - 2$, then both the binary expansion of n and the binary expansion of $(n-1)/3$ have a high percentage of 1's. Notably, if we repeat this division process and consider $((n-1)/3)/3$, then we will obtain a number with a nice binary expansion. Accordingly, we say that $2^{102} - 2^{100} - 2$ requires two iterations of division by 3.

Some numbers require numerous iterations of division by 3 before their binary expansions are nice. For example, $n = 2^{3000} - 2^{2975} - 2^{2807} - 1$ requires nine iterations. These sorts of counterexamples seem to follow some interesting patterns. Let n_i denote the number obtained after i iterations of division by 3 so that $n_0 = n$, $n_1 = (n_0 - (n_0 \bmod 3))/3$, etc. In general, it seems that the number of iterations that are necessary to produce a "nice" binary expansion is tied to the number of iterations for which $n \equiv 2 \bmod 3$. For example, when $n = 2^{3000} - 2^{2975} - 2^{2807} - 1$, then $n_0 \equiv n_1 \equiv n_2 \equiv \cdots \equiv n_7 \equiv 2 \bmod 3$, but $n_8 \equiv 0 \bmod 3$, and n_9 has the first "nice" binary expansion.

It should be noted that there is no reason to only employ division by 3. For example, when $n = 2^{3000} - 2^{2975} - 2^{2807} - 1$, $n \bmod 11 \equiv 5$, and $(n-6)/11$ has a nice binary expansion. It should be noted that $n \equiv 4 \bmod 5$ and $n \equiv 6 \bmod 7$, and the binary representations of $(n-4)/5$ and $(n-6)/7$ both contain a large percentage of 1's.

In general, then, performing this process of division by appropriate numbers before applying Guy's method is a promising strategy for obtaining an improvement on the unconditional upper bound on $f(n)$. We believe that it could be an interesting problem to make these vague heuristics precise and understand whether this could give rise to a new effective method of giving explicit constructions of n with sums and products that use few 1's.

Acknowledgments. We would like to thank Professor Arias de Reyna for generously sharing the code that he developed with Professor van de Lune. Thank you to the SMALL REU program, Williams College, and the Williams College Science Center where the bulk of this work took place. We would like to thank Professor Amanda Folsom for funding from NSF Grant DMS1449679 as well as SMALL REU for funding from NSF Grant DMS1347804, the Williams College Finnerty Fund, and the Clare Boothe Luce Program. The fourth listed author was supported by NSF grants DMS1265673 and DMS1561945, the fifth listed author was supported by Simons Foundation Grant #360560 and the seventh listed author was supported by NSF Grant DMS1763179 and the Alfred P. Sloan Foundation. Finally, we thank the anonymous referee for many helpful suggestions that significantly improved the paper.

References

[1] H. Altman, Integer complexity and well-ordering, *Michigan Math. J.* **64** (2015), no. 3, 509-538.

[2] H. Altman, Integer complexity: Representing numbers of bounded defect, *Theoret. Comput. Sci.* **652** (2016), 64-85.

[3] H. Altman, Integer complexity: Algorithms and computational results, *arXiv:1606.03635 [math.NT]*.

[4] H. Altman and J. Zelinsky, Numbers with integer complexity close to the lower bound, *Integers* **12** (2012), no. 6, 1093-1125.

[5] J. Arias de Reyna and J. van de Lune, Algorithms for determining integer complexity, *arXiv:1404.2183 [math.NT]*.

[6] R. Arratia and L. Gordon, Tutorial on large deviations for the binomial distribution, *Bull. Math. Biol.* **51** (1989), no. 1, 125-131.

[7] M. N. Fuller, *C-Program to Compute A005245*, Feburary 2008. http://oeis.org/A005245/a005245.c.txt.

[8] R. K. Guy, Unsolved problems: some suspiciously simple sequences, *Amer. Math. Monthly* **93** (1986), no. 3, 186-190.

[9] J. Iraids, K. Balodis, J.Cerenoks, M. Opmanis, R. Opmanis, and K. Podnieks, Integer complexity: experimental and analytical results, *Scientific Papers University of Latvia, Computer Science and Information Technologies* **787** (2012), 153-179.

[10] K. Mahler and J. Popken, On a maximum problem in arithmetic (Dutch), *Nieuw Arch. Wiskd.* **3** (1953), no. 1, 1-15.

[11] C. Shriver, Applications of Markov chain analysis to integer complexity, *arXiv:1511.07842 [math.NT].*

[12] S. Steinerberger, A short note on integer complexity, *Contrib. Discrete Math.* **9** (2014), no. 1.

[13] V. V. Srinivas and B. R. Shankar, Integer complexity: breaking the $\theta(n^2)$ barrier, *World Academy of Science, Engineering and Technology* **2** (2008), no. 5, 454-455.

11

TRIANGULAR RAMSEY NUMBERS

Zachary Chaney
Department of Applied Mathematics and Statistics, Colorado School of Mines, Golden, Colorado

Jed Menard
Department of Applied Mathematics and Statistics, Colorado School of Mines, Golden, Colorado

Connor Mattes
Department of Applied Mathematics and Statistics, Colorado School of Mines, Golden, Colorado

Timothy Trujillo
Department of Mathematics and Statistics, Sam Houston State University, Huntsville, Texas
trujillo@shsu.edu

Abstract

The purpose of this paper is to introduce the idea of triangular Ramsey numbers and provide values as well as upper and lower bounds for them. To do this, the combinatorial game Mines is introduced after some necessary theorems about triangular sets are proven. This game is easy enough that young children are able to play. The most basic variations of this game are analyzed, and theorems about winning strategies and the existence of draws are proven. The game of Mines is then used to define triangular Ramsey numbers. Lower bounds are found for these triangular Ramsey numbers using the probabilistic method and the theorems about triangular sets.

1. Combinatorial Games

Combinatorial games make it possible to easily explain the underlying workings of combinatorial problems, which in turn help build a deeper understanding of combinatorics. In fact, some combinatorial games are so simple that they where invented as tools to be used in grade school classrooms to introduce students to logical reasoning and mathematical concepts. For example, the game of Tri was introduced by Haggard and Schonberger in [9] with the goal of "developing logical skills of evaluating alternatives and their consequences." The game had the unintended learning outcome of developing the skill of visual disembedding, i.e., the skill of picking out simple figures from a more complex image. Haggard and Schonberger point out that this ability has been linked to success in solving mathematical problems. In this paper we introduce a new game in the spirit of Haggard and Schonberger. In theory, these games are simple enough that they can be used to help develop mathematical problem solving skills in primary school students.

This paper focuses on a new combinatorial game called Mines which we use to introduce the notion of a triangular Ramsey number. The game is called Mines because our original game boards were in the shape of a Reuleaux triangle, the logo of Colorado School of Mines (see Figure 1). Our main results concern theorems about this game, such as the existence of a winner and the possibility of a winning strategy. The existence of triangular Ramsey numbers follows from the work of Dobrinen and Todorcevic. The primary purpose of introducing the game of Mines is to provide a simplified presentation of the finite-dimensional Ramsey theory of the infinite-dimensional topological Ramsey space $\mathcal{R}_1$ introduced and studied [3]. The game provides a simplified method for defining triangular Ramsey numbers which are the direct analogue of the Ramsey numbers for the finite-dimensional Ramsey theory of $\mathcal{R}_1$.

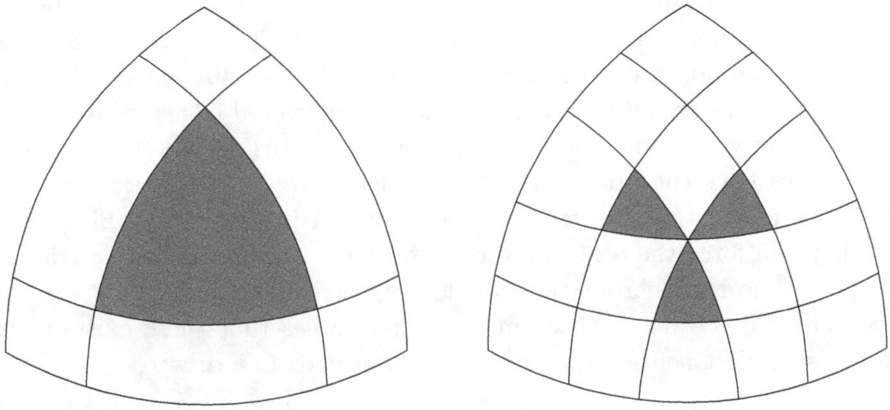

Figure 1: Two different sized Mines game boards

In Section 2, we introduce the games of Tri and Sim. Then, we describe how the games can be used to define Ramsey numbers. Near the end of the section we provide a short survey of some known Ramsey numbers and some bounds on unknown Ramsey numbers. The section concludes with a lemma about Tri needed later in the paper.

In Section 3, in order to help precisely describe the gameboards, we introduce the concept of a triangular set. The remainder of Section 3 is devoted to proving combinatorial results related to counting triangular sets.

In Section 4, we introduce the game of Mines and prove some theorems about its game play. For example, certain variations of Mines have the property that they can never end in a draw.

In Section 5, we use the game to define the notion of a triangular Ramsey number. Our main results in this section involve finding exact values of some triangular Ramsey numbers and bounds for other triangular Ramsey numbers. Section 5 ends by applying the combinatorial results from Section 3 and the probabilistic method as pioneered by Erdős to find lower bounds for triangular Ramsey numbers.

In Section 6, we collect together the main results of the paper in Table 11. We then discuss the connection between the triangular Ramsey numbers and the topological Ramsey space $\mathcal{R}_1$ introduced by Dobrinen and Todorcevic in [3]. We conclude with some open questions and problems related to Mines and triangular Ramsey numbers.

2. Tri, Sim, and Ramsey Numbers

Two games that have attracted attention in the literature are Sim and Tri. The game Sim_m was introduced by Simmons in 1969 [17], and Tri_m was introduced by Haggard and Schonberger in 1977 [9]. Both Sim_m and Tri_m are two player games played on a game board of $m > 2$ vertices with $\binom{m}{2}$ possible edges. Each player chooses a color; players alternate turns coloring uncolored edges using their color. Both games end when a monochromatic triangle is constructed (three vertices all of whose edges have the same color) or all edges have been colored. In Tri_m the winner is the player that constructs a monochromatic triangle. In Sim_m a player wins if they can force the other player to construct a monochromatic triangle. In either game, if no monochromatic triangle is constructed, we say the game ends in a draw. The finite Ramsey theorem for pairs implies that there exists a natural number m such that neither Tri_m nor Sim_m ever ends in a draw.

Problem 1. Find the smallest natural number $m > 2$ such that neither Tri_m nor Sim_m ever ends in a draw.

The solution to the problem is $m = 6$. For $n < m$, the two games have natural generalizations to $\text{Tri}_m(n)$ and $\text{Sim}_m(n)$. The only difference being that these versions end when a monochromatic complete graph with n vertices is constructed. In this notation Tri_m and Sim_m correspond to $\text{Tri}_m(3)$ and $\text{Sim}_m(3)$. The Finite Ramsey Theorem for pairs implies that for all natural numbers n there exists a natural number m such that neither $\text{Tri}_m(n)$ nor $\text{Sim}_m(n)$ ever ends in a draw.

Problem 2. Let n be a natural number greater than 2. Find the smallest natural number $m \geq n$ such that neither $\text{Tri}_m(n)$ nor $\text{Sim}_m(n)$ ever ends in a draw.

The solution to the second problem for the natural number m is called the *Ramsey number for* n, denoted by $R(n)$. It is known that $R(3) = 6$ and $R(4) = 18$; however, $R(5)$ still remains unknown. Figure 2 gives upper and lower bounds for some small Ramsey numbers. For example, from the table we have $43 \leq R(5) \leq 49$. In other words, there is a game of $\mathrm{Tri}_{43}(5)$ that ends in a draw but no game of $\mathrm{Tri}_{49}(5)$ can end in a draw. The lower bound in the second to last row of the table is the lower bound obtained by Erdős using the probabilistic method in [6]. The last row gives the best known upper and lower bounds.

n	Lower Bound	$R(n)$	Upper Bound	References
3	-	6	-	[8]
4	-	18	-	[8]
5	43	?	49	[7] [14]
6	102	?	165	[10] [13]
7	205	?	540	[16] [13]
n	$2^{n/2}$	?	4^{n-1}	[6][5]
n	$n2^{n/2}[\sqrt{2}/e + o(1)]$	?	$n^{-C\frac{\log n}{\log\log n}}4^n$	[18] [2]

Figure 2: Table of upper and lower bounds of some Ramsey numbers and their references.

The next lemma about Tri will be used later to obtain an upper bound for a small triangular Ramsey number. For natural numbers k the notation $R^k(n)$ denotes the Ramsey number $\underbrace{R(R(\cdots(R(n))}_{k-times}\cdots))$.

Lemma 1. *Let k be a natural number. If k games of $\mathrm{Tri}_{R^k(3)}(3)$ are played on the same game board, then there exists a complete graph with three vertices that is monochromatic for each of the k games.*

Proof. By the definition of Ramsey number there is a complete graph with $R^{k-1}(3)$ vertices that is monochromatic for the first game. Then, restrict the second game to this complete subgraph of the game board. Again by the definition of Ramsey number there is a complete subgraph of this graph with $R^{k-2}(3)$ vertices that is monochromatic for the first and second games. Continuing this way for k steps we obtain a complete graph with three vertices that is monochromatic for all n games. □

3. Combinatorics of Triangular Sets

A *triangular number* is a number that can be represented by a triangular arrangement of equally spaced points. For example, the number 15 can be arranged into a triangle with five levels (see Figure 3).

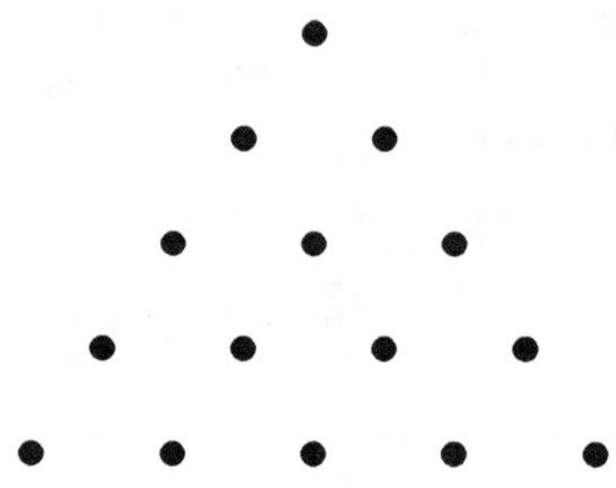

Figure 3: Triangular arrangement of 15 points.

For this reason, 15 is called a triangular number. The first four triangular numbers are 1, 3, 6, and 10 whose arrangements are given in Figure 4. If T_n denotes the nth triangular number, then by construction $T_{n+1} = T_n + n + 1$ with $T_1 = 1$. It is well known that these numbers can be represented as follows:

$$T_n = \sum_{i=1}^{n} i = \frac{n(n+1)}{2} = \binom{n+1}{2}.$$

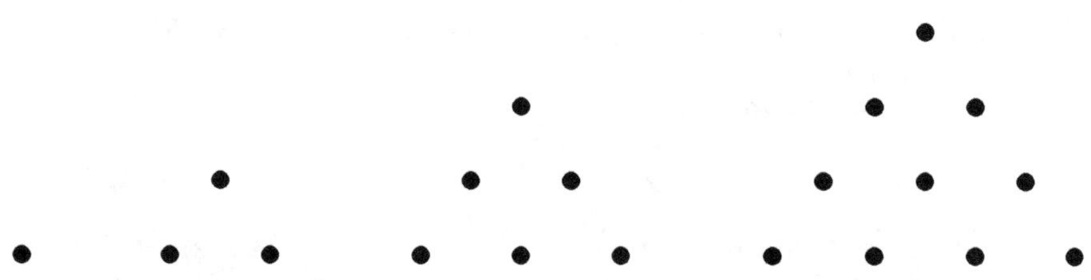

Figure 4: Triangular arrangement of 1, 3, 6, and 10 points

Let $\mathbb{N} = \{1, 2, 3, \ldots\}$ denote the set of natural numbers. A subset X of $\mathbb{N}$ is *triangular* if $|X|$ is a triangular number. Let $\{x_1, x_2, x_3, \ldots, x_{T_n}\}$ be an increasing enumeration of X. The numbers in X can be naturally arranged into a triangle with n levels as shown in Figure 5.

Let $\triangle$ denote the collection of all triangular subsets of $\mathbb{N}$. For $k \in \mathbb{N}$, let $\triangle_k$ denote the triangular sets with k levels, i.e., those subsets of $\mathbb{N}$ such that $|X| = T_k$.

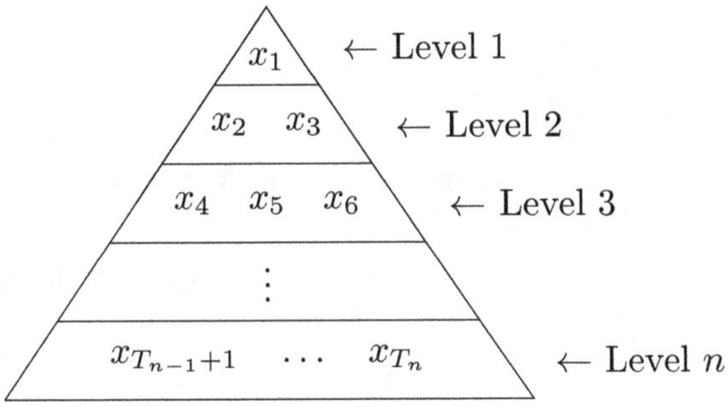

Figure 5: Triangular arrangement of $\{x_1, x_2, x_3, \ldots, x_{T_n}\}$

The next partial order is an adaptation of the order on $\mathcal{R}_1$ considered by Dobrinen and Todorcevic in [3] to our current setting. It can be seen as a restriction (to triangular sets) of the partial order used by Laflamme, which inspired the work in [3], to study complete combinatorics in [12].

Definition 1. For $X, Y \in \triangle$, $X \leqslant Y$ means that $X \subseteq Y$ and every level of X is contained in a single distinct level of Y.

For example, if we let $W = \{1, 2, 3, 4, 5, 6\}$, $X = \{2, 4, 5\}$, $Y = \{3, 5, 6\}$, and $Z = \{5\}$, then $W \in \triangle_3$, $X, Y \in \triangle_2$, and $Z \in \triangle_1$ such that $Z \subseteq X, Y$ and $X, Y \subseteq W$. Each level of X and Y are contained in a distinct single level of W and Z, being only one element, is contained in X and Y. Figure 6 displays this configuration and the associated Hasse diagram in the partial order $(\triangle, \leq)$.

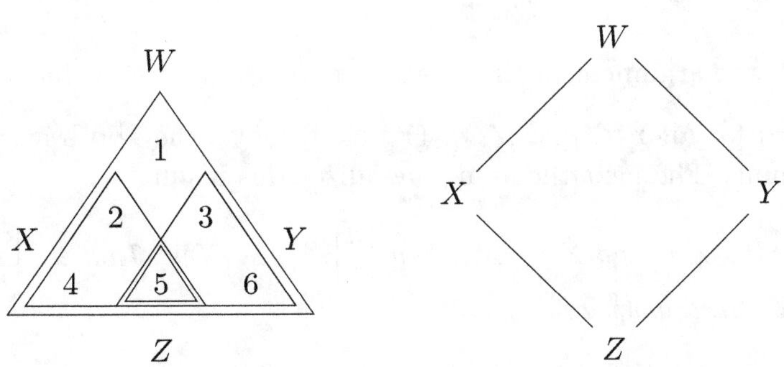

Figure 6: Hasse diagram and possible configuration for W, X, Y and Z

Definition 2. For $k \in \mathbb{N}$ and $X \in \triangle$, let $\triangle_k(X) = \{Y \in \triangle_k : Y \leq X\}$.

Suppose $X \in \triangle_3$ and let $\{x_1, x_2, x_3, x_4, x_5, x_6\}$ be an increasing enumeration of X. Then $\triangle_2(X)$ contains the elements $\{x_1, x_2, x_3\}$, $\{x_1, x_4, x_5\}$, $\{x_1, x_4, x_6\}$,

$\{x_1, x_5, x_6\}$, $\{x_2, x_4, x_5\}$, $\{x_2, x_4, x_6\}$, $\{x_2, x_5, x_6\}$, $\{x_3, x_4, x_5\}$, $\{x_3, x_4, x_6\}$, and $\{x_3, x_5, x_6\}$. Thus for any $X \in \triangle_3$, $|\triangle_2(X)| = 10$.

Note that if $Y \in \triangle_4$, then $|\triangle_3(Y)| = 41$. To see this, note that there are $\binom{4}{3} = 4$ ways to choose three elements from the last row of Y. Each one of these possibilities can be added onto any element of $\triangle_2(Y')$ where Y' is the element of $\triangle_3$ obtained by removing the last level of Y to obtain a distinct element of $\triangle_3(Y)$. In particular, there are $\binom{4}{3} \cdot |\triangle_2(X)| = 40$ elements of $\triangle_3(Y)$ obtained this way. The only other element of $\triangle_3(Y)$ is Y'. So $|\triangle_3(Y)| = \binom{4}{3} \cdot |\triangle_2(Y')| + 1 = 41$. By a similar argument, one can show that for all $k \in \mathbb{N}$ and for all $Y \in \triangle_k$ if $Y' \in \triangle_{k-1}$, then

$$|\triangle_{k-1}(Y)| = 1 + k|\triangle_{k-2}(Y')|.$$

To better express these types of combinatorial relationships we introduce a variant of the binomial coefficient $\binom{n}{m}$. The next definition should be contrasted with the recursive definition of the binomial coefficients using Pascal's Triangle.

Definition 3.

$$\begin{cases} \begin{bmatrix} n \\ k \end{bmatrix} = \begin{bmatrix} n-1 \\ k \end{bmatrix} + \binom{n}{k} \begin{bmatrix} n-1 \\ k-1 \end{bmatrix} \\ \\ \begin{bmatrix} n \\ 0 \end{bmatrix} = 1, \begin{bmatrix} n \\ n \end{bmatrix} = 1 \end{cases}$$

With this definition, note that for all natural numbers k

$$\begin{bmatrix} k \\ k-1 \end{bmatrix} = \begin{bmatrix} k-1 \\ k-1 \end{bmatrix} + \binom{k}{k-1} \begin{bmatrix} k-1 \\ k-2 \end{bmatrix} = 1 + k \begin{bmatrix} k-1 \\ k-2 \end{bmatrix}.$$

Since $\begin{bmatrix} 2 \\ 1 \end{bmatrix} = 1$, the argument in the previous paragraph implies that for all natural numbers k and for all $Y \in \triangle_k$, $|\triangle_{k-1}(Y)| = \left[\begin{smallmatrix} k \\ k-1 \end{smallmatrix}\right]$ as they both satisfy the same recursive formula. The next theorem generalizes this result.

Theorem 1. *If $k < n$ and $X \in \triangle_n$, then $\begin{bmatrix} n \\ k \end{bmatrix} = |\triangle_k(X)|$. That is, $\begin{bmatrix} n \\ k \end{bmatrix}$ counts the number of $\triangle_k$'s in a given $\triangle_n$.*

Proof. Let $k < n$ and $X \in \triangle_n$. We show that $|\triangle_k(X)|$ satisfies the same recursive definition as $\left[\begin{smallmatrix} n \\ k \end{smallmatrix}\right]$. It is clear that $|\triangle_n(X)| = 1$. If we consider $\emptyset$ to be the only element of $\triangle_0$ then $|\triangle_0(X)| = 1$. Thus, the base cases of the recursions are the same. We complete the proof by verifying that $|\triangle_k(X)| = |\triangle_k(X')| + \binom{n}{k} |\triangle_{k-1}(X')|$ where X' is the triangular set in $\triangle_{n-1}(X)$ obtained by removing the last level of X.

Note that $|\triangle_k(X)|$ can be thought of as the number of $\triangle_k$'s in a $\triangle_n$. It should be clear that $|\triangle_k(X)| = |\triangle_k(X')| + c$, where c is the amount of new $\triangle_k' s$ formed when the last level of X is added back to X'. Each $\triangle_k$ contributing to c must have its last level in the last level of X. There are $\binom{n}{k}$ possibilities for those k points in the final level of X. For each collection of k points in the last level of X there are $|\triangle_{k-1}(X')|$ possibilities for triangular sets in $\triangle_k(X)$ whose last level is the given k points. Therefore, $c = \binom{n}{k}|\triangle_{k-1}(X')|$ and $|\triangle_k(X)| = |\triangle_k(X')| + \binom{n}{k}|\triangle_{k-1}(X')|$. □

Later in the paper we use the next corollary to obtain estimates for upper and lower bounds on $\left[{n \atop k}\right]$. These estimates are needed to apply the probabilistic method to our combinatorial game and obtain lower bounds on triangular Ramsey numbers.

Corollary 1. *For* $0 < k < n$, $\displaystyle \left[{n \atop k}\right] = \sum_{0<i_1<\cdots<i_k\leq n} \left(\prod_{j=1}^{k} \binom{i_j}{j} \right).$

Proof. We show that $\sum\limits_{0<i_1<\cdots<i_k\leq n} \left(\prod\limits_{j=1}^{k} \binom{i_j}{j} \right)$ satisfies the same recursive definition as $\left[{n \atop k}\right]$. By the previous theorem, $\sum\limits_{0<i_1\leq n} \binom{i_1}{1} = T_n = \left[{n \atop 1}\right]$. Further,

$$\sum_{0<i_1<\cdots<i_n\leq n} \left(\prod_{j=1}^{n} \binom{i_j}{j} \right) = \binom{1}{1}\binom{2}{2}\cdots\binom{n}{n} = 1 = \left[{n \atop n}\right].$$

Thus, the base case of the two recursions are the same. Note that

$$\sum_{0<i_1<\cdots<i_k\leq n-1} \left(\prod_{j=1}^{k} \binom{i_j}{j} \right) + \binom{n}{k} \cdot \sum_{0<i_1<\cdots<i_{k-1}\leq n-1} \left(\prod_{j=1}^{k-1} \binom{i_j}{j} \right)$$
$$= \sum_{0<i_1<\cdots<i_k\leq n-1} \left(\prod_{j=1}^{k} \binom{i_j}{j} \right) + \sum_{0<i_1<\cdots<i_{k-1}\leq n-1, i_k=n} \left(\prod_{j=1}^{k} \binom{i_j}{j} \right).$$

The right hand side of the previous equation is just the sum $\sum\limits_{0<i_1<\ldots<i_k\leq n} \left(\prod\limits_{j=1}^{k} \binom{i_j}{j} \right)$ broken into the parts where $i_k = n$ in the second sum and where it isn't in the first sum, so the equality holds. In particular, the formula satisfies the same recursion formula as $\left[{n \atop k}\right]$. □

4. The Game of Mines

The most basic variant of the game of Mines, denoted by Mines_3, is played on a game board of size $\triangle_3$ with two players each assigned one of two markings and/or colors. Two example game boards are given in Figure 7 where the numbers represent the positions that can be marked. The players have alternating turns in which they may choose to mark one position on the game board or none at all. The game ends when all positions have been played.

Definition 4. Let X denote the set of moves made by Player 1 and Y the set of moves made by Player 2. Let x_m denote a move by Player 1 on position m and y_m a move by Player 2 on position m.

At any point in the game we have $X \cap Y = \emptyset$. If both players always choose to mark a position, then given that Player 1 goes first, we also have $|X| = |Y| - 1$ or $|X| = |Y|$, and $|X| + |Y| = T_3$ when all positions have been played.

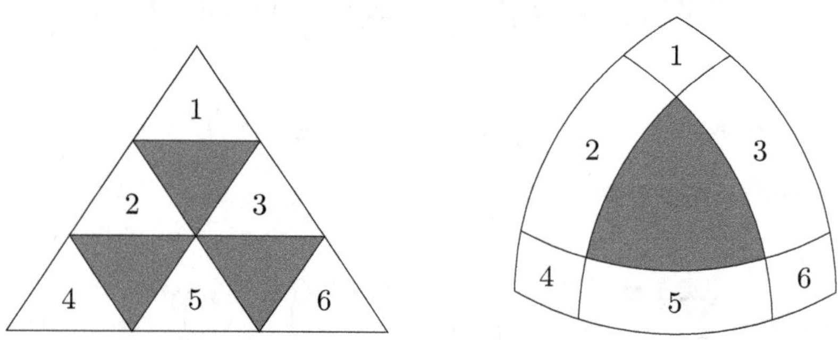

Figure 7: Two alternative gameboards for Mines$_3$.

Player 1 *wins* if $\triangle_2(X) \neq \emptyset$. Player 2 *wins* if $\triangle_2(Y) \neq \emptyset$. A *draw* occurs if all positions have been played but neither player has won the game. That is, $\triangle_2(X) = \triangle_2(Y) = \emptyset$ and $|X| + |Y| = T_3$.

Figure 8 describes a possible game such that $X = \{x_2, x_3, x_4\}$ and $Y = \{y_1, y_5, y_6\}$. Here $\triangle_2(Y) \neq \emptyset$ since $\{y_1, y_5, y_6\} \in \triangle_2$ indicating that Player 2 has won the game.

Theorem 2. *Both players cannot construct a* $\triangle_2$. *That is, it is impossible for both* $\triangle_2(X) \neq \emptyset$ *and* $\triangle_2(Y) \neq \emptyset$.

Proof. Assume this is not the case. In other words, there exists a situation such that $\triangle_2(X) \neq \emptyset$ and $\triangle_2(Y) \neq \emptyset$. Without loss of generality, we may assume that all positions have been played.

The largest row of the game board has 3 positions, and the largest level of a $\triangle_2$ has 2 positions. Therefore, both X and Y cannot construct their $\triangle_2$'s largest level in the same row of the game board. By the pigeon hole principle there exists $a, b \in \{4, 5, 6\}$ such that $x_a, x_b \in X$ or $y_a, y_b \in Y$. If $x_a, x_b \in X$, then $y_2, y_3 \in Y$. In order for $\triangle_2(X) \neq \emptyset$ we must have $x_1 \in X$ and we find $\triangle_2(Y) = \emptyset$, a contradiction. If instead we had $y_a, y_b \in Y$, then $x_2, x_3 \in X$. Once again, in order for $\triangle_2(X) \neq \emptyset$

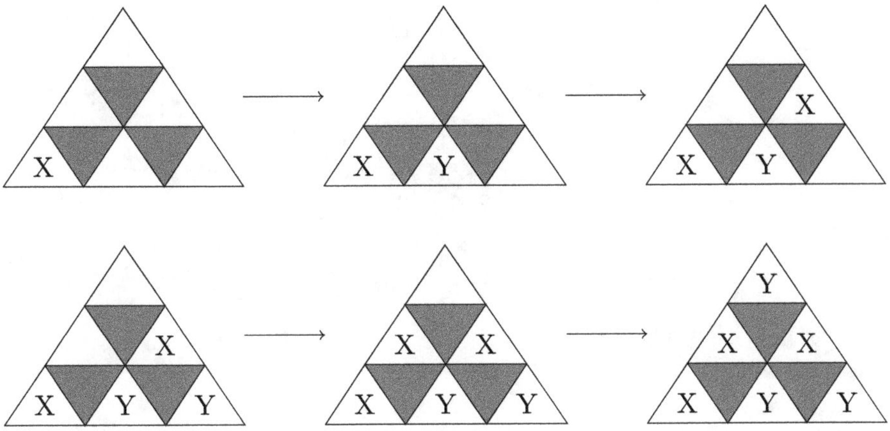

Figure 8: An example game of Mines_3.

we must have $x_1 \in X$ and we find $\triangle_3(Y) = \emptyset$, a contradiction. Therefore, it is impossible for both players to construct a $\triangle_2$. □

Theorem 3. *Mines_3 never ends in a draw.*

Proof. Assume a full game of Mines_3 has been played. By the pigeonhole principle, at least two of the bottom three elements must be of the same color. If any of the three elements above the bottom row are of the same color as the two on the bottom, then a $\triangle_2$ has been constructed in that color. For this not to happen, the three elements in the top two rows must all be in the opposite color. If this is the case, then a $\triangle_2$ has been constructed in the opposite color. Therefore, it is impossible for a game of Mines_3 to be played in which neither player constructs a $\triangle_2$, and no game can be played in which both players construct a $\triangle_2$. □

An interesting variation of the game play exploits the fact that the game board has 120° rotational symmetry about its center. There are three directions to the game board given by the perpendicular from any of the three edges to its adjacent vertex. We let D_n denote the orientation of the game board in the n direction (see Figure 9). We use the notation $\triangle_{k,n}(X)$ to denote all $\triangle_k$'s on the game board in the D_n direction. For example, $\{1,3,4\} \in \triangle_{2,2}(X)$ but $\{1,3,4\} \notin \triangle_{2,1}(X)$. Note that if $\triangle_{k,n}(X) \neq \emptyset$, then there exists a $\triangle_k$ in the direction of D_n contained in X. Likewise, if $\triangle_{k,n}(Y) \neq \emptyset$, then there exists a $\triangle_k$ in the direction of D_n contained in Y. The player that wins in two of the three directions wins this variation of the game.

By Theorem 2 and Theorem 3 we see that there can never be a draw in a single direction. In addition, given that there are three directions one player must win in at least two directions. Therefore this variation of Mines_3 can never end in a draw.

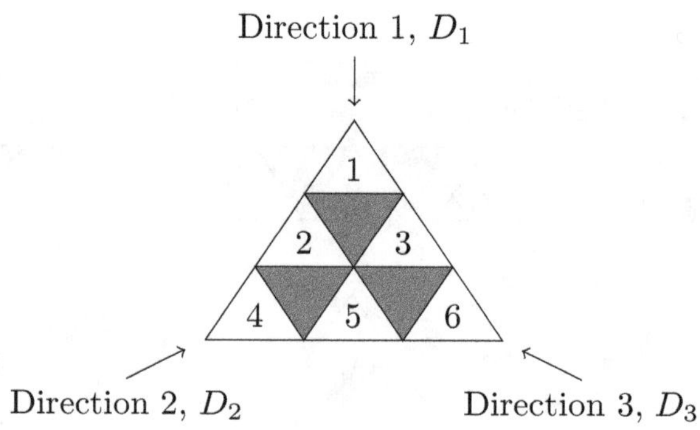

Figure 9: A game board for Mines_3 with directions D_1, D_2 and D_3 labeled.

The next theorem is true for both variations of Mines_3. We give the proof for the omnidirectional case as it is more informative.

Theorem 4. *Player 1 has a winning strategy for* Mines_3.

Proof. In the first three moves of the game we have $|X| = 2$ and $|Y| = 1$. Therefore, Player 1 can guarantee there exists $q, r \in \{1, 4, 6\}$ such that $x_q, x_r \in X$.

By rotating the game board 120° to the left or right, we can, without loss of generality, assume $x_4, x_6 \in X$. At this point in the game $\{y_1, y_2, y_3\} \not\subseteq Y$ and $\{x_1, x_2, x_3\} \not\subseteq X$ since $|X| + |Y| = 3$. On the next move Player 1 plays position 1, 2 or 3, whichever is available. Player 1 wins in direction D_1 since $\triangle_{3,1}(X) \neq \emptyset$. Thus, Player 1 only needs to win in one other direction to win the game. If $x_1 \in X$ then, $\{x_1, x_4, x_6\} = X$ and $\triangle_{3,2}(X) \neq \emptyset$. If $x_2 \in X$, then $\{x_2, x_4, x_6\} = X$ and $\triangle_{3,3}(X) \neq \emptyset$. If $x_3 \in X$, then $\{x_3, x_4, x_6\} = X$ and $\triangle_{3,2}(X) \neq \emptyset$. Thus, in any case, Player 1 wins in at least two out of the three directions. □

4.1. The Game of $\text{Mines}_m(p, q, k)$

Let p, q, m and k be positive integers with $k \leq p, q \leq m$. The game $\text{Mines}_m(p, q, k)$ is played on a game board of size $\triangle_m$ with two players each assigned one of two markings and/or colors. Players have alternating turns in which they may choose to mark one position, in this case positions are the $\triangle_k$'s on the game board, or not mark a position. If $p = q = n$ then we use the notation $\text{Mines}_m(n, k)$. If $p = q = n$ and $k = 1$ then we use the notation $\text{Mines}_m(n)$. In this notation, $\text{Mines}_3 = \text{Mines}_3(2) = \text{Mines}_3(2, 1) = \text{Mines}_3(2, 2, 1)$ since the positions played by both players in Mines_3 are the $\triangle_1$'s on the game board.

We again let $X \subseteq \triangle_k$ denote the set of moves made by Player 1 and $Y \subseteq \triangle_k$ the set of moves made by Player 2. At any point in the game we have $X \cap Y = \emptyset$. If both players always choose to mark a position then given that Player 1 goes first we also have $|X| = |Y| - 1$ or $|X| = |Y|$, and $|X| + |Y| = \begin{bmatrix} m \\ k \end{bmatrix}$ when all positions have been played. Player 1 *wins* if they construct a $Z \in \triangle_p$ such that $\triangle_k(Z) \subseteq X$ before the second player is able to construct a $Z \in \triangle_q$ such that $\triangle_k(Z) \subseteq Y$. Player 2 *wins* if they construct a $Z \in \triangle_q$ such that $\triangle_k(Z) \subseteq Y$ before the first player is able to construct a $Z \in \triangle_p$ such that $\triangle_k(Z) \subseteq X$. A *draw* occurs if all positions have been played but neither player has won the game. That is, $|X| + |Y| = \begin{bmatrix} m \\ k \end{bmatrix}$ and for all $Z \in \triangle_n$, $\triangle_k(Z) \nsubseteq X$ and $\triangle_k(Z) \nsubseteq Y$. The game ends when all positions have been played or one of the players wins.

For some small game boards it is unnecessary to keep track of which player first constructs the winning triangular set. Instead the players can simply fill out the game board completely, then check to see who wins. For example, $\text{Mines}_5(3) = \text{Mines}_5(3,3,1)$ has this property. Figure 10 gives two examples of game boards for $\text{Mines}_5(n)$.

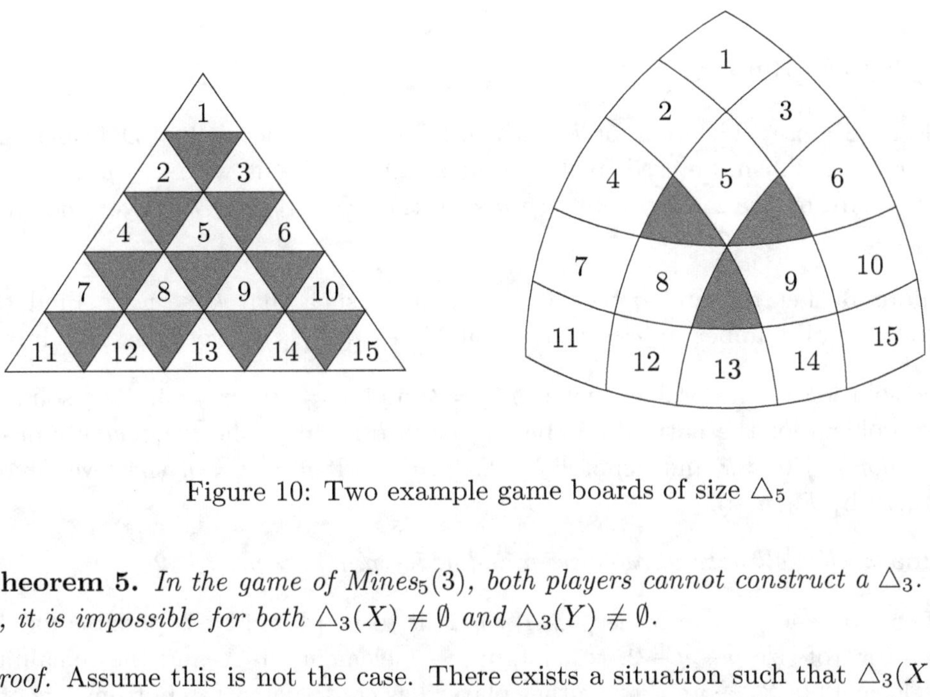

Figure 10: Two example game boards of size $\triangle_5$

Theorem 5. *In the game of* $\text{Mines}_5(3)$*, both players cannot construct a* $\triangle_3$*. That is, it is impossible for both* $\triangle_3(X) \neq \emptyset$ *and* $\triangle_3(Y) \neq \emptyset$.

Proof. Assume this is not the case. There exists a situation such that $\triangle_3(X) \neq \emptyset$ and $\triangle_3(Y) \neq \emptyset$. The largest level of a $\triangle_5$ has 5 positions, and the largest level of a $\triangle_3$ has 3 positions. Therefore, both X and Y cannot construct their $\triangle_3$'s largest level in the same row of the game board. By the pigeonhole principle there exist $a, b, c \in \{11, 12, 13, 14, 15\}$ such that $x_a, x_b, x_c \in X$ or $y_a, y_b, y_c \in Y$. Suppose that $x_a, x_b, x_c \in X$. We have the following two cases:

Case 1: There exist $d, e, f \in \{7, 8, 9, 10\}$ such that $y_d, y_e, y_f \in Y$. Since $\triangle_3(Y) \neq \emptyset$, either there exist $g, h \in \{4, 5, 6\}$ such that $y_g, y_h \in Y$ or $y_2, y_3 \in Y$. If $g, h \in \{4, 5, 6\}$, then $x_2, x_3 \in X$. If $y_2, y_3 \in Y$, then there exist $i, j \in \{4, 5, 6\}$ such that

$x_i, x_j \in X$. In order for $\triangle_3(X) \neq \emptyset$ we must have $x_1 \in X$ which causes $\triangle_3(Y) = \emptyset$, a contradiction.

Case 2: There exist $d, e, f \in \{4, 5, 6\}$ such that $y_d, y_e, y_f \in Y$. Then either there exist $g, h \in \{7, 8, 9, 10\}$ such that $x_g, x_h \in X$ or $x_2, x_3 \in X$. If $x_2, x_3 \in X$, then clearly $\triangle_3(Y) \neq \emptyset$. If there exist $g, h \in \{7, 8, 9, 10\}$ such that $x_g, x_h \in X$, then we must have $y_2, y_3 \in Y$. In order for $\triangle_3(X) \neq \emptyset$ we must have $x_1 \in X$ which causes $\triangle_3(Y) = \emptyset$, a contradiction.

If instead $y_a, y_b, y_c \in Y$, then a similar argument gives a contradiction. So it is impossible for both $\triangle_3(X) \neq \emptyset$ and $\triangle_3(Y) \neq \emptyset$. □

The previous theorem can be extended, in the game of $\text{Mines}_{2n-1}(n)$ both players cannot construct a $\triangle_n$. That is, it is impossible for both $\triangle_n(X) \neq \emptyset$ and $\triangle_n(Y) \neq \emptyset$. We leave the proof to the interested reader.

5. Triangular Ramsey Numbers

The Finite Ramsey Theorem for $\mathcal{R}_1$, which follows from the work of Dobrinen and Todorcevic in [3], can be used to show that for all $p, q, k \in \mathbb{N}$ with $k \leq p, q$ there is a game board of size $\triangle_m$ with $m \geq p, q$ such that $\text{Mines}_m(p, q, k)$ never ends in a draw.

Problem 3. Let p, q and k be natural numbers such that $k \leq p, q$. Find the smallest natural number $m \geq p, q$ such that $\text{Mines}_m(p, q, k)$ never ends in a draw.

The solution to the problem when $p, q = 2$ and $k = 1$ is $m = 3$. The solution to the problem for the natural numbers p, q and k is called the *triangular Ramsey number for* p, q *and* k and denoted by $\mathcal{R}_1(p, q, k)$. If $p = q = n$ then we denote $\mathcal{R}_1(p, q, k)$ by $\mathcal{R}_1(n, k)$.

Lemma 2. *For all natural numbers* p *and* q, $\mathcal{R}_1(p, q, 1) > p + q - 2$.

Proof. Suppose a game of $\text{Mines}_{p+q-2}(p, q, 1)$ is to be played. It is possible for the bottom row, of size $p + q - 2$, to contain $p - 1$ elements in X and the remaining $q - 1$ elements in Y. Note that neither player has constructed the bottom row of a winning triangle in the bottom row. Since the rows decrease in size as players move up the triangle, it is possible for both players to fail to construct the bottom row of their winning triangle anywhere on the board. Therefore, $\mathcal{R}_1(p, q, 1) > p+q-2$. □

Theorem 6. *For all natural numbers* p *and* q, $\mathcal{R}_1(p, q, 1) = p+q-1$. *In particular, for all numbers* n, $\mathcal{R}_1(n, 1) = 2n - 1$.

Proof. By the previous lemma, $\mathcal{R}_1(p, q, 1) > p + q - 2$. The result follows by showing via induction on $p + q$ that $\mathcal{R}_1(p, q, 1) \leq p + q - 1$. The base case occurs when $p = q = 1$, i.e., $p + q = 2$. The base case is trivial; the first person to color an element creates a complete $\triangle_1$ and wins. In other words, $\mathcal{R}_1(1, 1, 1) = 1$.

Now, suppose $p + q = n + 1$ and the results holds when $p + q = n$. Note that $(p - 1) + q = n$ and $p + (q - 1) = n$. Thus by the inductive hypothesis, $\mathcal{R}_1(p - 1, q) = \mathcal{R}_1(p, q - 1) \leq p + q - 2$. We can now prove by contradiction that no game of $\text{Mines}_{p+q-1}(p, q, 1)$ ends in a draw. Towards a contradiction, suppose a full game of $\text{Mines}_{p+q-1}(p, q, 1)$ has been played on a board of height n and ends in a draw, i.e., $\triangle_p(X) = \emptyset$ and $\triangle_q(Y) = \emptyset$. Since $p + q - 1 = n$ and the bottom level of the game board contains n positions, the pigeon hole principle mandates that the bottom row of the game board must contain either p elements in X, considered *Case 1*, or q elements in Y, considered *Case 2*.

Case 1: The bottom row of the game board contains Z', a set of p elements in X. By the equation above, $\mathcal{R}_1(p - 1, q) = p + q - 1$, and the inductive hypothesis, we can see that there exists either $Z \in \triangle_{p-1}(X) \neq \emptyset$ or $\zeta \in \triangle_q(Y) \neq \emptyset$. If the first is the case, then $Z \cup Z' \in \triangle_p(X)$ and the game is won by Player 1. If the second is the case, then the game is won by Player 2. In either, we have a contradiction.

Case 2: The bottom row of the game board contains q elements in Y. By a similar method to that used above, we can show that either $Z \in \triangle_{q-1}(Y) \neq \emptyset$ or $\zeta \in \triangle_p(X)$ exists. In either situation, a fully colored triangle is made, and we have a contradiction.

In either case, we obtain a contradiction. Therefore, $\mathcal{R}_1(p, q, 1) \leq p + q - 1$ when $p + q = n + 1$. □

Next, we define a sequence $(M_{n,k})$ which we use to establish upper bounds for triangular Ramsey numbers. We let $R(n, k)$ denote the smallest number of vertices of a complete graph such that, for any coloring of its complete subgraphs with k vertices with two colors, there exists a complete subgraph with n vertices where the coloring is monochromatic. The existence of these Ramsey numbers also follows from Ramsey's theorem and could also be introduced by generalizing the game of Tri. Here we let $R^l(n, k)$ denote $\underbrace{R(R(\cdots(R}_{l-times}(n, k), k)\cdots), k)$.

Theorem 7. *Let $(M_{n,k})$ be the sequence recursively defined by*

$$\begin{cases} M_{n,k} = n & \text{if } k = 1, \\ M_{n,k} = n & \text{if } n = k, \\ M_{n+1,k} = R^{\left\lceil \frac{\mathcal{R}_1(M_{n,k}, k-1)}{k-1} \right\rceil}(n+1, k) & \text{if } n > k > 1. \end{cases}$$

If $n \geq k$, then $\mathcal{R}_1(n, k) \leq M_{2n-1,k}$.

Proof. We begin by establishing a simpler result by induction on n.

Claim 1. *Suppose that a game of* $\text{Mines}_{M_{n,k}}(n,k)$ *is played to completion. There exists a* $Z \in \triangle_n$ *on the game board such that given any level* i *of* Z*, either all* $\triangle_k$*'s contained in* Z *whose last level is contained in the* i^{th} *level of* Z *are played by Player 1 or all* $\triangle_k$*'s contained in* Z *whose last level is contained in the* i^{th} *level of* Z *are played by Player 2.*

Proof of Claim 1. First note that by the previous theorem $\mathcal{R}_1(n,1) = 2n-1 = M_{2n-1,1}$. Thus, the claim holds when $k=1$. Next, fix $k \geq 2$. Consider the base case when $n = k$ and $M_{n,k} = n$. If a game of $\text{Mines}_n(n,k)$ has been played to completion, then since there is only one playable position on the game board, the base case holds trivially.

Assume that the claim holds for $M_{n,k}$. Suppose that a game of $\text{Mines}_{M_{n+1,k}}(n+1,k)$ has been played to completion. As usual, let X denote the moves made by Player 1 and Y denote those made by Player 2. For each element $Z \in \triangle_{k-1}$ in the first $\mathcal{R}_1(M_{n,k}, k-1)$ levels of the game board, we play a game of $\text{Tri}_{M_{n+1,k}}(n+1,k)$ on the final level of the game board as follows: Player 1 plays the k-element set $\{i_1, i_2, \dots i_k\}$ if $Z \cup \{i_1, i_2, \dots, i_k\} \in X$ and Player 2 plays the k-element set $\{i_1, i_2, \dots i_k\}$ if $Z \cup \{i_1, i_2, \dots, i_k\} \in Y$. By an argument similar to the proof of Lemma 1, there exists an $(n+1)$-element set in the last level of the game board $\{z_1, z_2, \dots z_{n+1}\}$ such that for all $Z \in \triangle_{k-1}$ in the first $\mathcal{R}_1(M_{n,k}, k-1)$ levels of the game board either (†) for all $i_1, i_2, \dots i_k \in \{z_1, z_2, \dots z_{n+1}\}$, $Z \cup \{i_1, i_2, \dots i_k\} \in X$ or (‡) for all $i_1, i_2, \dots i_k \in \{z_1, z_2, \dots z_{n+1}\}$, $Z \cup \{i_1, i_2, \dots i_k\} \in Y$.

Next, consider the following hypothetical game of $\text{Mines}_{\mathcal{R}_1(M_{n,k},k-1)}(M_{n,k}, k-1)$ played on the first $\mathcal{R}_1(M_{n,k}, k-1)$ levels of the our original game board. Let $\bar{X}$ denote the moves made by Player 1 and $\bar{Y}$ denote those made by Player 2. In this game, Player 1 plays position $Z \in \triangle_{k-1}$ if (†) holds, and Player 2 plays position $Z \in \triangle_{k-1}$ if (‡) holds. By definition, this game does not end in a draw. If Player 1 wins this game, then there exists $W \in \triangle_{M_{n,k}}$ such that all $\triangle_k$'s whose first $k-1$ levels are in W and whose last level is contained in $\{z_1, z_2, \dots z_{n+1}\}$ are played by Player 1. If Player 2 wins this game, then there exists $W \in \triangle_{M_{n,k}}$ such that all $\triangle_k$'s whose first $k-1$ levels are in W and whose last level is contained in $\{z_1, z_2, \dots z_{n+1}\}$ are played by Player 2.

By the induction hypothesis there exists a $Z' \in \triangle_n(W)$ such that given any level i of Z', either all $\triangle_k$'s contained in Z' whose last level is contained in the ith level of Z' are played by Player 1, or all $\triangle_k$'s contained in Z' whose last level is contained in the ith level of Z' are played by Player 2.

Let $Z'' = Z' \cup \{z_1, z_2, \dots z_{n+1}\} \in \triangle_{n+1}$. Then, either all $\triangle_2$'s contained in Z'' whose last level is contained in the ith level of Z'' are played by Player 1, or all $\triangle_2$'s contained in Z'' whose last level is contained in the ith level of Z'' are played by Player 2. Therefore, the claim holds by induction. □

To prove the inequality, assume toward a contradiction that a game of $\text{Mines}_{M_{2n-1,k}}$

ends in a draw. By the previous claim, there exists a $Z \in \triangle_{2n-1}$ on the game board such that given any level i of Z, either all $\triangle_k$'s contained in Z whose last level is contained in the ith level of Z are played by Player 1 or all $\triangle_k$'s contained in Z whose last level is contained in the ith level of Z are played by Player 2. By the pigeonhole principle there are either at least n levels where Player 1 plays all $\triangle_k$'s or at least n levels where Player 2 plays all $\triangle_k$'s. If there are at least n levels where Player 1 wins, then any $W \in \triangle_n$ whose levels come from these n levels witnesses a win for Player 1, a contradiction. Similarly, if there are at least n levels where Player 2 wins, then any $W \in \triangle_n$ whose levels come from these n levels witnesses a win for Player 2, a contradiction. Therefore, this game could not have ended in a draw. □

When $n = k+1$ the previous proof can be simplified and we obtain smaller upper bounds. In fact, in this special case, induction on n is unnecessary.

Theorem 8. *Suppose that $k \geq 2$. Then*

$$\mathcal{R}_1(k+1,k) \leq R^{\lceil \frac{\mathcal{R}_1(k+1,k-1)}{k-1} \rceil}(k+1,k).$$

Proof. Let $M = R^{\lceil \frac{\mathcal{R}_1(k+1,k-1)}{k-1} \rceil}(k+1,k)$. Toward a contradiction, suppose that a game of $\text{Mines}_M(k+1,k)$ ends in a draw. As usual, let X denote the moves made by Player 1 and Y denote those made by Player 2. For each element $Z \in \triangle_{k-1}$ in the first $\mathcal{R}_1(k+1,k-1)$ levels of the game board, we play a game of $\text{Tri}_M(k+1,k)$ on the final level of the game board as follows: Player 1 plays the k-element set $\{i_1, i_2, \dots i_k\}$ if $Z \cup \{i_1, i_2, \dots, i_k\} \in X$ and Player 2 plays the k-element set $\{i_1, i_2, \dots i_k\}$ if $Z \cup \{i_1, i_2, \dots, i_k\} \in Y$. By Lemma 1, there exists a $(k+1)$-element set in the last level of the game board $\{z_1, z_2, \dots z_{k+1}\}$ such that for all $Z \in \triangle_{k-1}$ in the first $\mathcal{R}_1(k+1,k-1)$ levels of the game board, either (†) for all $i_1, i_2, \dots i_k \in \{z_1, z_2, \dots z_{k+1}\}$, $Z \cup \{i_1, i_2, \dots i_k\} \in X$ or (‡) for all $i_1, i_2, \dots i_k \in \{z_1, z_2, \dots z_{k+1}\}$, $Z \cup \{i_1, i_2, \dots i_k\} \in Y$.

Next, consider the following hypothetical game of $\text{Mines}_{\mathcal{R}_1(k+1,k-1)}(k+1,k-1))$. Let $\bar{X}$ denote the moves made by Player 1 and $\bar{Y}$ denote those made by Player 2. In this game, Player 1 plays position Z if (†) holds and Player 2 plays position Z if (‡) holds. In other words, $\bar{X} = \{Z \in \triangle_{k-1} : Z \cup \{i_1, i_2, \dots i_k\} \in X\}$ and $\bar{Y} = \{Z \in \triangle_{k-1} : Z \cup \{i_1, i_2, \dots i_k\} \in Y\}$. By definition, this game does not end in a draw.

Suppose that Player 1 wins and let W be some element of $\triangle_{k+1}$ witnessing a win for Player 1 (in our hypothetical game of $\text{Mines}_{\mathcal{R}_1(k+1,k-1)}(k+1,k-1))$. Note that not all $\triangle_k's$ in W are played by Player 2 (in our original game) because otherwise the game would not have ended in a draw. So, without loss of generality, we can assume that there is at least one $W' \in \triangle_k(W)$ played by Player 1 (in the original game). However, this is a contradiction because $W' \cup \{z_1, z_2, \dots z_{k+1}\}$ then witnesses a win for Player 1 (in our original game). If instead Player 2 wins, we can let $\bar{Z}$ be some element of $\triangle_3(\bar{Y})$ and we obtain a similar contradiction. □

5.1. The Probabilistic Method

The work in this section follows closely from the probabilistic method described by Erdös [6]. In our case, we apply it to a randomly played game of Mines.

Theorem 9. *Let p, q, k and m be natural numbers such that $k \leq p \leq q \leq m$. If $\begin{bmatrix} m \\ p \end{bmatrix} \cdot 2^{-\left[\begin{smallmatrix} p \\ k \end{smallmatrix}\right]} + \begin{bmatrix} m \\ q \end{bmatrix} \cdot 2^{-\left[\begin{smallmatrix} q \\ k \end{smallmatrix}\right]} < 1$, then $\mathcal{R}_1(p,q,k) > m$.*

Proof. Let p, q, k and m be given. To use the probabilistic method, we consider two players playing a game of $\text{Mines}_m(p,q,k)$ with a fair coin. The players pick a position and flip the coin to see if they will play that triangular set or skip their turn. The game is then played to completion using these random moves.

Consider the random variables X and Y where X counts the number of winning $\triangle_p$'s colored by Player 1 and Y the number of winning $\triangle_q$'s colored by Player 2. By Theorem 1, the probability that a randomly chosen $\triangle_p$ from the game board witnesses a win for Player 1 is $2^{-\left[\begin{smallmatrix} p \\ k \end{smallmatrix}\right]}$ and for a randomly chosen $\triangle_q$ the probability that it witnesses a win for Player 2 is $2^{-\left[\begin{smallmatrix} q \\ k \end{smallmatrix}\right]}$. By Theorem 1, there are exactly $\begin{bmatrix} m \\ p \end{bmatrix}$ possible $\triangle_p$'s on the game board and $\begin{bmatrix} m \\ q \end{bmatrix}$ possible $\triangle_q$'s. Thus, $\begin{bmatrix} m \\ p \end{bmatrix} \cdot 2^{1-\left[\begin{smallmatrix} p \\ k \end{smallmatrix}\right]} = E[X]$ and $\begin{bmatrix} m \\ q \end{bmatrix} \cdot 2^{1-\left[\begin{smallmatrix} q \\ k \end{smallmatrix}\right]} = E[Y]$. Hence, $E[X+Y] = E[X]+E[Y] = \begin{bmatrix} m \\ p \end{bmatrix} \cdot 2^{-\left[\begin{smallmatrix} p \\ k \end{smallmatrix}\right]} + \begin{bmatrix} m \\ q \end{bmatrix} \cdot 2^{-\left[\begin{smallmatrix} q \\ k \end{smallmatrix}\right]} < 1$. Since the expected value of $X+Y$ is less than one there exists some game of $\text{Mines}_m(p,q,k)$ that ends in a draw. Therefore, $m < \mathcal{R}_1(p,q,k)$. □

The previous theorem can be used to find lower bounds for small values of p, q and k by searching for the largest value of m that satisfies the inequality. A summary of these values for small p, q and k can be found in Figure 11. The next theorem provides asymptotic estimates for large values of p, q and k with $p = q = n$ which don't require the computation of $\begin{bmatrix} n \\ k \end{bmatrix}$ nor $\begin{bmatrix} m \\ n \end{bmatrix}$ for any m.

Theorem 10. *For all natural numbers n and k with $n \leq k$,*

$$\mathcal{R}_1(n,k) \geq (2\pi T_n)^{\frac{1}{4T_n}} \cdot \sqrt{\frac{2T_n}{e}} \cdot 2^{\frac{n^k - k^k}{2k^k T_n}} - 1.$$

Proof. Let $m = \mathcal{R}_1(n,k)$. By the previous theorem, we must have $\begin{bmatrix} m \\ n \end{bmatrix} 2^{-\left[\begin{smallmatrix} n \\ k \end{smallmatrix}\right]+1} \geq 1$. Now we also know that

$$\begin{bmatrix} m \\ n \end{bmatrix} \leq \binom{T_m}{T_n}.$$

This is because if we are counting $\triangle_n$'s (which has T_n points) in a $\triangle_m$ (which has T_m points) we cannot exceed the number of ways of choosing T_n points arbitrarily from T_m points. Therefore, we are counting some T_n sized subset of T_m points with certain properties. We know that $\binom{m}{n} \leq \frac{m^n}{n!}$. Substituting in these inequalities gives,

$$\frac{T_m^{T_n}}{T_n!} \cdot 2^{-\left[{n \atop k}\right]+1} \geq 1$$

$$\frac{(m(m+1))^{T_n}}{T_n!} \cdot 2^{-\left[{n \atop k}\right]+1-T_n} \geq 1.$$

Then by Sterling's formula since $n! > \sqrt{2\pi n} \cdot \left(\frac{n}{e}\right)^n$ we have

$$\frac{(m(m+1))^{T_n}}{\sqrt{2\pi T_n} \cdot T_n^{T_n} \cdot e^{-T_n}} \cdot 2^{-\left[{n \atop k}\right]+1-T_n} \geq 1$$

$$\frac{(m+1)^{2T_n}}{\sqrt{2\pi T_n} \cdot T_n^{T_n} \cdot e^{-T_n}} \cdot 2^{-\left[{n \atop k}\right]+1-T_n} \geq 1$$

$$(m+1)^{2T_n} \geq \sqrt{2\pi T_n} \cdot \left(\frac{T_n}{e}\right)^{T_n} \cdot 2^{\left[{n \atop k}\right]+T_n-1}.$$

From Corollary 1, $\left[{n \atop k}\right] = \sum_{0<i_1<\ldots<i_k\leq n} \left(\prod_{j=1}^{k} \binom{i_j}{j}\right)$. The smallest product in this summation is $\binom{1}{1}\binom{2}{2}\cdots\binom{n}{n} = 1$ since there are $\binom{n}{k}$ terms in the sum $\left[{n \atop k}\right] \geq \binom{n}{k} \geq \left(\frac{n}{k}\right)^k$. So

$$(m+1)^{2T_n} \geq \sqrt{2\pi T_n} \cdot \left(\frac{T_n}{e}\right)^{T_n} \cdot 2^{\left(\frac{n}{k}\right)^k+T_n-1}.$$

Isolating m in the previous inequality and simplifying gives the result. □

6. Conclusion

Recall that the existence of triangular Ramsey numbers follows from the work of Dobrinen and Todorcevic. Our primary purpose for introducing the game of Mines was to provide a simplified presentation of the finite-dimensional Ramsey theory of the infinite-dimensional topological Ramsey space $\mathcal{R}_1$ introduced and studied by Dobrinen and Todorcevic. The next table summarizes the main results for small values of p, q and k. Note that $\mathcal{R}_1(p,q,k) = \mathcal{R}_1(q,p,k)$ so we only give values for $p \leq q$.

These upper and lower bounds have applications to characterizing the Dedekind cuts in nonstandard models of arithmetic that arise from ultrafilter mapping that are associated to the space $\mathcal{R}_1$ introduced by Dobrinen and Todorcevic in [3]. In [1], Blass uses upper and lower bounds for Ramsey numbers to characterize, under the continuum hypothesis, the Dedekind cuts that can be associated to ultrafilter mappings from Ramsey and weakly-Ramsey ultrafilters. Trujillo in [19], characterizes

p	q	k	Lower Bound	$\mathcal{R}_1(p,q,k)$	Upper Bound	Result
1	2	1	–	2	–	Thm. 6
2	2	1	–	3	–	Thm. 6
2	3	1	–	4	–	Thm. 6
3	3	1	–	5	–	Thm. 6
2	2	2	–	2	–	trivial
2	3	2	–	3	–	trivial
3	3	2	6	?	$R^{15}(3)$	Thm. 8
3	4	2	6	?	$M_{7,2}$	Thm. 7&Thm. 9
4	4	2	25	?	$M_{7,2}$	Thm. 7&Thm. 9
3	3	3	–	3	–	trivial
3	4	3	–	4	–	trivial
4	4	3	20	?	$R^{\lceil \frac{M_{7,2}}{2} \rceil}(4,3)$	Thm. 8&Thm. 9
4	5	3	20	?	$M_{9,3}$	Thm. 7&Thm. 9
5	5	3	9.39×10^7	?	$M_{9,3}$	Thm. 7&Thm. 9
4	4	4	–	4	–	trivial
4	5	4	–	5	–	trivial
5	5	4	3425	?	$R^{\lceil \frac{M_{9,3}}{3} \rceil}(5,4)$	Thm. 8&Thm. 9
30	30	20	221	?	$M_{59,20}$	Thm. 7&Thm. 10
35	35	20	4.70^{18}	?	$M_{69,20}$	Thm. 7&Thm. 10
40	40	20	7.29×10^{193}	?	$M_{79,20}$	Thm. 7&Thm. 10

Figure 11: Table of some triangular Ramsey numbers.

the Dedekind cuts that can be associated to ultrafilter mappings among ultrafilters within the Tukey-type of a Ramsey for $\mathcal{R}_1$ ultrafilter. This motivates the following open problems:

Problem 4. Use upper and lower bounds for triangular Ramsey numbers to characterize the Dedekind cuts that can be associated to ultrafilter mappings from Ramsey for $\mathcal{R}_1$ ultrafilters.

In a followup paper, Dobrinen and Todorcevic [4] introduce a hierarchy of spaces $\mathcal{R}_\alpha$ for $\alpha < \omega$ that extend the space $\mathcal{R}_1$.

Problem 5. Introduce a combinatorial game, similar to Mines, that provides a simplified presentation of the finite-dimensional Ramsey theory of the infinite-dimensional topological Ramsey spaces $\mathcal{R}_\alpha$ for $\alpha < \omega_1$ defined and studied by Dobrinen and Todorcevic in [4]. Then find upper and lower bounds for Ramsey numbers based on these games.

Upper and lower bounds for the Ramsey numbers associated to the spaces $\mathcal{R}_\alpha$, $\alpha > 1$, also have similar applications to characterizing the Dedekind cuts that can be associated to ultrafilter mappings from Ramsey for $\mathcal{R}_\alpha$ ultrafilters. In addition to these open problems, there are problems still open related to playing the game of Mines.

Theorem 4 provides an explicit description of a winning strategy for Mines_3. The proof of Zermelo's theorem in [11] can be adapted to show that either Player 1 or Player 2 must have a winning strategy for any game of $\text{Mines}_m(n, k)$ where $m \geq \mathcal{R}_1(n, k)$. A standard strategy stealing argument can then be used to show that Player 1 must have a winning strategy for the game. However, the proof of Zermelo's theorem does not provide for an explicit description of how Player 1 should play to win the game. In the game of $\text{Mines}_5(3, 1)$, Player 1 has the opening move allowing them to guarantee two of the three corners (assuming this is still the optimal strategy with larger game boards) which could give them the win. However, Player 2 has the ability to react to Player 1's moves and possibly prevent them from forming a $\triangle_3$. In addition, some game boards will have an odd number of positions giving Player 1 an additional position over Player 2; perhaps this gives them an even bigger advantage. Regardless, giving an explicit description of the winning strategy for a game board of size $\triangle_m$ seems like a difficult problem.

Problem 6. Find an explicit description of the winning strategy for Player 1 in a game of $\text{Mines}_m(n, k)$ where $m \geq \mathcal{R}_1(n, k)$. In particular, describe the winning strategy for Player 1 in a game of $\text{Mines}_5(3, 1)$.

The complexity of Mines increases as the number of players is increased from 2 players to k players. In this variation, we have a different set of Ramsey numbers. It is evident that as the number of players increases, the size of the associated triangular Ramsey numbers also increase.

Problem 7. How does the game of Mines change when adding more players? In particular, find upper and lower bounds for triangular Ramsey numbers for variations of Mines with more than two players.

In the off-diagonal case of the game $\text{Mines}_m(p, q, 1)$, if $p < q$, then Player 1 has an advantage as they are now constructing a smaller triangular set. Player 2 can be given an advantage in this game by allowing them to play more than one $\triangle_1$ on each turn. The question then becomes how many $\triangle_1$s should Player 2 be allowed to color per turn such that the game is fair? Note that, if $m = p + q - 1$ and we let Player 2 play q $\triangle_1$s per turn, then they clearly have a winning strategy provided that $p > 1$.

Problem 8. Suppose p and q are given with $1 < p < q$. What is the smallest number of $\triangle_1$s Player 2 can be allowed to color per turn such that Player 2 has the winning strategy?

In addition to simplifying the finite Ramsey theory of the Ramsey space $\mathcal{R}_1$, a secondary goal was to introduce a game that was simple enough to played by young children. In this way, the game can be played with the intention of developing logical skill. In [9], the authors give evidence that the game of Tri can be used to develop logical thinking skills in young children. Our final problem will be of interest mainly to researchers in math education.

Problem 9. Give concrete evidence that the game Mines can be played by young children and can be used to develop their visual disembedding skills.

References

[1] A. Blass, Ultrafilter mappings and their Dedekind cuts, *Trans. Amer. Math. Soc.*, 327-340, (1974), 327-340.

[2] D. Conlon, A new upper bound for diagonal Ramsey numbers, *Ann. of Math.*, (2009), 941-960.

[3] N. Dobrinen and S. Todorcevic, A new class of Ramsey-classification theorems and their application in the Tukey theory of ultrafilters, Part 1, *Trans. Amer. Math. Soc.*, **7** (2015), 4627–4659.

[4] N. Dobrinen and S. Todorcevic, A new class of Ramsey-classification theorems and their application in the Tukey theory of ultrafilters, Part 2, *Trans. Amer. Math. Soc.*, **367(7)** (2015), 4627-4659.

[5] P. Erdős and G. Szekeres, A combinatorial problem in geometry, *Compo. Math.*, **2** (1935), 463-470.

[6] P. Erdős, Some remarks on the theory of graphs, *Bull. Amer. Math. Soc.*, **53(4)** (1947), 292-294.

[7] G. Exoo, A lower bound for $R(5,5)$, *J. Graph Theory*, **13(1)** (1989), 97-8.

[8] R. Greenwood and A. Gleason, Combinatorial relations and chromatic graphs, *Canad. J. of Math.*, **7(1)** (1955), 7.

[9] G. Haggard and A. Schonberger, The game of Tri, *Arithmetic Teacher*, (1977), 318-320.

[10] J. Kalbfleisch, Construction of special edge-chromatic graphs, *Canad. Math. Bull.*, **8** (1965), 575-584.

[11] Y. Khomskii, *Infinite Games*, Summer course at the University of Sofia, Bulgaria, 2010.

[12] C. Laflamme, Forcing with filters and complete combinatorics, *Ann. of Pure Appl. Logic*, **42(2)** (1989), 125-163.

[13] J. Mackey, *Combinatorial Remedies*, Ph.D. thesis, Department of Mathematics, University of Hawaii, 1994.

[14] B. McKay and S. Radziszowski, Subgraph counting identities and Ramsey numbers, *J. Comb. Theory Ser. B*, **69(2)** (1997), 193-209.

[15] F. Ramsey, On a problem of formal logic, *Proc. Lond. Math. Soc. (2)* , **1** (1930), 264-286.

[16] J. Shearer, Lower bounds for small diagonal Ramsey numbers, *J. Comb. Theory Ser. A*, **42(2)** (1986), 302-304.

[17] G. Simmons, The game of SIM, *Mathematical Solitaires & Games*, (1980), 50.

[18] J. Spencer, Ramsey's theorem - a new lower bound, *J. Combin. Theory Ser. A*, **18(1)** (1975), 108-115.

[19] T. Trujillo, Ramsey for R1 ultrafilter mappings and their Dedekind cuts, *MLQ Math. Log. Q.*, **61(4-5)** (2015), 263-273.

12

AN ALTERNATIVE PROOF OF MELHAM'S IDENTITIES

George Humby
University of Exeter, Exeter, United Kingdom
g.humby2@exeter.ac.uk

Abstract

Melham conjectures 21 identities, all of which are analogous to Jacobi's two-square theorem. A small number of these have already been proved in various ways by other authors, and in 2013 Toh proved all of them using similar, known identities. In this paper we offer an alternative and straightforward method to proving all of them using the theory of modular forms, in particular, Sturm's bound, which we explain in detail.

1. Introduction

Define

$$G_3(q) := \sum_{n=0}^{\infty} q^{\frac{n(n+1)}{2}}, \qquad G_5(q) := \sum_{n=-\infty}^{\infty} q^{\frac{n(3n-1)}{2}}, \qquad G_7(q) := \sum_{n=-\infty}^{\infty} q^{\frac{n(5n-3)}{2}},$$

the generating functions of the triangular, pentagonal, and heptagonal numbers, respectively, with $q = e^{2\pi i\tau}$, and τ in the upper half plane, H. Melham's 21 identities (all analogous to Jacobi's two-square theorem) [9] are all of the form

$$G_k(q^\alpha)G_k(q^\beta) = \text{an explicit } q\text{-series}$$

where $k = 3, 5$ or 7, $\alpha, \beta \in \mathbb{N}$. For example, using the labeling in his paper, his Identity (6) is

$$G_3(q)G_3(q^5) = \sum_{n=0}^{\infty} \left[\frac{q^{3n} + q^{7n+1}}{1 - q^{20n+5}} - \frac{q^{13n+9} + q^{17n+12}}{1 - q^{20n+15}} \right].$$

We will frequently refer to the $G_k(q^\alpha)G_k(q^\beta)$ part as the left-hand side of the identity, and the explicit q-series as the right-hand side.

The method we shall use relies on the theory of *modular forms*. We provide a brief description here, but for a more thorough reading, the reader should look to such sources as [2], [8], [10]. A *congruence subgroup*, Γ, of $SL_2(\mathbb{Z})$ is a subgroup of $SL_2(\mathbb{Z})$ that contains

$$\Gamma(N) := \mathrm{Ker}\Big(SL_2(\mathbb{Z}) \to SL_2(\mathbb{Z}/N\mathbb{Z})\Big), \tag{1}$$

where $\Gamma(N)$ is called the *principal congruence subgroup*. The smallest such N is the *level* of Γ. Three such particular congruence subgroups of interest are:

$$\Gamma_0(N) := \left\{\begin{pmatrix} a & b \\ c & d \end{pmatrix} \in SL_2(\mathbb{Z}) : \begin{pmatrix} a & b \\ c & d \end{pmatrix} \equiv \begin{pmatrix} * & * \\ 0 & * \end{pmatrix} \pmod N\right\}, \tag{2}$$

$$\begin{aligned}\Gamma^0(N) &:= \begin{pmatrix} 0 & 1 \\ 1 & 0 \end{pmatrix} \Gamma_0(N) \begin{pmatrix} 0 & 1 \\ 1 & 0 \end{pmatrix} \\ &= \left\{\begin{pmatrix} a & b \\ c & d \end{pmatrix} \in SL_2(\mathbb{Z}) : \begin{pmatrix} a & b \\ c & d \end{pmatrix} \equiv \begin{pmatrix} * & 0 \\ * & * \end{pmatrix} \pmod N\right\},\end{aligned} \tag{3}$$

$$\Gamma_1(N) := \left\{\begin{pmatrix} a & b \\ c & d \end{pmatrix} \in SL_2(\mathbb{Z}) : \begin{pmatrix} a & b \\ c & d \end{pmatrix} \equiv \begin{pmatrix} 1 & * \\ 0 & 1 \end{pmatrix} \pmod N\right\}, \tag{4}$$

where $*$ represents any number. All three groups have level N. Let $A = \begin{pmatrix} a & b \\ c & d \end{pmatrix}$, and for a complex function f, define

$$f|_A(z) := (cz+d)^{-k} f\left(\frac{az+b}{cz+d}\right). \tag{5}$$

A function f is a modular form of weight k, with k a positive integer, over a congruence subgroup Γ if f is holomorphic on H, $f|_\gamma$ remains bounded as $\Im(z) \to \infty$ for any $\gamma \in SL_2(\mathbb{Z})$, and verifies, for all $\gamma \in \Gamma$, $f|_\gamma = f$.

Our method is to show that the left-hand side, or LHS, (raised to an even power) of each identity is a type of theta function, a modular form over a certain congruence subgroup. We then show that the right-hand side, or RHS, (raised to the same even power) is also a modular form for the same weight over this subgroup, by showing that it can be written in terms of the Weierstrass zeta function. It is then a simple case of checking a small number of coefficients to show that each identity satisfies Sturm's bound, and thus that both sides must be equivalent. Melham mentions in his paper that a small number of these have already been proved in various ways by Hirschhorn [5], Sun [12], and Dickson [3] (combined with work from Adiga, Cooper, and Han [1]). In 2013 Toh [13] proved all of them using similar, known identities. The method demonstrated in this paper is easily expandable to similar identities, and a textbook demonstration of how to apply Sturm's bound.

2. The LHS as Theta Functions

We begin this section by defining a *lattice* as in [7, p. 149]. Let V be an n-dimensional real vector space. A lattice in V is a subgroup of V of the form

$$\Gamma = \mathbb{Z}v_1 + \mathbb{Z}v_2 + \cdots + \mathbb{Z}v_n, \tag{6}$$

where $\{v_1, v_2, \ldots, v_n\}$ is a linearly independent set of vectors in V. An *integral lattice* is a lattice where the inner product of any two elements in the lattice is integral. An *even lattice* is an integral lattice where the inner product of an element with itself (or norm) is always even. The *dual lattice*, denoted Λ^*, of a lattice Λ, is the lattice of vectors having integral inner products with all the elements of Λ. Define the *discriminant* of a lattice to be $\text{disc}(\Lambda) = |\Lambda^*/\Lambda|$. Finally, the *level* of a lattice Λ is the minimum $N \in \mathbb{N}$ with Nx^2 even for all $x \in \Lambda^*$ [4, p. 91].

Ebeling [4, p. 86] defines a *generalized Theta function* for a lattice $\Lambda \subset \mathbb{R}^n$, a point $z \in \mathbb{R}^n$, the variable $\tau \in H$, and a spherical polynomial P of degree r, as

$$\vartheta_{z+\Lambda,P}(\tau) := \sum_{x \in z+\Lambda} P(x)e^{\pi i \tau x^2} = \sum_{x \in z+\Lambda} P(x)q^{\frac{x^2}{2}}. \tag{7}$$

We have no need to discuss spherical polynomials; for our matters it is sufficient to state that the constant polynomial $P(x) = 1$ is spherical, of degree 0. We take $P(x) = 1$ from here on, and will drop it from our notation.

Lemma 1. *Let $\Lambda \subset \mathbb{R}^n$ (n even) be an even lattice of level N, $\rho \in \Lambda^*$, $A = \begin{pmatrix} a & b \\ c & d \end{pmatrix} \in \Gamma_0(N)$. Then for odd $d > 0$, $c \neq 0$,*

$$\vartheta_{\rho+\Lambda}(A\tau) = (c\tau + d)^{\frac{n}{2}} \left(\frac{\Delta}{d}\right) e^{\pi i a b \rho^2} \vartheta_{a\rho+\Lambda}(\tau) \tag{8}$$

with $\Delta := (-1)^{\frac{n}{2}} \text{disc}(\Lambda)$, $\left(\frac{\Delta}{d}\right)$ the Jacobi symbol, and for $c = 0$,

$$\vartheta_{\rho+\Lambda}(A\tau) = e^{\pi i a b \rho^2} \vartheta_{a\rho+\Lambda}(\tau). \tag{9}$$

Proof. The proof of this is Corollary 3.1 combined with the remarks before Theorem 3.2 of [4, pp. 92-94]. □

Now, as $\rho \in \Lambda^*$, $\rho^2 \in \mathbb{Q}$. Let $\rho^2 = \frac{u}{v}$, with $\gcd(u, v) = 1$. We have

$$\vartheta^{2v}_{\rho+\Lambda}|A = e^{2\pi i a b u} \vartheta^{2v}_{a\rho+\Lambda} = \vartheta^{2v}_{a\rho+\Lambda}.$$

We now take $n = 2$, and consider lattices in $\mathbb{R}^2$.

2.1. Triangular Numbers, $G_3(q)$

Let $\Lambda = \left\langle \begin{pmatrix} \sqrt{2\alpha} \\ 0 \end{pmatrix}, \begin{pmatrix} 0 \\ \sqrt{2\beta} \end{pmatrix} \right\rangle$, with α, β integers. Then Λ is even, and has dual

$$\Lambda^* = \left\langle \begin{pmatrix} \frac{1}{\sqrt{2\alpha}} \\ 0 \end{pmatrix}, \begin{pmatrix} 0 \\ \frac{1}{\sqrt{2\beta}} \end{pmatrix} \right\rangle. \tag{10}$$

Calculating the level N of Λ is straightforward, and gives $N = 4 \cdot \operatorname{lcm}(\alpha, \beta)$. Let $\rho = \begin{pmatrix} \sqrt{\frac{\alpha}{2}} \\ \sqrt{\frac{\beta}{2}} \end{pmatrix} \in \Lambda^*$. We have, with $q = e^{2\pi i\tau}$,

$$\begin{aligned} \vartheta_{\rho+\Lambda}(\tau) &= \sum_{n,m\in\mathbb{Z}} q^{\alpha n^2+\alpha n+\frac{\alpha}{4}+\beta m^2+\beta m+\frac{\beta}{4}} \\ &= q^{\frac{\alpha+\beta}{4}} \sum_{n,m\in\mathbb{Z}} q^{2\alpha\frac{n(n+1)}{2}+2\beta\frac{m(m+1)}{2}}. \end{aligned} \tag{11}$$

Hence

$$\vartheta_{\rho+\Lambda}\left(\frac{\tau}{2}\right) = q^{\frac{\alpha+\beta}{8}} \sum_{n\in\mathbb{Z}} q^{\alpha\frac{n(n+1)}{2}} \sum_{m\in\mathbb{Z}} q^{\beta\frac{m(m+1)}{2}}, \tag{12}$$

so

$$\frac{1}{4} q^{-\frac{\alpha+\beta}{8}} \vartheta_{\rho+\Lambda}\left(\frac{\tau}{2}\right) = G_3(q^\alpha)G_3(q^\beta). \tag{13}$$

Now, it is easy to see that $\rho + \Lambda = a\rho + \Lambda$ whenever $a \equiv 1 \pmod 2$. Further, as N is clearly even, for $A = \begin{pmatrix} a & b \\ c & d \end{pmatrix} \in \Gamma_0(N)$ we can see that a must be odd. Now, $\rho^2 = \frac{\alpha+\beta}{2}$, hence if $\alpha+\beta$ is even then $\vartheta_{\rho+\Lambda}(\tau)^2$ is a weight 2 modular form for $\Gamma_0(N)$, and if $\alpha+\beta$ is odd then $\vartheta_{\rho+\Lambda}(\tau)^4$ is a weight 4 modular form for $\Gamma_0(N)$. Therefore $\vartheta_{\rho+\Lambda}(\frac{\tau}{2})^{2v}$ is a weight $2v$ modular form for $\Gamma' = \begin{pmatrix} 1 & 0 \\ 0 & \frac{1}{2} \end{pmatrix} \Gamma_0(N) \begin{pmatrix} 1 & 0 \\ 0 & 2 \end{pmatrix}$, where v is defined as

$$v = \begin{cases} 1 & \alpha+\beta \equiv 0 \pmod 2 \\ 2 & \alpha+\beta \equiv 1 \pmod 2. \end{cases} \tag{14}$$

In other words, depending on the values of α and β, we need to raise $\vartheta_{\rho+\Lambda}(\tau)$ to an appropriate power $(2v)$, in order to ensure modularity. This turns out to be exactly analogous in the next two cases.

We see that for $A \in \Gamma'$ and $A = \begin{pmatrix} a & b \\ c & d \end{pmatrix}$, we have $b \equiv 0 \pmod 2$ and $c \equiv 0 \pmod{\frac{N}{2}}$. In other words,

$$\Gamma' = \left\{ A = \begin{pmatrix} a & b \\ c & d \end{pmatrix} \in \Gamma_0\left(\frac{N}{2}\right) \quad : \quad b \text{ even} \right\} = \Gamma_0\left(\frac{N}{2}\right) \cap \Gamma^0(2). \tag{15}$$

Hence we can conclude that

$$\left(q^{\frac{\alpha+\beta}{8}} G_3(q^\alpha) G_3(q^\beta)\right)^{2v} \tag{16}$$

is a modular form for Γ' as well.

2.2. Pentagonal Numbers, $G_5(q)$

Let $\Lambda = \left\langle \begin{pmatrix} \sqrt{6\alpha} \\ 0 \end{pmatrix}, \begin{pmatrix} 0 \\ \sqrt{6\beta} \end{pmatrix} \right\rangle$, with α, β integers. Then Λ is even and has dual

$$\Lambda^* = \left\langle \begin{pmatrix} \frac{1}{\sqrt{6\alpha}} \\ 0 \end{pmatrix}, \begin{pmatrix} 0 \\ \frac{1}{\sqrt{6\beta}} \end{pmatrix} \right\rangle. \tag{17}$$

The level of Λ is $N = 12 \cdot \mathrm{lcm}(\alpha, \beta)$. Let $\rho = \begin{pmatrix} -\sqrt{\frac{\alpha}{6}} \\ -\sqrt{\frac{\beta}{6}} \end{pmatrix} \in \Lambda^*$. We find

$$q^{-\frac{\alpha+\beta}{24}} \vartheta_{\rho+\Lambda}\left(\frac{\tau}{2}\right) = G_5(q^\alpha) G_5(q^\beta). \tag{18}$$

Again, it is easy to see that $\rho + \Lambda = a\rho + \Lambda$ whenever $a \equiv 1 \pmod 6$. When $a \equiv -1 \pmod 6$ we see $a\rho + \Lambda = -(\rho + \Lambda)$, but as we are taking the norm of each shifted lattice point this is also acceptable. Further, as N is clearly even, for $A = \begin{pmatrix} a & b \\ c & d \end{pmatrix} \in \Gamma_0(N)$ we can see that a must be odd and coprime to 3.

Now, $\rho^2 = -\frac{\alpha+\beta}{6}$, hence $\vartheta_{\rho+\Lambda}(\tau)^{2v}$ is a weight $2v$ modular form for the congruence subgroup $\Gamma' = \{A = \begin{pmatrix} a & b \\ c & d \end{pmatrix} \in \Gamma_0(N) \quad : \quad a \equiv \pm 1 \pmod 6\}$ where

$$v = \begin{cases} 1 & \alpha+\beta \equiv 0 \pmod 6 \\ 6 & \alpha+\beta \equiv \pm 1 \pmod 6 \\ 3 & \alpha+\beta \equiv \pm 2 \pmod 6 \\ 2 & \alpha+\beta \equiv 3 \pmod 6. \end{cases} \tag{19}$$

Therefore $\vartheta_{\rho+\Lambda}(\frac{\tau}{2})^{2v}$ is a weight $2v$ modular form for $\Gamma'' = \begin{pmatrix} 1 & 0 \\ 0 & \frac{1}{2} \end{pmatrix} \Gamma' \begin{pmatrix} 1 & 0 \\ 0 & 2 \end{pmatrix}$.

We see for $A \in \Gamma''$, $A = \begin{pmatrix} a & b \\ c & d \end{pmatrix}$, we have $b \equiv 0 \pmod 2$ and $c \equiv 0 \pmod{\frac{N}{2}}$. In other words,

$$\begin{aligned} \Gamma'' &= \left\{ A = \begin{pmatrix} a & b \\ c & d \end{pmatrix} \in \Gamma_0\left(\frac{N}{2}\right) \quad : \quad b \text{ even}, \ a \equiv \pm 1 \pmod 6 \right\} \\ &= \Gamma_0\left(\frac{N}{2}\right) \cap \Gamma^0(2) \end{aligned} \tag{20}$$

as $1 = ad - bc \equiv d \pmod 6$. Hence we can conclude that

$$\left(q^{\frac{\alpha+\beta}{24}} G_5(q^\alpha) G_5(q^\beta)\right)^{2v} \tag{21}$$

is a modular form for Γ'' as well.

2.3. Heptagonal Numbers, $G_7(q)$

Let $\Lambda = \left\langle \begin{pmatrix} \sqrt{10\alpha} \\ 0 \end{pmatrix}, \begin{pmatrix} 0 \\ \sqrt{10\beta} \end{pmatrix} \right\rangle$, with α, β integers. Then Λ is even, and has dual

$$\Lambda^* = \left\langle \begin{pmatrix} \frac{1}{\sqrt{10\alpha}} \\ 0 \end{pmatrix}, \begin{pmatrix} 0 \\ \frac{1}{\sqrt{10\beta}} \end{pmatrix} \right\rangle. \tag{22}$$

The level of Λ is $N = 20 \cdot \operatorname{lcm}(\alpha, \beta)$. Let $\rho = \begin{pmatrix} -3\sqrt{\frac{\alpha}{10}} \\ -3\sqrt{\frac{\beta}{10}} \end{pmatrix} \in \Lambda^*$. We find

$$q^{-\frac{9(\alpha+\beta)}{40}} \vartheta_{\rho+\Lambda}\left(\frac{\tau}{2}\right) = G_7(q^\alpha) G_7(q^\beta). \tag{23}$$

Once more, we have $\rho + \Lambda = a\rho + \Lambda$ whenever $a \equiv 1 \pmod{10}$. Further, as N is clearly even, for $A = \begin{pmatrix} a & b \\ c & d \end{pmatrix} \in \Gamma_0(N)$ we can see that a must be odd.

Now, $\rho^2 = \frac{\alpha+\beta}{10}$, hence $\vartheta_{\rho+\Lambda}(\tau)^{2v}$ is a weight $2v$ modular form for the congruence subgroup $\Gamma' = \{A = \begin{pmatrix} a & b \\ c & d \end{pmatrix} \in \Gamma_0(N) \quad : \quad a \equiv 1 \pmod{10}\}$ where

$$v = \begin{cases} 1 & \alpha+\beta \equiv 0 \pmod{10} \\ 10 & \alpha+\beta \equiv \pm 1 \pmod{10} \\ 5 & \alpha+\beta \equiv \pm 2 \pmod{10} \\ 10 & \alpha+\beta \equiv \pm 3 \pmod{10} \\ 5 & \alpha+\beta \equiv \pm 4 \pmod{10} \\ 2 & \alpha+\beta \equiv 5 \pmod{10}. \end{cases} \tag{24}$$

Therefore $\vartheta_{\rho+\Lambda}(\frac{\tau}{2})^{2v}$ is a weight $2v$ modular form for $\Gamma'' = \begin{pmatrix} 1 & 0 \\ 0 & \frac{1}{2} \end{pmatrix} \Gamma' \begin{pmatrix} 1 & 0 \\ 0 & 2 \end{pmatrix}$. We see for $A \in \Gamma''$, $A = \begin{pmatrix} a & b \\ c & d \end{pmatrix}$, we have $b \equiv 0 \pmod 2$ and $c \equiv 0 \pmod{\frac{N}{2}}$. In other words,

$$\begin{aligned} \Gamma' &= \left\{A = \begin{pmatrix} a & b \\ c & d \end{pmatrix} \in \Gamma_0\left(\frac{N}{2}\right) \quad : \quad b \text{ even}, a \equiv \pm 1 \pmod{10}\right\} \\ &= \Gamma_0\left(\frac{N}{2}\right) \cap \Gamma^0(2) \cap \Gamma_1(10). \end{aligned} \tag{25}$$

3. The RHS as the Weierstrass Zeta Function

The *Weierstrass zeta function* is a function that naturally leads to an Eisenstein series of weight 1 [2, p. 138]. We write it as Z to avoid confusion and define it as

$$Z_\Lambda(z) := \frac{1}{z} + \sum_{\substack{w\in\Lambda \\ w\neq 0 \\ w\neq z}} \left(\frac{1}{z-w} + \frac{1}{w} + \frac{z}{w^2} \right) \tag{26}$$

for a 2 dimensional lattice $\Lambda := \omega_1\mathbb{Z} \oplus \omega_2\mathbb{Z}$, and $z \in \mathbb{C}$. This function is not quite periodic, but rather with each step along the lattice increases by a lattice constant, $\eta_1(\Lambda)$ or $\eta_2(\Lambda)$, defined by:

$$\eta_1(\Lambda) := Z_\Lambda(z+\omega_1) - Z_\Lambda(z) \quad \text{and} \quad \eta_2(\Lambda) := Z_\Lambda(z+\omega_2) - Z_\Lambda(z). \tag{27}$$

If we assume without loss of generality that $\Im\left(\frac{\omega_1}{\omega_2}\right) > 0$, then these lattice constants satisfy the *Legendre relation*

$$\eta_2(\Lambda)\omega_1 - \eta_1(\Lambda)\omega_2 = 2\pi i. \tag{28}$$

We can thus define a periodic function, $Z^*_{\Lambda_\tau}$, for $u, v \in \mathbb{R}$, and $\Lambda_\tau := \mathbb{Z} + \tau\mathbb{Z}$, as follows,

$$Z^*_{\Lambda_\tau}(u\tau + v) := Z_{\Lambda_\tau}(u\tau+v) - u\eta_1(\Lambda_\tau) - v\eta_2(\Lambda_\tau). \tag{29}$$

For some scalar m, and writing $Z(z \mid \omega_1, \omega_2) := Z_\Lambda(z)$ as is common, we see

$$\begin{aligned} Z(mz \mid m\omega_1, m\omega_2) &= \frac{1}{mz} + \sum_{\substack{w\in m\Lambda \\ w\neq 0 \\ w\neq mz}} \left(\frac{1}{mz-w} + \frac{1}{w} + \frac{mz}{w^2} \right) \\ &= \frac{1}{m}\left(\frac{1}{z} + \sum_{\substack{w\in\Lambda \\ w\neq 0 \\ w\neq z}} \left(\frac{1}{z-w} + \frac{1}{w} + \frac{z}{w^2} \right) \right) \\ &= \frac{1}{m} Z(z \mid \omega_1, \omega_2). \end{aligned} \tag{30}$$

Applying this to $Z(z \mid 1, \tau) = Z_{\Lambda_\tau}(z)$ transformed by a matrix in $SL_2(\mathbb{Z})$, we find

$$\begin{aligned} Z\left(z \mid 1, \frac{a\tau+b}{c\tau+d}\right) &= (c\tau+d)Z(z(c\tau+d) \mid c\tau+d, a\tau+b) \\ &= (c\tau+d)Z(z(c\tau+d) \mid 1, \tau) \end{aligned} \tag{31}$$

as $\begin{pmatrix} a & b \\ c & d \end{pmatrix} \in SL_2(\mathbb{Z})$. We define

$$f_{r,s}(\tau) := Z\left(\frac{r\tau+s}{N} \mid 1, \tau\right). \tag{32}$$

Hence

$$f_{r,s}\left(\frac{a\tau+b}{c\tau+d}\right)=Z\left(\frac{1}{N}\left(s+r\left(\frac{a\tau+b}{c\tau+d}\right)\right)\mid 1,\frac{a\tau+b}{c\tau+d}\right)$$
$$=(c\tau+d)Z\left(\frac{1}{N}(s(c\tau+d)+r(a\tau+b))\mid 1,\tau\right)$$
$$=(c\tau+d)f_{ar+cs,br+ds}(\tau). \tag{33}$$

Diamond and Shurman [2, p. 138] further state that the Weierstrass zeta function can be expressed as

$$Z_{\Lambda_\tau}(z)=\eta_2(\Lambda_\tau)z-\pi i\frac{1+e^{2\pi iz}}{1-e^{2\pi iz}}-2\pi i\sum_{n=1}^{\infty}\left[\frac{e^{2\pi iz}q^n}{1-e^{2\pi iz}q^n}-\frac{e^{-2\pi iz}q^n}{1-e^{-2\pi iz}q^n}\right]. \tag{34}$$

We consider the holomorphic function of τ,

$$Z^*_{\Lambda_\tau}\left(\frac{a\tau+b}{N}\right)=Z_{\Lambda_\tau}\left(\frac{a\tau+b}{N}\right)-\frac{a\eta_1(\Lambda_\tau)+b\eta_2(\Lambda_\tau)}{N} \tag{35}$$

which we call an Eisenstein series of weight 1.

3.1. $E_{a,b,N}$ When $a \not\equiv 0 \pmod N$

If we have $a\not\equiv 0(N)$ we can expand the denominators, so we take $0<a<N$ and, using the fact that $e^{2\pi iz}=e^{2\pi i\left(\frac{a\tau+b}{N}\right)}=e^{2\pi i\frac{a\tau}{N}}e^{2\pi i\frac{b}{N}}=q^{\frac{a}{N}}\zeta_N^b$ (with $\zeta_N:=e^{\frac{2\pi i}{N}}$ as always), simplify using the Legendre relation:

$$Z^*_{\Lambda_\tau}\left(\frac{a\tau+b}{N}\right)=\eta_2(\Lambda_\tau)\left(\frac{a\tau+b}{N}\right)-\pi i\frac{1+q^{\frac{a}{N}}\zeta_N^b}{1-q^{\frac{a}{N}}\zeta_N^b}-\frac{a\eta_1(\Lambda_\tau)+b\eta_2(\Lambda_\tau)}{N}$$
$$-2\pi i\sum_{n=1}^{\infty}\left[\frac{q^{\frac{a}{N}}\zeta_N^b q^n}{1-q^{\frac{a}{N}}\zeta_N^b q^n}-\frac{q^{-\frac{a}{N}}\zeta_N^{-b}q^n}{1-q^{-\frac{a}{N}}\zeta_N^{-b}q^n}\right]$$
$$=\frac{a}{N}(\tau\eta_2(\Lambda_\tau)-\eta_1(\Lambda_\tau))-\pi i\left(1+2\sum_{m=1}^{\infty}q^{\frac{am}{N}}\zeta_N^{bm}\right)$$
$$-2\pi i\sum_{n=1}^{\infty}\sum_{m=1}^{\infty}\left[q^{\frac{m(a+nN)}{N}}\zeta_N^{bm}-q^{\frac{m(-a+nN)}{N}}\zeta_N^{-bm}\right]$$
$$=\frac{2\pi ia}{N}-\pi i-2\pi i\sum_{m=1}^{\infty}q^{\frac{am}{N}}\zeta_N^{bm}$$
$$-2\pi i\sum_{n=1}^{\infty}\sum_{m=1}^{\infty}\left[q^{m(\frac{a}{N}+n)}\zeta_N^{bm}-q^{\frac{m(-a+nN)}{N}}\zeta_N^{-bm}\right]$$

$$= 2\pi i\left(\frac{a}{N}-\frac{1}{2}-\sum_{m=1}^{\infty}\left[\sum_{n=0}^{\infty}q^{\frac{m}{N}(a+nN)}\zeta_N^{bm}-\sum_{n=1}^{\infty}q^{\frac{m}{N}(-a+nN)}\zeta_N^{-bm}\right]\right)$$

$$= 2\pi i\left(\frac{a}{N}-\frac{1}{2}-\sum_{r=1}^{\infty}q^{\frac{r}{N}}\left[\sum_{\substack{s|r\\ s\equiv a\ (N)\\ s>0}}\zeta_N^{\frac{br}{s}}-\sum_{\substack{s|r\\ s\equiv -a\ (N)\\ s>0}}\zeta_N^{-\frac{br}{s}}\right]\right)$$

$$:= E_{a,b,N}(\tau). \tag{36}$$

Hence we have replaced $Z^*_{\Lambda_\tau}$ with $E_{a,b,N}$, to make our choices of a, b, N explicit. Let $M = \begin{pmatrix} a & b \\ c & d \end{pmatrix} \in SL_2(\mathbb{Z})$. Then by our earlier result involving the transformation of $f_{r,s}(\tau)$ (Equation (33)),

$$E_{\alpha,\beta,N}|_M = E_{a\alpha+c\beta,b\alpha+d\beta,N}. \tag{37}$$

The first two indices can be reduced modulo N, hence we see that $E_{a,b,N}$ remains invariant under transformation by $\Gamma(N)$.

3.2. $E_{a,b,N}$ When $a \equiv 0 \pmod N$

For the case $a \equiv 0 \pmod N$ we start with

$$Z^*_{\Lambda_\tau}\left(\frac{a\tau+b}{N}\right) = \eta_2(\Lambda)\left(\frac{a\tau+b}{N}\right) - \pi i\frac{1+q^{\frac{a}{N}}\zeta_N^b}{1-q^{\frac{a}{N}}\zeta_N^b} - 2\pi i\sum_{n=1}^{\infty}\left[\frac{q^{\frac{a}{N}}\zeta_N^b q^n}{1-q^{\frac{a}{N}}\zeta_N^b q^n}-\frac{q^{-\frac{a}{N}}\zeta_N^{-b}q^n}{1-q^{-\frac{a}{N}}\zeta_N^{-b}q^n}\right] - \frac{a\eta_1(\Lambda)+b\eta_2(\Lambda)}{N}, \tag{38}$$

with $a = tN$, with $t \in \mathbb{Z}$. We consider $t = 0$, i.e., $a = 0$, and notice that the result will hold for any $a \equiv 0 \pmod N$ due to periodicity:

$$Z^*_{\Lambda_\tau}\left(\frac{b}{N}\right) = \eta_2(\Lambda)\left(\frac{b-b}{N}\right) - \pi i\frac{1+\zeta_N^b}{1-\zeta_N^b} - 2\pi i\sum_{n=1}^{\infty}\left[\frac{\zeta_N^b q^n}{1-\zeta_N^b q^n}-\frac{\zeta_N^{-b}q^n}{1-\zeta_N^{-b}q^n}\right]$$

$$= -\pi i\frac{1+\zeta_N^b}{1-\zeta_N^b} - 2\pi i\sum_{m=1}^{\infty}\sum_{n=1}^{\infty}\left[\zeta_N^{mb}q^{mn}-\zeta_N^{-mb}q^{mn}\right]$$

$$= 2\pi i\left(-\frac{1}{2}\cdot\frac{1+\zeta_N^b}{1-\zeta_N^b}-\sum_{m=1}^{\infty}\sum_{n=1}^{\infty}q^{nm}\left[\zeta_N^{mb}-\zeta_N^{-mb}\right]\right)$$

$$= 2\pi i\left(\frac{1}{2} + \frac{1}{\zeta_N^b - 1} - \sum_{t=1}^{\infty}\sum_{\substack{s|t\\ s>0}} q^t\left[\zeta_N^{\frac{bt}{s}} - \zeta_N^{-\frac{bt}{s}}\right]\right)$$

$$= 2\pi i\left(\frac{1}{2} + \frac{1}{\zeta_N^b - 1} - \sum_{r=1}^{\infty}\sum_{\substack{s|r\\ s\equiv 0\ (N)\\ s>0}} q^{\frac{r}{N}}\left[\zeta_N^{\frac{br}{s}} - \zeta_N^{-\frac{br}{s}}\right]\right)$$

$$:= E_{0,b,N}(\tau). \tag{39}$$

3.3. Defining $F_{a,c,N}$ for $a, c \not\equiv 0 \pmod N$

Now we consider the following linear combination, for $0 < a < N$:

$$\frac{1}{2\pi i}F_{a,c,N}(\tau) := \frac{1}{2\pi i}\sum_{b=0}^{N-1}\zeta_N^{-bc}E_{a,b,N}(\tau)$$

$$= \sum_{b=0}^{N-1}\zeta_N^{-bc}\left(\frac{a}{N} - \frac{1}{2}\right)$$

$$- \sum_{b=0}^{N-1}\zeta_N^{-bc}\sum_{r=1}^{\infty}\left[\sum_{\substack{s|r\\ s\equiv a\ (N)\\ s>0}} q^{\frac{r}{N}}\zeta_N^{\frac{br}{s}} - \sum_{\substack{s|r\\ s\equiv -a\ (N)\\ s>0}} q^{\frac{r}{N}}\zeta_N^{\frac{-br}{s}}\right]$$

$$= \left(\frac{a}{N} - \frac{1}{2}\right)\sum_{b=0}^{N-1}\zeta_N^{-bc}$$

$$- \sum_{r=1}^{\infty}\left[\sum_{\substack{s|r\\ s\equiv a\ (N)\\ s>0}} q^{\frac{r}{N}}\sum_{b=0}^{N-1}\zeta_N^{b\left(\frac{r}{s}-c\right)} - \sum_{\substack{s|r\\ s\equiv -a\ (N)\\ s>0}} q^{\frac{r}{N}}\sum_{b=0}^{N-1}\zeta_N^{b\left(-\frac{r}{s}-c\right)}\right]$$

$$= \left(\frac{a}{N} - \frac{1}{2}\right)\sum_{b=0}^{N-1}\zeta_N^{-bc} - \sum_{r=1}^{\infty}\left[\sum_{\substack{s|r\\ s\equiv a\ (N)\\ \frac{r}{s}\equiv c\ (N)\\ s>0}} q^{\frac{r}{N}}N - \sum_{\substack{s|r\\ s\equiv -a\ (N)\\ \frac{r}{s}\equiv -c\ (N)\\ s>0}} q^{\frac{r}{N}}N\right]$$

$$= \left(\frac{a}{N} - \frac{1}{2}\right)\sum_{b=0}^{N-1}\zeta_N^{-bc} - N\left(\sum_{\substack{s\equiv a\ (N)\\ t\equiv c\ (N)\\ s,t>0}} q^{\frac{st}{N}} - \sum_{\substack{s\equiv -a\ (N)\\ t\equiv -c\ (N)\\ s,t>0}} q^{\frac{st}{N}}\right). \tag{40}$$

So if we pick c such that $N \nmid c$, the sum on the left disappears, and we have

$$F_{a,c,N}(\tau) = -2\pi i N \left(\sum_{\substack{s \equiv a \ (N) \\ t \equiv c \ (N) \\ s,t>0}} q^{\frac{st}{N}} - \sum_{\substack{s \equiv -a \ (N) \\ t \equiv -c \ (N) \\ s,t>0}} q^{\frac{st}{N}} \right). \tag{41}$$

For simplicity we define

$$F^*_{a,c,N}(\tau) := \sum_{\substack{s \equiv a \ (N) \\ t \equiv c \ (N) \\ s,t>0}} q^{\frac{st}{N}} - \sum_{\substack{s \equiv -a \ (N) \\ t \equiv -c \ (N) \\ s,t>0}} q^{\frac{st}{N}} = -\frac{1}{2\pi i N} F_{a,c,N}(\tau). \tag{42}$$

We write

$$K_{a,c,N} := \sum_{\substack{s \equiv a \ (N) \\ t \equiv c \ (N) \\ s,t>0}} q^{\frac{st}{N}} \tag{43}$$

so that $F^*_{a,c,N} = K_{a,c,N} - K_{-a,-c,N}$. Suppose $\gcd(a,N) = d > 1$. If we take the sum over $c \equiv r \pmod{\frac{N}{d}}$ of the $F^*_{a,c,N}$ for some r, we observe

$$\begin{aligned} \sum_{\substack{0<c<N \\ c \equiv r \ (\frac{N}{d})}} F^*_{a,c,N} &= \sum_{\substack{0<c<N \\ c \equiv r \ (\frac{N}{d})}} K_{a,c,N} - \sum_{\substack{0<c<N \\ c \equiv -r \ (\frac{N}{d})}} K_{-a,c,N} \\ &= K_{\frac{a}{d},r,\frac{N}{d}} - K_{-\frac{a}{d},-r,\frac{N}{d}} \\ &= F^*_{\frac{a}{d},r,\frac{N}{d}}. \end{aligned} \tag{44}$$

Of course, if $\frac{a}{d} \equiv -r \pmod{\frac{N}{d}}$ then this is 0.

4. The Identities

In this section, we use what we have covered in the previous sections to rearrange the identities conjectured by Melham to find equivalent identities which can then be easily proven. We continue to use the labeling from Melham's paper to refer to each of the identities.

4.1. Triangular Numbers

4.1.1. Identity (6)

As mentioned before, Melham's Identity (6) is

$$G_3(q)G_3(q^5) = \sum_{n=0}^{\infty} \left[\frac{q^{3n} + q^{7n+1}}{1 - q^{20n+5}} - \frac{q^{13n+9} + q^{17n+12}}{1 - q^{20n+15}} \right].$$

We expand the denominators in the RHS to give

$$\sum_{n=0}^{\infty}\sum_{m=0}^{\infty}\left[q^{3n+(20n+5)m}+q^{7n+1+(20n+5)m}-q^{13n+9+(20n+15)m}-q^{17n+12+(20n+15)m}\right]$$
$$=\sum_{n=0}^{\infty}\sum_{m=0}^{\infty}\left[q^{\frac{(20n+5)(20m+3)}{20}-\frac{3}{4}}+q^{\frac{(20n+5)(20m+7)}{20}-\frac{3}{4}}\right.$$
$$\left.-q^{\frac{(20n+15)(20m+13)}{20}-\frac{3}{4}}-q^{\frac{(20n+15)(20m+17)}{20}-\frac{3}{4}}\right]$$
$$=q^{-\frac{3}{4}}\left(\sum_{\substack{a,b>0\\ a\equiv 5\ (20)\\ b\equiv 3\ (20)}}q^{\frac{ab}{20}}-\sum_{\substack{a,b>0\\ a\equiv -5\ (20)\\ b\equiv -3\ (20)}}q^{\frac{ab}{20}}+\sum_{\substack{a,b>0\\ a\equiv 5\ (20)\\ b\equiv 7\ (20)}}q^{\frac{ab}{20}}-\sum_{\substack{a,b>0\\ a\equiv -5\ (20)\\ b\equiv -7\ (20)}}q^{\frac{ab}{20}}\right)$$
$$=q^{-\frac{3}{4}}\left(F^*_{5,3,20}+F^*_{5,7,20}\right). \tag{45}$$

There exists a similar result for all the identities, that is, each identity is equivalent to an identity of the form

$$G_k(q^{\alpha})G_k(q^{\beta})=q^{\frac{\alpha+\beta}{8(2-k)}}\sum_{(a_i,c_i)}F^*_{a_i,c_i,4(k-2)}, \tag{46}$$

with the sum going over a number of pairs (a_i, c_i). We will use these pairs later to calculate coefficients. Continuing with Identity (6), we recall by our earlier result that

$$\sum_{\substack{0<c<20\\ c\equiv 3\ (4)}}F^*_{5,c,20}=F^*_{1,3,4}=K_{1,3,4}-K_{3,1,4}=0. \tag{47}$$

Hence

$$F^*_{5,3,20}+F^*_{5,7,20}=-F^*_{5,11,20}-F^*_{5,19,20} \tag{48}$$

as

$$F^*_{5,15,20}=0. \tag{49}$$

Noting that $1,4$ are quadratic residues modulo 5, where as $2,3$ are not, we see the RHS is equal to, where $\left(\frac{c}{p}\right)$ denotes the Legendre symbol,

$$-\frac{q^{-\frac{3}{4}}}{2}\sum_{\substack{0<c<20\\ c\equiv 3\ (4)}}\left(\frac{c}{5}\right)F^*_{5,c,20}, \tag{50}$$

or equivalently,

$$\frac{q^{-\frac{3}{4}}}{2}\sum_{\substack{0<c<20\\ c\equiv 1\ (4)}}\left(\frac{c}{5}\right)F^*_{15,c,20}=-\frac{q^{-\frac{3}{4}}}{80\pi i}\sum_{b=0}^{19}E_{15,b,20}\sum_{\substack{0<c<20\\ c\equiv 1\ (4)}}\left(\frac{c}{5}\right)\zeta_{20}^{-bc}. \tag{51}$$

Here we have used the fact that

$$F^*_{a,c,N} = -\frac{1}{2\pi i N}\sum_{b=0}^{N-1}\zeta_N^{-bc}E_{a,b,N}. \tag{52}$$

We will encounter sums of the form

$$\sum_{\substack{0<c<N\\ c\equiv\alpha\ (\beta)}}\left(\frac{c}{p}\right)\zeta_N^{-bc}$$

frequently. To deal with these we use a small lemma.

Lemma 2. *Let $\alpha,\beta,p\in\mathbb{N}$, $\alpha<\beta$, p an odd prime, $(\beta,p)=1$. Define $N=\beta p$, and let $\gamma\equiv\beta^{-1}$ modulo p. We have*

$$\sum_{\substack{0<c<N\\ c\equiv\alpha\ (\beta)}}\left(\frac{c}{p}\right)\zeta_N^{-bc} = \begin{cases}\zeta_N^{\alpha b(\gamma\beta-1)}\left(\frac{b\gamma}{p}\right)\sqrt{p} & p\equiv 1\ (mod\ 4)\\ -\zeta_N^{\alpha b(\gamma\beta-1)}\left(\frac{b\gamma}{p}\right)i\sqrt{p} & p\equiv 3\ (mod\ 4).\end{cases} \tag{53}$$

Proof. We define S as follows:

$$\begin{aligned} S &:= \sum_{\substack{0<c<N\\ c\equiv\alpha\ (\beta)}}\left(\frac{c}{p}\right)\zeta_N^{-bc}\\ &= \sum_{k=0}^{\left\lfloor\frac{N-\alpha}{\beta}\right\rfloor}\left(\frac{\alpha+k\beta}{p}\right)\zeta_N^{-b(\alpha+k\beta)}\\ &= \zeta_N^{-b\alpha}\sum_{k=0}^{\left\lfloor\frac{N-\alpha}{\beta}\right\rfloor}\left(\frac{\alpha+k\beta}{p}\right)\zeta_p^{-bk}. \end{aligned}\tag{54}$$

As $\alpha<\beta$ we have $\left\lfloor\frac{N-\alpha}{\beta}\right\rfloor = p-1$. Letting $n=\alpha+k\beta$ (which is of course a bijection modulo p), we therefore have $k=(n-\alpha)\gamma$ modulo p, so

$$\begin{aligned} S &= \zeta_N^{-\alpha b}\sum_{n=0}^{p-1}\left(\frac{n}{p}\right)\zeta_p^{-b(n-\alpha)\gamma}\\ &= \zeta_N^{-\alpha b}\zeta_p^{b\alpha\gamma}\sum_{n=0}^{p-1}\left(\frac{n}{p}\right)\zeta_N^{-b\gamma n}\\ &= \zeta_N^{\alpha b(\gamma\beta-1)}\sum_{n=0}^{p-1}\left(\frac{n}{p}\right)\zeta_N^{-b\gamma n}. \end{aligned}\tag{55}$$

This is a Gauss sum. Gauss proved [6] that

$$\sum_{n=0}^{p-1}\left(\frac{n}{p}\right)\zeta_p^{-bn}=\begin{cases}\left(\frac{b}{p}\right)\sqrt{p} & p\equiv 1 \pmod 4\\ -\left(\frac{b}{p}\right)i\sqrt{p} & p\equiv 3 \pmod 4\end{cases} \tag{56}$$

for p an odd prime. We can therefore conclude with the statement of the lemma. □

Now, returning to the RHS of Identity (6), we have

$$\sum_{\substack{0<c<20\\ c\equiv 1\ (4)}}\left(\frac{c}{5}\right)\zeta_{20}^{-bc}=\left(\frac{b}{5}\right)\sqrt{5}. \tag{57}$$

Hence we see the RHS is

$$-\frac{\sqrt{5}}{80\pi i}q^{-\frac{3}{4}}\sum_{b=0}^{19}\left(\frac{b}{5}\right)\zeta_{20}^{15b}E_{15,b,20}(\tau). \tag{58}$$

From before, we have

$$\frac{1}{4}q^{-\frac{\alpha+\beta}{8}}\vartheta_{\rho+\Gamma}\left(\frac{\tau}{2}\right)=G_3(q^{\alpha})G_3(q^{\beta}) \tag{59}$$

with $\Gamma=\left\langle\begin{pmatrix}\sqrt{2\alpha}\\0\end{pmatrix},\begin{pmatrix}0\\\sqrt{2\beta}\end{pmatrix}\right\rangle$, and $\rho=\begin{pmatrix}\sqrt{\frac{\alpha}{2}}\\\sqrt{\frac{\beta}{2}}\end{pmatrix}\in\Gamma^*$. Hence Identity (6) is equivalent to

$$\vartheta_{\rho+\Gamma}\left(\frac{\tau}{2}\right)=-\frac{\sqrt{5}}{20\pi i}\sum_{b=0}^{19}\left(\frac{b}{5}\right)\zeta_{20}^{15b}E_{15,b,20}(\tau). \tag{60}$$

Raising both sides to the power of 2, as $\alpha+\beta=1+5=6$ is even, we see the LHS is a modular form for the congruence subgroup $\Gamma_0(10)\cap\Gamma^0(2)$, as detailed in Section 2.1. Thus we aim to show that the RHS (ignoring the constant term),

$$H(\tau):=\left(\sum_{b=0}^{19}\left(\frac{b}{5}\right)\zeta_{20}^{15b}E_{15,b,20}(\tau)\right)^2, \tag{61}$$

is also a modular form for this subgroup. Before we do this, we must transform the other identities into a similar form. Our method above works perfectly for the majority of the identities, but there are exceptions to this. The first such identity is Identity (13), where we must use a slight variation. We detail this method next.

Table 1: Triangular and Pentagonal Identities

Id.	$\binom{\alpha}{\beta}$	N	p	$\binom{k_1}{a_1}$	$\binom{k_2}{a_2}$	$\binom{k_3}{a_3}$	$\binom{k_4}{a_4}$	$2v$
6, T	$\binom{1}{5}$	20	5	$\binom{15}{15}$				2
11, T	$\binom{1}{13}$	52	13	$\binom{39}{39}$				2
14, T	$\binom{1}{37}$	148	37	$\binom{111}{148}$				2
7, T	$\binom{1}{6}$	24	3	$\binom{9}{3}$	$\binom{21}{15}$			4
8, T	$\binom{2}{3}$			$\binom{21}{9}$	$\binom{9}{21}$			4
9, T	$\binom{1}{10}$	40	5	$\binom{5}{5}$	$\binom{25}{25}$			4
10, T	$\binom{2}{5}$			$\binom{5}{25}$	$\binom{25}{5}$			4
12, T	$\binom{1}{22}$	88	11	$\binom{33}{11}$	$\binom{77}{55}$			4
13*, T	$\binom{2}{11}$			$\binom{33}{33}$	$\binom{77}{77}$			4
15*, T	$\binom{1}{58}$	232	29	$\binom{29}{29}$	$\binom{145}{145}$			4
16, T	$\binom{2}{29}$			$\binom{145}{29}$	$\binom{29}{145}$			4
17, P	$\binom{1}{5}$	60	5	$\binom{35}{15}$	$\binom{5}{45}$			2
20, P	$\binom{1}{13}$	156	13	$\binom{143}{91}$	$\binom{65}{13}$			6
23*, P	$\binom{1}{37}$	444	37	$\binom{407}{259}$	$\binom{185}{37}$			6
18*, P	$\binom{1}{10}$	120	5	$\binom{105}{25}$	$\binom{45}{85}$	$\binom{105}{65}$	$\binom{45}{5}$	12
19*, P	$\binom{2}{5}$			$\binom{105}{5}$	$\binom{45}{65}$	$\binom{105}{85}$	$\binom{45}{25}$	12
21, P	$\binom{1}{22}$	264	11	$\binom{121}{11}$	$\binom{253}{143}$	$\binom{209}{187}$	$\binom{187}{209}$	12
22, P	$\binom{2}{11}$			$\binom{121}{121}$	$\binom{187}{187}$	$\binom{209}{209}$	$\binom{253}{253}$	12
24*, P	$\binom{1}{58}$	696	29	$\binom{493}{29}$	$\binom{145}{377}$	$\binom{319}{551}$	$\binom{667}{203}$	12
25*, P	$\binom{2}{29}$			$\binom{319}{667}$	$\binom{667}{319}$	$\binom{493}{145}$	$\binom{145}{493}$	12

Table 2: Heptagonal Identity

Id.	$\binom{\alpha}{\beta}$	N	p	$\binom{k_1}{a_1}$	$\binom{k_2}{a_2}$	$\binom{k_3}{a_3}$	$\binom{k_4}{a_4}$	$\binom{k_5}{a_5}$	$\binom{k_6}{a_6}$	$\binom{k_7}{a_7}$	$\binom{k_8}{a_8}$	$2v$
26, H	$\binom{1}{6}$	120	3	$\binom{111}{21}$	$\binom{51}{81}$	$\binom{39}{69}$	$\binom{99}{9}$	$\binom{63}{117}$	$\binom{3}{57}$	$\binom{87}{93}$	$\binom{27}{33}$	20

4.1.2. Identity (13)

This identity is slightly resistant to our previous method, and requires a little more effort. Recalling that $K_{a,c,N} - K_{-a,-c,N} = -\frac{1}{2\pi i N}F_{a,c,N} = F^*_{a,c,N}$, we expand the denominators, and find the identity is equivalent to

$$\begin{aligned}G_3(q^2)G_3(q^{11}) = q^{-\frac{13}{8}}\Big(&- K_{11,5,88} - K_{11,37,88} - K_{11,45,88} - K_{11,53,88} - K_{11,69,88}\\ &- K_{77,3,88} - K_{77,11,88} - K_{77,27,88} - K_{77,59,88} - K_{77,67,88} - K_{77,75,88}\\ &+ K_{33,15,88} + K_{33,23,88} + K_{33,31,88} + K_{33,47,88} + K_{33,71,88}\\ &+ K_{55,1,88} + K_{55,9,88} + K_{55,25,88} + K_{55,33,88} + K_{55,49,88} + K_{55,81,88}\Big). \quad (62)\end{aligned}$$

Now, noting that we can write $-K_{-a,-c,N} = F^*_{a,c,N} - K_{a,c,N}$, and also $K_{a,c,N} = F^*_{a,c,N} + K_{-a,-c,N}$, we find this is equivalent to (multiplying by $q^{\frac{13}{8}}$ for simplicity)

$$\begin{aligned}q^{\frac{13}{8}}G_3(q^2)G_3(q^{11}) =& -K_{11,5,88} - K_{11,37,88} - K_{11,45,88} - K_{11,53,88} - K_{11,69,88}\\ &+ (F^*_{11,85,88} - K_{11,85,88}) + (F^*_{11,77,88} - K_{11,77,88}) + (F^*_{11,61,88} - K_{11,61,88})\\ &+ (F^*_{11,29,88} - K_{11,29,88}) + (F^*_{11,21,88} - K_{11,21,88}) + (F^*_{11,13,88} - K_{11,13,88})\\ &+ (F^*_{33,15,88} + K_{55,73,88}) + (F^*_{33,23,88} + K_{55,65,88}) + (F^*_{33,31,88} + K_{55,57,88})\\ &+ (F^*_{33,47,88} + K_{55,41,88}) + (F^*_{33,71,88} + K_{55,17,88})\\ &+ K_{55,1,88} + K_{55,9,88} + K_{55,25,88} + K_{55,33,88} + K_{55,49,88} + K_{55,81,88}\\ =& F^*_{11,85,88} + F^*_{11,77,88} + F^*_{11,61,88} + F^*_{11,29,88} + F^*_{11,21,88} + F^*_{11,13,88}\\ &+ F^*_{33,15,88} + F^*_{33,23,88} + F^*_{33,31,88} + F^*_{33,47,88} + F^*_{33,71,88}\\ &- K_{11,5,88} - K_{11,13,88} - K_{11,21,88} - K_{11,29,88} - K_{11,37,88} - K_{11,45,88} - K_{11,53,88}\\ &- K_{11,61,88} - K_{11,69,88} - K_{11,77,88} - K_{11,85,88}\\ &+ K_{55,1,88} + K_{55,9,88} + K_{55,17,88} + K_{55,25,88} + K_{55,33,88} + K_{55,41,88} + K_{55,49,88}\\ &+ K_{55,57,88} + K_{55,65,88} + K_{55,73,88} + K_{55,81,88}\\ =& F^*_{11,85,88} + F^*_{11,77,88} + F^*_{11,61,88} + F^*_{11,29,88} + F^*_{11,21,88} + F^*_{11,13,88} + F^*_{33,15,88}\\ &+ F^*_{33,23,88} + F^*_{33,31,88} + F^*_{33,47,88} + F^*_{33,71,88} - \sum_{r=0}^{10} K_{11,5+8r,88} + \sum_{r=0}^{10} K_{55,1+8r,88}\\ =& F^*_{11,85,88} + F^*_{11,77,88} + F^*_{11,61,88} + F^*_{11,29,88} + F^*_{11,21,88} + F^*_{11,13,88} + F^*_{33,15,88}\\ &+ F^*_{33,23,88} + F^*_{33,31,88} + F^*_{33,47,88} + F^*_{33,71,88} - K_{1,5,8} + K_{5,1,8}\\ =& F^*_{11,85,88} + F^*_{11,77,88} + F^*_{11,61,88} + F^*_{11,29,88} + F^*_{11,21,88} + F^*_{11,13,88} + F^*_{33,15,88}\\ &+ F^*_{33,23,88} + F^*_{33,31,88} + F^*_{33,47,88} + F^*_{33,71,88}. \quad (63)\end{aligned}$$

Hence Identity (13) is equivalent to (note this is the form we mentioned during Identity (6), giving us the pairs (a_i, c_i), required for the calculation of coefficients later)

$$\begin{aligned} G_3(q^2)G_3(q^{11}) =& q^{-\frac{13}{8}}\Big(F^*_{11,13,88} + F^*_{11,21,88} + F^*_{11,29,88} + F^*_{11,61,88} + F^*_{11,85,88} \\ &+ F^*_{33,15,88} + F^*_{33,23,88} + F^*_{33,31,88} + F^*_{33,47,88} + F^*_{33,71,88}\Big) \\ =& \frac{1}{2}q^{-\frac{13}{8}}\left(\sum_{\substack{0<c<88 \\ c\equiv 7\ (8)}}\left(\frac{c}{11}\right)F^*_{33,c,88} - \sum_{\substack{0<c<88 \\ c\equiv 3\ (8)}}\left(\frac{c}{11}\right)F^*_{77,c,88}\right) \end{aligned} \tag{64}$$

as -1 is not a quadratic residue modulo 11. This is therefore the same as

$$G_3(q^2)G_3(q^{11}) = \frac{\sqrt{11}}{352\pi}q^{-\frac{13}{8}}\sum_{b=0}^{87}\left(\frac{b}{11}\right)\left(\zeta_{88}^{33b}E_{33,b,88} - \zeta_{88}^{77b}E_{77,b,88}\right), \tag{65}$$

which becomes

$$\vartheta_{\rho+\Gamma}\left(\frac{\tau}{2}\right) = \frac{\sqrt{11}}{88\pi}\sum_{b=0}^{87}\left(\frac{b}{11}\right)\left(\zeta_{88}^{33b}E_{33,b,88} - \zeta_{88}^{77b}E_{77,b,88}\right). \tag{66}$$

Hence

$$H(\tau) := \left(\sum_{b=0}^{87}\left(\frac{b}{11}\right)\left(\zeta_{88}^{33b}E_{33,b,88} - \zeta_{88}^{77b}E_{77,b,88}\right)\right)^4. \tag{67}$$

4.2. Table of Identities

We find that all identities are equivalent to an identity of the form

$$\vartheta_{\rho+\Gamma}\left(\frac{\tau}{2}\right) = \frac{\sqrt{p}}{N\pi Q}\sum_{b=0}^{N-1}\left(\frac{b}{p}\right)\sum_{(k_i,a_i)}(-1)^{i+1}\zeta_N^{k_i b}E_{a_i,b,N}(\tau) \tag{68}$$

with $Q = 1$ if $p \equiv 1 \pmod 4$, and $Q = i$ if $p \equiv 3 \pmod 4$. The second sum runs over the pairs (k_i, a_i), $i = 1, 2, \ldots$. For our triangular identities, there are either 1 or 2 pairs, for our pentagonal, 2 or 4 pairs. For the single heptagonal identity, there are 8 pairs. Therefore each of Melham's identities (once transformed) is defined simply by α, β (recall from Sections 2.1 - 2.3 that N is defined by α, β), p, the pairs (k_i, a_i), and whether the identity is triangular, pentagonal, or heptagonal. We display this information in Table 1 and Table 2, with the understanding that these equivalent identities can be obtained by simply applying the method used for Identity (6) or, in some cases, Identity (13) (which we mark with an asterisk) to Melham's original identities in his paper.

Each identity has a corresponding $H(\tau)$, obtained by ignoring the constant term and raising to a certain even power, of the form

$$H(\tau) = \left(\sum_{b=0}^{N-1} \left(\frac{b}{p}\right) \sum_{(k_i,a_i)} (-1)^{i+1} \zeta_N^{k_i b} E_{a_i,b,N}(\tau) \right)^{2v} \tag{69}$$

for some integer v, as discussed in Sections 2.1 - 2.3. This is what we wish to show is a modular form for the relevant subgroup.

5. Identities Under Transformation by the Congruence Subgroup

For all our identities, we have already shown that the LHS (as a theta function) raised to some even power is modular for Γ', where

$$\Gamma' = \Gamma_0\left(\frac{N}{2}\right) \cap \Gamma^0(2), \tag{70}$$

further intersected with $\Gamma_1(10)$ for the heptagonal case. As we have also seen, the RHS of each identity is an Eisenstein series, and is therefore modular with respect to $\Gamma(N)$, where thankfully the N of the LHS is equal to the N of the RHS. In order to reduce the amount of coefficients needed to be calculated, we would like to show that the RHS (raised to the same even power as the LHS) is also modular for Γ'. As the RHS is already modular for $\Gamma(N)$, we only need to consider elements in $\Gamma'' = \Gamma'/\Gamma(N)$. A set of generators for Γ'' is

$$A = \begin{pmatrix} 1 & 2 \\ 0 & 1 \end{pmatrix}, \quad B = \begin{pmatrix} 1 & 0 \\ \frac{N}{2} & 1 \end{pmatrix},$$

$$C_0 = -I, \quad C_1 = \begin{pmatrix} c_1 & 0 \\ 0 & c_1^{-1} \end{pmatrix}, \quad C_2 = \begin{pmatrix} c_2 & 0 \\ 0 & c_2^{-1} \end{pmatrix}, \quad \dots$$

where the C_j continue to include all c_j necessary to generate $(\mathbb{Z}/N\mathbb{Z})^*$ (noticing we have already included -1).

5.1. Invariance Under Transformation by Matrix A

We begin this section with a lemma we will require.

Lemma 3. *Let p be an odd prime, $p \mid N$. Let $\gcd(d, N) = g$, and $kg = d$, $Mg = N$,* *all integers. Then*

$$\sum_{b=0}^{N-1}\left(\frac{b}{p}\right)\zeta_N^{-db}=\sum_{b=0}^{N-1}\left(\frac{b}{p}\right)\zeta_M^{-kb}=\begin{cases}0 & p\neq M\\ \left(\frac{k}{p}\right)\frac{N}{p}\sqrt{p} & p=M\equiv 1\ (mod\ 4)\\ -\left(\frac{k}{p}\right)\frac{iN}{p}\sqrt{p} & p=M\equiv 3\ (mod\ 4).\end{cases} \tag{71}$$

Proof. Define

$$\begin{aligned}S:&=\sum_{b=0}^{p-1}\left(\frac{b}{p}\right)\sum_{r=0}^{\frac{N}{p}-1}\zeta_M^{-k(b+pr)}\\ &=\sum_{b=0}^{p-1}\left(\frac{b}{p}\right)\zeta_M^{-kb}\sum_{r=0}^{\frac{N}{p}-1}\zeta_M^{-kpr}\\ &=\sum_{b=0}^{p-1}\left(\frac{b}{p}\right)\zeta_M^{-kb}\sum_{r=0}^{\frac{N}{p}-1}\zeta_{\frac{N}{p}}^{-kgr}.\end{aligned} \tag{72}$$

Now,

$$\sum_{r=0}^{\frac{N}{p}-1}\zeta_{\frac{N}{p}}^{-kgr}=\begin{cases}0 & kg\not\equiv 0\ (\text{mod } \frac{N}{p})\\ \frac{N}{p} & kg\equiv 0\ (\text{mod } \frac{N}{p}).\end{cases} \tag{73}$$

For some $t\in\mathbb{Z}$,

$$kg\equiv 0\ (\text{mod } \frac{N}{p})\iff k\frac{N}{M}=\frac{N}{p}t\iff kp=Mt\iff M\mid p. \tag{74}$$

So $S=0$ when $p\not\equiv 0$ (mod M), and as p an odd prime, $p\neq M$. When $p\equiv 0$ (mod M), i.e., $p=M$, we have

$$S=\frac{N}{p}\sum_{b=0}^{p-1}\left(\frac{b}{p}\right)\zeta_p^{-kb}=\begin{cases}\left(\frac{k}{p}\right)\frac{N}{p}\sqrt{p} & p\equiv 1\ (\text{mod } 4)\\ -\left(\frac{k}{p}\right)\frac{N}{p}i\sqrt{p} & p\equiv 3\ (\text{mod } 4),\end{cases} \tag{75}$$

which completes the proof. □

To show that $H(\tau)$ is invariant under the matrix A, we recall that $E_{a,b,N}$ is an Eisenstein series, and make use of the transformation formula in Equation (37). We also write

$$\sum_{j\geq 0}s_j(\alpha,\beta,N)q^{\frac{j}{N}}:=E_{\alpha,\beta,N}(\tau) \tag{76}$$

so that $s_j(\alpha,\beta,N)$ is the $q^{\frac{j}{N}}$ coefficient of $E_{\alpha,\beta,N}(\tau)$. We spot

$$\begin{aligned} E_{a,b+a,N}(\tau) &= E_{a,b,N}(\tau+1) \\ &= 2\pi i\left(\frac{a}{N}-\frac{1}{2}-\sum_{r=1}^{\infty} q^{\frac{r}{N}}\zeta_N^r\left[\sum_{\substack{s\mid r\\ s\equiv a\ (N)\\ s>0}}\zeta_N^{\frac{br}{s}}-\sum_{\substack{s\mid r\\ s\equiv -a\ (N)\\ s>0}}\zeta_N^{-\frac{br}{s}}\right]\right) \\ &= \sum_{j\geq 0} s_j(a,b,N)\zeta_N^j q^{\frac{j}{N}}. \end{aligned} \tag{77}$$

Thus

$$\begin{aligned} H|_A &= \left(\sum_{\beta=0}^{N-1}\left(\frac{\beta}{p}\right)\sum_{(k_i,a_i)}(-1)^{i+1}\zeta_N^{k_i\beta}E_{a_i,\beta+2a_i,N}\right)^{2v} \\ &= \left(\sum_{\beta=0}^{N-1}\left(\frac{\beta}{p}\right)\sum_{(k_i,a_i)}(-1)^{i+1}\zeta_N^{k_i\beta}\sum_{j\geq 0}s_j(a_i,\beta,N)\zeta_N^{2j}q^{\frac{j}{N}}\right)^{2v} \\ &= \left(\sum_{j\geq 0}\zeta_N^{2j}q^{\frac{j}{N}}\sum_{(k_i,a_i)}(-1)^{i+1}\sum_{\beta=0}^{N-1}\left(\frac{\beta}{p}\right)\zeta_N^{k_i\beta}s_j(a_i,\beta,N)\right)^{2v}. \end{aligned} \tag{78}$$

By Lemma 3, we have

$$\sum_{\beta=0}^{N-1}\left(\frac{\beta}{p}\right)\zeta_N^{k_i\beta}s_j(a_i,\beta,N)=0 \tag{79}$$

unless $s_j(a_i,\beta,N)$ has a $\zeta_N^{m\beta}$ term with $\gamma:=(m+k_i,N)=\frac{N}{p}$, hence we need

$$m=-k_i+d_i\gamma \tag{80}$$

with $d_i\not\equiv 0$ modulo p. For $j>0$,

$$s_j(a_i,\beta,N)=-2\pi i\left[\sum_{\substack{s\mid j\\ s\equiv a_i\ (N)\\ s>0}}\zeta_N^{\frac{\beta j}{s}}-\sum_{\substack{s\mid j\\ s\equiv -a_i\ (N)\\ s>0}}\zeta_N^{-\frac{\beta j}{s}}\right]. \tag{81}$$

Therefore we have $j=\pm(a_i+lN)m$ for some $l\in\mathbb{Z}$, hence $j\equiv\pm a_i m \pmod N$. We note that for all our identities $p\mid a_i$, hence $j\equiv\pm a_i m\equiv\pm a_i(-k_i+d_i\gamma)\equiv\pm a_i k_i \pmod N$.

For triangular identities of the form that have just one pair of (a_i,k_i), we notice that $a_ik_i\equiv\frac{N}{4} \pmod N$. Hence $\zeta_N^{2j}=\pm 1$ for all j of nonvanishing terms. These identities were raised to the power of 2, hence $H|_A = H$.

For triangular identities that have two pairs of (a_i, k_i), we see that $a_ik_i \equiv \frac{N}{8}$ or $\pm\frac{3N}{8} \pmod N$. Hence $\zeta_N^{2j} = \pm i$ for all j of nonvanishing terms. These identities were raised to the power of 4, hence $H|_A = H$.

For pentagonal identities that have two pairs of (a_i, k_i), we see that $a_ik_i \equiv \frac{3N}{4}$ (Identity (17)) or $\frac{5N}{12}$ (Identities (20), (23)) modulo N. Hence for Identity (17) $\zeta_N^{2j} = -1$, and for the others, $\zeta_N^{2j} = \zeta_6^5$ for all j of nonvanishing terms. Identity (17) was raised to the power of 2, and Identities (20) and (23) were raised to the power of 6, hence in both cases $H|_A = H$.

For pentagonal identities that have four pairs of (a_i, k_i), we see that $a_ik_i \equiv \frac{7N}{8}$ or $\frac{3N}{8}$ (Identities (18), (19)) or $\frac{N}{24}, \frac{11N}{24}, \frac{13N}{24}, \frac{17N}{24}$ (Identities (21), (22), (24), (25)) modulo N. Hence for Identities (18), (19) $\zeta_N^{2j} = -i$, and for the others, $\zeta_N^{2j} = \zeta_{12}$, ζ_{12}^{11}, or ζ_{12}^5 for all j of nonvanishing terms. All of these identities were raised to the power of 12, hence in both cases $H|_A = H$.

For the one heptagonal identity we have $a_ik_i \equiv 51 \equiv \frac{17N}{40} \pmod N$ for all 8 pairs. Thus $\zeta_N^{2j} = \zeta_{20}^{17}$, and as the identity is raised to the power of 20, $H|_A = H$.

5.2. Invariance Under Transformation by Matrix B

To show that $H(\tau)$ is invariant under the matrix $B = \begin{pmatrix} 1 & 0 \\ \frac{N}{2} & 1 \end{pmatrix}$ we notice

$$H|_B = H|_{B_1^{-1}}|_{B_2}|_{B_1} \tag{82}$$

with

$$B_1^{-1} = \begin{pmatrix} 0 & 1 \\ -1 & 0 \end{pmatrix}, \quad B_2 = \begin{pmatrix} 1 & -\frac{N}{2} \\ 0 & 1 \end{pmatrix}, \quad B_1 = \begin{pmatrix} 0 & -1 \\ 1 & 0 \end{pmatrix},$$

because $B = B_1^{-1}B_2B_1$. If $H|_{B_1}$ remains fixed under transformation by B_2, then H will be fixed by B. Therefore we just need to check that $H|_{B_1}$ is a sum of powers of $q^{\frac{2}{N}}$. We have

$$\begin{aligned} H|_{B_1} &= \left(\sum_{\beta=0}^{N-1}\left(\frac{\beta}{p}\right)\sum_{(k_i,a_i)}(-1)^{i+1}\zeta_N^{k_i\beta}E_{\beta,-a_i,N}\right)^{2v} \qquad (83)\\ &= \left(2\pi i\sum_{(k_i,a_i)}(-1)^{i+1}\sum_{\beta=0}^{N-1}\left(\frac{\beta}{p}\right)\zeta_N^{k_i\beta}\left[\frac{\beta}{N}-\frac{1}{2}+\frac{1}{2\pi i}\sum_{r=1}^{\infty}q^{\frac{r}{N}}s_{-r}(\beta,a_i,N)\right]\right)^{2v}\\ &= \left(S+\sum_{(k_i,a_i)}(-1)^{i+1}\sum_{\beta=0}^{N-1}\left(\frac{\beta}{p}\right)\zeta_N^{k_i\beta}\sum_{r=1}^{\infty}q^{\frac{r}{N}}s_{-r}(\beta,a_i,N)\right)^{2v} \end{aligned}$$

with S some constant. Rearranging the RHS, we get

$$
\begin{aligned}
H|_{B_1} &= \Bigg(S - 2\pi i \sum_{(k_i,a_i)} (-1)^{i+1} \sum_{r=1}^{\infty} q^{\frac{r}{N}} \Bigg[\sum_{\beta=0}^{N-1} \sum_{\substack{s|r \\ s\equiv\beta\ (N) \\ s>0}} \left[\left(\frac{\beta}{p}\right) \zeta_N^{k_i\beta} \zeta_N^{-\frac{a_i r}{s}}\right] \\
&\quad - \sum_{\beta=0}^{N-1} \sum_{\substack{s|r \\ s\equiv-\beta\ (N) \\ s>0}} \left[\left(\frac{\beta}{p}\right) \zeta_N^{k_i\beta} \zeta_N^{\frac{a_i r}{s}}\right]\Bigg]\Bigg)^{2v} \qquad (84) \\
&= \Bigg(S - 2\pi i \sum_{(k_i,a_i)} (-1)^{i+1} \sum_{r=1}^{\infty} q^{\frac{r}{N}} \Bigg[\sum_{\substack{s|r \\ s>0}} \left[\left(\frac{s}{p}\right) \zeta_N^{k_i s} \zeta_N^{-\frac{a_i r}{s}}\right] \\
&\quad - \sum_{\substack{s|r \\ s>0}} \left[\left(\frac{-s}{p}\right) \zeta_N^{-k_i s} \zeta_N^{\frac{a_i r}{s}}\right]\Bigg]\Bigg)^{2v} \\
&= \Bigg(S - 2\pi i \sum_{r=1}^{\infty} q^{\frac{r}{N}} \Bigg[\sum_{\substack{s|r \\ s>0}} \left(\frac{s}{p}\right) \sum_{(k_i,a_i)} (-1)^{i+1} \Bigg[\zeta_N^{k_i s - a_i \frac{r}{s}} \\
&\quad - \left(\frac{-1}{p}\right) \zeta_N^{-k_i s + a_i \frac{r}{s}}\Bigg]\Bigg]\Bigg)^{2v}.
\end{aligned}
$$

Define

$$
G := \sum_{(k_i,a_i)} (-1)^{i+1} \left[\zeta_N^{k_i s - a_i \frac{r}{s}} - \left(\frac{-1}{p}\right) \zeta_N^{-k_i s + a_i \frac{r}{s}}\right]. \qquad (85)
$$

We aim to show $G = 0$ whenever r is odd. Suppose first that $p \equiv 1 \pmod 4$, then, as we always have an even number of pairs of (a_i, k_i), using a simple trigonometric identity,

$$
\begin{aligned}
G &= \sum_{(k_i,a_i)} (-1)^{i+1} \left[\zeta_N^{k_i s - a_i \frac{r}{s}} - \zeta_N^{-k_i s + a_i \frac{r}{s}}\right] \\
&= 2i \sum_{(k_i,a_i)} (-1)^{i+1} \sin\left[\frac{2\pi}{N}\left(k_i s - a_i \frac{r}{s}\right)\right] \\
&= 2i \sum_{\text{odd } i} \left(\sin\left[\frac{2\pi}{N}\left(k_i s - a_i \frac{r}{s}\right)\right] - \sin\left[\frac{2\pi}{N}\left(k_{i+1} s - a_{i+1} \frac{r}{s}\right)\right]\right) \\
&= 4i \sum_{\text{odd } i} \cos[\theta_1] \sin[\theta_2], \qquad (86)
\end{aligned}
$$

with $\theta_1 = \frac{\pi}{N}\left(s(k_i + k_{i+1}) - \frac{r}{s}(a_i + a_{i+1})\right)$ and $\theta_2 = \frac{\pi}{N}\left(s(k_i - k_{i+1}) - \frac{r}{s}(a_i - a_{i+1})\right)$. Similarly, if $p \equiv 3 \pmod 4$ we have

$$
\begin{aligned}
G &= \sum_{(k_i,a_i)} (-1)^{i+1}\left[\zeta_N^{k_i s - a_i \frac{r}{s}} + \zeta_N^{-k_i s + a_i \frac{r}{s}}\right] \\
&= 2\sum_{(k_i,a_i)} (-1)^{i+1}\cos\left[\frac{2\pi}{N}\left(k_i s - a_i \frac{r}{s}\right)\right] \\
&= -4\sum_{\text{odd } i} \sin\left[\theta_1\right]\sin\left[\theta_2\right]. \qquad (87)
\end{aligned}
$$

For identities that have $p \equiv 1 \pmod 4$, we note that for odd i, $k_i - k_{i+1} \equiv a_i - a_{i+1} \equiv \frac{N}{2}$ modulo N. Thus G vanishes when r is odd, as s and $\frac{r}{s}$ must share the same parity.

For identities that have $p \equiv 3 \pmod 4$, we are not as restricted, and we can have either (for odd i) $k_i - k_{i+1} \equiv a_i - a_{i+1} \equiv \frac{N}{2}$ or $k_i + k_{i+1} \equiv a_i + a_{i+1} \equiv \frac{N}{2}$ modulo N. For all of these identities we have this requirement, hence G vanishes for odd r, as required.

5.3. Invariance Under Transformation by Matrices C_j

To show these identities remain fixed under matrices of the form (disregarding the subscript for now) $C = \begin{pmatrix} c & 0 \\ 0 & c^{-1} \end{pmatrix}$ modulo N, we apply the transformation formula again:

$$
H|_C = \left(\sum_{\beta=0}^{N-1}\left(\frac{\beta}{p}\right)\sum_{(k_i,a_i)}(-1)^{i+1}\zeta_N^{k_i\beta}E_{ca_i,c^{-1}\beta,N}\right)^{2v}. \qquad (88)
$$

We let $\beta' = c^{-1}\beta$ and find, as we must have $(c, N) = 1$ and $p \mid N$,

$$
\begin{aligned}
H|_C &= \left(\sum_{\beta'=0}^{N-1}\left(\frac{c\beta'}{p}\right)\sum_{(k_i,a_i)}(-1)^{i+1}\zeta_N^{k_i c\beta'}E_{ca_i,c^{-1}\beta,N}\right)^{2v} \\
&= \left(\frac{c}{p}\right)^{2v}\left(\sum_{\beta'=0}^{N-1}\left(\frac{\beta'}{p}\right)\sum_{(k_i,a_i)}(-1)^{i+1}\zeta_N^{k_i c\beta'}E_{ca_i,\beta',N}\right)^{2v}. \qquad (89)
\end{aligned}
$$

This will be equal to H if c acts on the pairs (k_i, a_i), by multiplication modulo N, by mapping each to $\pm(k_j, a_j)$, with $j = 1, 2, \ldots$ and so on. To see this we consider the effect of one mapping. We allow for each (k_i, a_i) to be mapped to either itself, its negative, a different (k_j, a_j), or the negative of that. If (k_i, a_i) is mapped to itself or (k_j, a_j), there are no issues. If it is mapped to the negative of one of these we notice, using $E_{-a,-b,N} = -E_{a,b,N}$,

$$\sum_{\beta=0}^{N-1}\left(\frac{\beta}{p}\right)\zeta_N^{-k_i\beta}E_{-a_i,\beta,N} = -\sum_{\beta=0}^{N-1}\left(\frac{\beta}{p}\right)\zeta_N^{-k_i\beta}E_{a_i,-\beta,N}$$
$$= -\left(\frac{-1}{p}\right)\sum_{\beta=0}^{N-1}\left(\frac{\beta}{p}\right)\zeta_N^{k_i\beta}E_{a_i,\beta,N}. \tag{90}$$

If $p \equiv 3 \pmod 4$ then the negatives cancel, and we are done. If however $p \equiv 1 \pmod 4$, we need to have either all the pairs (k_i, a_i) map to negatives, or for each pair that maps to a negative match to a pair (k_j, a_j) where j has a different parity than i, which will be accounted for by the $(-1)^{i+1}$ term.

For example, Identity (7) has RHS

$$H = \left(\sum_{b=0}^{23}\left(\frac{b}{3}\right)\left(\zeta_{24}^{9b}E_{3,b,24} - \zeta_{24}^{21b}E_{15,b,24}\right)\right)^4. \tag{91}$$

A set of generators for $(\mathbb{Z}/24\mathbb{Z})^*$ is 5, 7, 23. Of course $23 \equiv -1$ is trivial, so we focus first on 5. We get $5\cdot(k_1, a_1) = 5\cdot(9,3) = (21,15) \equiv (k_2, a_2)$ modulo 24, and $5\cdot(k_2,a_2) = 5\cdot(21,15) \equiv (9,3) \equiv (k_1,a_1)$ modulo 24. Thus H transforms to

$$\left(\sum_{b=0}^{23}\left(\frac{b}{3}\right)\left(\zeta_{24}^{21b}E_{15,b,24} - \zeta_{24}^{9b}E_{3,b,24}\right)\right)^4 = H. \tag{92}$$

Similarly, for 7 we find $7\cdot(k_1,a_1) \equiv -(k_1,a_1)$, and $7\cdot(k_2,a_2) \equiv -(k_2,a_2)$ modulo 24. So H transforms to

$$\left(\sum_{b=0}^{23}\left(\frac{b}{3}\right)\left(\zeta_{24}^{-9b}E_{-3,b,24} - \zeta_{24}^{-21b}E_{-15,b,24}\right)\right)^4$$
$$= \left(-\left(\frac{-1}{3}\right)\sum_{b=0}^{23}\left(\frac{b}{3}\right)\left(\zeta_{24}^{9b}E_{3,b,24} - \zeta_{24}^{21b}E_{15,b,24}\right)\right)^4$$
$$= H. \tag{93}$$

This is shown in Tables 3 and 4 below. Listed is the identity number, the value of N, a list of the c_i required, and the values of the pairs (k_i, a_i). The table then shows how the c_i permute the pairs. All permutations are in fact involutions, and we denote the fact that c_i swaps (k_i, a_i) to (k_j, a_j) as simply (i, j), whereas if c_i takes (k_i, a_i) to $-(k_j, a_j)$ and (k_j, a_j) to $-(k_i, a_i)$, we write $(i, -j)$. If (k_i, a_i) maps to itself, or to $-(k_i, a_i)$, we write (i), or $(-i)$, respectively. Of course, -1 just takes the pair (k_i, a_i) to $-(k_i, a_i)$.

Table 3: Identities with 2 pairs of (k_i, a_i)

Id.	N	c_1, c_2, c_3	$\begin{pmatrix} k_1 \\ a_1 \end{pmatrix}$	$\begin{pmatrix} k_2 \\ a_2 \end{pmatrix}$	c_1 **acts**	c_2 **acts**
7	24	$5, 7, 23$	$\begin{pmatrix} 9 \\ 3 \end{pmatrix}$	$\begin{pmatrix} 21 \\ 15 \end{pmatrix}$	$(1, 2)$	$(1, -2)$
8			$\begin{pmatrix} 21 \\ 9 \end{pmatrix}$	$\begin{pmatrix} 9 \\ 21 \end{pmatrix}$	$(1, 2)$	$(1, -2)$
9	40	$3, 11, 39$	$\begin{pmatrix} 5 \\ 5 \end{pmatrix}$	$\begin{pmatrix} 25 \\ 25 \end{pmatrix}$	$(1, -2)$	$(1, -2)$
10			$\begin{pmatrix} 5 \\ 25 \end{pmatrix}$	$\begin{pmatrix} 25 \\ 5 \end{pmatrix}$	$(1, -2)$	$(1, -2)$
12	88	$3, 5, 87$	$\begin{pmatrix} 33 \\ 11 \end{pmatrix}$	$\begin{pmatrix} 77 \\ 55 \end{pmatrix}$	$(1, -2)$	$(1, 2)$
13			$\begin{pmatrix} 33 \\ 33 \end{pmatrix}$	$\begin{pmatrix} 77 \\ 77 \end{pmatrix}$	$(1, -2)$	$(1, 2)$
15	232	$3, 5, 231$	$\begin{pmatrix} 29 \\ 29 \end{pmatrix}$	$\begin{pmatrix} 145 \\ 145 \end{pmatrix}$	$(1, -2)$	$(1, 2)$
16			$\begin{pmatrix} 145 \\ 29 \end{pmatrix}$	$\begin{pmatrix} 29 \\ 145 \end{pmatrix}$	$(1, -2)$	$(1, 2)$
17	60	7,13,59	$\begin{pmatrix} 35 \\ 15 \end{pmatrix}$	$\begin{pmatrix} 5 \\ 45 \end{pmatrix}$	$(1, -2)$	$(1), (2)$
20	156	7,11,155	$\begin{pmatrix} 143 \\ 91 \end{pmatrix}$	$\begin{pmatrix} 65 \\ 13 \end{pmatrix}$	$(1, 2)$	$(-1), (-2)$
23	444	5,7,443	$\begin{pmatrix} 407 \\ 259 \end{pmatrix}$	$\begin{pmatrix} 185 \\ 37 \end{pmatrix}$	$(1, -2)$	$(1, 2)$

For the three Identities (6), (11), (14), that have just a single pair, it is easy to find a set of generators that, for example, maps the pair to its negative. For the single heptagonal identity, we need to avoid matrices with first entry congruent to ± 3 modulo 10. So, as $N = 120$ we take $c_0 = 119$, $c_1 = 11$, $c_2 = 19$, and $c_3 = 29$.

As before, c_0 is immediate. Next, c_1 takes (k_1, a_1) to $-(k_4, a_4)$ and vice versa, (k_2, a_2) to $-(k_3, a_3)$ and vice versa, (k_5, a_5) to $-(k_8, a_8)$ and vice versa, and (k_6, a_6) to $-(k_7, a_7)$ and vice versa. The value 19 swaps the pair with index 1 with the negative of the pair with index 2, 3 with -4, 5 with -6, 7 with -8. Finally, 29 swaps 1 and 4, 2 and 3, 5 and 8, 6 and 7.

6. Equivalence Using Sturm's Bound

Let $f = \sum_{k\in\mathbb{Z}} \alpha_k q^k$. We define $ord(f)$ to be the smallest such n that $\alpha_n \neq 0$. We first state a simplified version of Sturm's bound [11].

Theorem 1. *Define* $\Gamma := SL_2(\mathbb{Z})$, *and let* f, g *be modular forms on* $\Gamma' \supseteq \Gamma(N)$ *of weight* k, *a positive integer. Suppose* $ord(f-g) > \frac{k[\Gamma:\Gamma']}{12}$. *Then* $f = g$.

In other words, Sturm's bound says that given two modular forms over the same congruence subgroup and of the same weight, then they are equivalent if their q-expansions agree up to the $q^{\frac{kM}{12}}$ coefficient, where $k = 2v$ is the weight, and M is the index of the congruence subgroup in Γ.

Recall that we have defined for the triangular and pentagonal identities

$$\Gamma' = \Gamma_0\left(\frac{N}{2}\right) \cap \Gamma^0(2), \tag{94}$$

and for the heptagonal identity,

$$\Gamma' = \Gamma_0\left(\frac{N}{2}\right) \cap \Gamma^0(2) \cap \Gamma_1(10). \tag{95}$$

We have now shown that each side of our transformed identities are modular of weight $2v$ for $\Gamma'' = \Gamma'/\Gamma(N)$. We can now apply Sturm's bound. We have [2, p. 14], for the triangular and pentagonal cases,

$$M = [\Gamma : \Gamma'] = [\Gamma : \Gamma_0(N)] = N \prod_{p|N}\left(1 + \frac{1}{p}\right) \tag{96}$$

as, despite the fact that $\Gamma' \neq \Gamma_0(N)$, they have the same index in Γ. To see this, we

Table 4: Identities with 4 pairs of (k_i, a_i)

Id.	N	c_1, c_2, c_3, c_4	$\binom{k_1}{a_1}$	$\binom{k_2}{a_2}$	$\binom{k_3}{a_3}$	$\binom{k_4}{a_4}$	c_1 acts	c_2 acts	c_3 acts
18	120	$7, 11, 19, 119$	$\binom{105}{25}$	$\binom{45}{85}$	$\binom{105}{65}$	$\binom{45}{5}$	$(1, -3), (2, -4)$	$(1, -2), (3, -4)$	$(1, -4), (2, -3)$
19			$\binom{105}{5}$	$\binom{45}{65}$	$\binom{105}{85}$	$\binom{45}{25}$	$(1, -3), (2, -4)$	$(1, -2), (3, -4)$	$(1, -4), (2, -3)$
21	264	$5, 7, 13, 263$	$\binom{121}{11}$	$\binom{253}{143}$	$\binom{209}{187}$	$\binom{187}{209}$	$(1, -4), (2, 3)$	$(1, -3), (2, 4)$	$(1, 2), (3, -4)$
22			$\binom{121}{121}$	$\binom{187}{187}$	$\binom{209}{209}$	$\binom{253}{253}$	$(1, -2), (3, 4)$	$(1, -3), (2, 4)$	$(1, 4), (2, -3)$
24	696	$5, 7, 13, 695$	$\binom{493}{29}$	$\binom{145}{377}$	$\binom{319}{551}$	$\binom{667}{203}$	$(1, -3), (2, -4)$	$(1, 4), (2, 3)$	$(1, 2), (3, 4)$
25			$\binom{319}{667}$	$\binom{667}{319}$	$\binom{493}{145}$	$\binom{145}{493}$	$(1, -3), (2, -4)$	$(1, 4), (2, 3)$	$(1, 2), (3, 4)$

simply notice that Γ' and $\Gamma_0(N)$ both share a common subgroup, $\Gamma_0(N) \cap \Gamma^0(2)$, and this common subgroup clearly has index 2 in both. For the heptagonal case, we simply have to double M. We need to check $\frac{kM}{12} + 1$ coefficients. Starting with Identity (6), we have $M = 36$, and we need to check $\frac{kM}{12} + 1 = 7$ coefficients. We use the form of

$$q^{\frac{\alpha+\beta}{8(k-2)}} G_k(q^\alpha) G_k(q^\beta) = \sum_{(a_i,c_i)} F^*_{a_i,c_i,4(k-2)} \tag{97}$$

for ease, which as we mentioned before each identity can be written as. For Identity (6), this is

$$q^{\frac{3}{4}} G_3(q) G_3(q^5) = F^*_{5,3,20} + F^*_{5,7,20} \tag{98}$$

and find both sides start with

$$1 + 1 \cdot q + 0 \cdot q^2 + 1 \cdot q^3 + 0 \cdot q^4 + 1 \cdot q^5 + 2 \cdot q^6 + 0 \cdot q^7.$$

As the coefficients of this form agree, the coefficients of $H(\tau)$ and $\vartheta_{\rho+\Gamma}\left(\frac{\tau}{2}\right)^{2v}$ must also agree, and we are finished.

Our computations show that both the LHS and RHS have the same coefficients for all triangular, pentagonal, and heptagonal identities. Notice that while we always knew both sides were modular forms for $\Gamma(N)$, using that would have made the number of coefficients required for Sturm's Bound much larger, hence the efforts to show both sides were modular forms for the larger congruence subgroups. As an example, Identity (15) has $N = 232$, and $2v = 4$. With our efforts we only need to check just 121 coefficients, a relatively easy task. If we had used $\Gamma(N)$, we would have utilized the well known formula [2, p. 13],

$$M = [\Gamma : \Gamma(N)] = N^3 \prod_{p|N} \left(1 - \frac{1}{p^2}\right) \tag{99}$$

and found that we had to check 1,559,041 coefficients, which is somewhat more daunting. Worse still, when we finally got round to checking Identities (24) and (25), with $N = 696$ and $2v = 12$, we would have had to check 224,501,761 coefficients. With what we now know we of course only have to check 1,441.

Acknowledgments This paper is based on work undertaken during an MRes at the University of Exeter, supervised by Robin Chapman, who was immensely invaluable throughout, both patient and motivating. The author would also like to thank the reviewer for providing such helpful feedback and guidance.

References

[1] C. Adiga, S. Cooper, and J. H. Han, A general relation between sums of squares and sums of triangular numbers, *Int. J. Number Theory* **1** (2005), 175-182.

[2] F. Diamond and J. Shurman, *A First Course in Modular Forms*, Springer, 2005.

[3] L. E. Dickson, *Introduction to the Theory of Numbers*, University of Chicago Press, 1929. Reprinted by Dover, 1957.

[4] W. Ebeling, *Lattices and Codes*, Vieweg, 1994.

[5] M. Hirschhorn, *The Power of q*, Springer, 2017.

[6] K. Ireland and M. Rosen, *A Classical Introduction to Modern Number Theory, 2nd Edition*, Springer, 1990.

[7] F. Jarvis, *Algebraic Number Theory*, Springer, 2014.

[8] N. Koblitz, *Introduction to Elliptic Curves and Modular Forms, 2nd Edition.* Springer, 1993.

[9] R. S. Melham, Analogues of Jacobis two-square theorem: an informal account, *Integers* **10** (2010), 83-100.

[10] J.-P. Serre, *A Course in Arithmetic*, Springer, 1973.

[11] J. Sturm, On the Congruence of Modular Forms. In Number Theory, D. V. Chudnovsky, G. V. Chudnovsky, H. Cohn, and M. B. Nathanson (eds.), *Lectures Notes in Math.* **1240** (1987), 275-280.

[12] Z.-H. Sun, On the number of representations of n by $ax(x-1)/2 + by(y-1)/2$, *J. Number Theory* **129** (2009), 971-989.

[13] P. C. Toh, On representations by figurate numbers: a uniform approach to the conjectures of Melham, *Int. J. Number Theory* **9** (2013), 1055-1071.

Permissions

 Every chapter published in this book has been scrutinized by our experts. Their significance has been extensively debated. The topics covered herein carry significant findings which will fuel the growth of the discipline. They may even be implemented as practical applications or may be referred to as a beginning point for another development.

The contributors of this book come from diverse backgrounds, making this book a truly international effort. This book will bring forth new frontiers with its revolutionizing research information and detailed analysis of the nascent developments around the world.

We would like to thank all the contributing authors for lending their expertise to make the book truly unique. They have played a crucial role in the development of this book. Without their invaluable contributions this book wouldn't have been possible. They have made vital efforts to compile up to date information on the varied aspects of this subject to make this book a valuable addition to the collection of many professionals and students.

This book was conceptualized with the vision of imparting up-to-date information and advanced data in this field. To ensure the same, a matchless editorial board was set up. Every individual on the board went through rigorous rounds of assessment to prove their worth. After which they invested a large part of their time researching and compiling the most relevant data for our readers.

The editorial board has been involved in producing this book since its inception. They have spent rigorous hours researching and exploring the diverse topics which have resulted in the successful publishing of this book. They have passed on their knowledge of decades through this book. To expedite this challenging task, the publisher supported the team at every step. A small team of assistant editors was also appointed to further simplify the editing procedure and attain best results for the readers.

Apart from the editorial board, the designing team has also invested a significant amount of their time in understanding the subject and creating the most relevant covers. They scrutinized every image to scout for the most suitable representation of the subject and create an appropriate cover for the book.

The publishing team has been an ardent support to the editorial, designing and production team. Their endless efforts to recruit the best for this project, has resulted in the accomplishment of this book. They are a veteran in the field of academics and their pool of knowledge is as vast as their experience in printing. Their expertise and guidance has proved useful at every step. Their uncompromising quality standards have made this book an exceptional effort. Their encouragement from time to time has been an inspiration for everyone.

The publisher and the editorial board hope that this book will prove to be a valuable piece of knowledge for researchers, students, practitioners and scholars across the globe.

List of Contributors

Ivica Martinjak
Faculty of Science, University of Zagreb, Zagreb, Croatia

Helmut Prodinger
Department of Mathematics, University of Stellenbosch, Stellenbosch, South Africa

Fritz Schweiger
Department of Mathematics, University of Salzburg, Salzburg, Austria

Takashi Agoh
Department of Mathematics, Tokyo University of Science, Noda, Chiba, Japan

Maria Axenovich
Karlsruhe Institute of Technology, Karlsruhe, Germany

John Goldwasser
West Virginia University, Morgantown, West Virginia

Bernard Lidický, Michael Young and Ryan R. Martin
Iowa State University, Ames, Iowa

David Offner
Westminster College, New Wilmington, Pennsylvania

John Talbot
University College London, London, UK

H. Belbachir and S. Hadj-Brahim
Department of Mathematics, USTHB, RECITS Laboratory, Algiers, Algeria

István Mező and Chen-Ying Wang
School of Mathematics and Statistics, Nanjing University of Information Science & Technology, Nanjing, P.R. China

José L. Ramírez
Departamento de Matem´aticas, Universidad Nacional de Colombia, Bogotá, Colombia

Gregory A. Freiman
The Raymond and Beverly Sackler Faculty of Exact Sciences, School of Mathematical Sciences, Tel Aviv University, Tel Aviv, Israel

Oriol Serra and Christoph Spiegel
Department of Mathematics, Universitat Politécnica de Catalunya and Barcelona Graduate School of Mathematics, Barcelona, Spain

John Asplund
Dept. of Technology and Mathematics, Dalton State College, Dalton, Georgia

N. Bradley Fox
Department of Mathematics and Statistics, Austin Peay State University, Clarksville, Tennessee

Brian D. Sittinger
Department of Mathematics, CSU Channel Islands, Camarillo, California

Marina C. Morales
ACE Charter High School, Camarillo, California

Katherine Cordwell
Department of Mathematics, University of Maryland, College Park, Maryland Computer Science Dept., Carnegie Mellon University, Pittsburgh, Pennsylvania

Alyssa Epstein and Anand Hemmady
Dept. of Math. and Statistics, Williams College, Williamstown, Massachusetts Stanford Law School, Stanford, California

Steven J. Miller
Dept. of Math. and Statistics, Williams College, Williamstown, Massachusetts
Dept. of Math. Sciences, Carnegie Mellon University, Pittsburgh, Pennsylvania

Eyvindur Palsson
Dept. of Math., Virginia Polytechnic Institute and State University, Virginia

Aaditya Sharma
Dept. of Math. and Statistics, Williams College, Williamstown, Massachusetts DPMMS, University of Cambridge

Stefan Steinerberger
Department of Mathematics, Yale University, Connecticut

Yen Nhi Truong Vu
Dept. of Mathematics and Statistics, Amherst College, Amherst, Massachusetts

Zachary Chaney, Jed Menard and Connor Mattes
Department of Applied Mathematics and Statistics, Colorado School of Mines, Golden, Colorado

Timothy Trujillo
Department of Mathematics and Statistics, Sam Houston State University, Huntsville, Texas

George Humby
University of Exeter, Exeter, United Kingdom

Index

www.ingramcontent.com/pod-product-compliance
Lightning Source LLC
LaVergne TN
LVHW080613210926
848485LV00003B/153